STRESS WAVES IN NON-ELASTIC SOLIDS

Other Pergamon Titles of Interest

BARTENEV, G. M. and ZUYEV, Yu. S.
Strength and Failure of Visco-elastic Materials

BELYAYEV, N. M.
Problems in Strength of Materials

BRITVEC, S. J.
The Stability of Elastic Systems

CALLADINE, C. R.
Engineering Plasticity

CHEUNG, Y. K.
Finite Strip Method in Structural Analysis

CHRISTIAN, J. W.
The Theory of Transformations in Metals and Alloys, 2nd edition

DUGDALE, D. S.
Elements of Elasticity

HENDRY, A. W.
Elements of Experimental Stress Analysis (SI edition)

HERRMANN, G. and PERRONE, N.
Dynamic Response of Structures

HULL, D.
Introduction to Dislocations, 2nd edition

ILYUSHIN, A. A. and LENSKII, V. S.
Strength of Materials

JAEGER, L. G.
Cartesian Tensors in Engineering Science

JAEGER, L. G.
Elementary Theory of Elastic Plates

KOVACS, I. and ZSOLDOS, L.
Dislocations and Plastic Deformation

SAADA, A. S.
Elasticity: Theory and Applications

WASHIZU, K.
Variational Methods in Elasticity and Plasticity, 2nd edition

STRESS WAVES IN NON-ELASTIC SOLIDS

by

W. K. NOWACKI

Translated by

ZBIGNIEW OLESIAK

PERGAMON PRESS
OXFORD · NEW YORK · TORONTO · SYDNEY
PARIS · FRANKFURT

U. K.	Pergamon Press Ltd., Headington Hill Hall, Oxford OX3 0BW, England
U. S. A.	Pergamon Press Inc., Maxwell House, Fairview Park, Elmsford, New York 10523, U.S.A.
CANADA	Pergamon of Canada Ltd., 75 The East Mall, Toronto, Ontario, Canada
AUSTRALIA	Pergamon Press (Aust.) Pty. Ltd., 19a Boundary Street, Rushcutters Bay, N.S.W. 2011, Australia
FRANCE	Pergamon Press SARL, 24 rue des Ecoles, 75240 Paris, Cedex 05, France
FEDERAL REPUBLIC OF GERMANY	Pergamon Press GmbH, 6242 Kronberg Taunus, Pferdstrasse 1, Federal Republic of Germany

First English Edition 1978

Library of Congress Cataloging in Publication Data

Nowacki, Wojciech Krzysztof.
Stress waves in non-elastic solids.

Bibliography: p.
1. Elastic solids. 2. Stress waves. 3. Visco-plasticity. I. Title.
QC191.N68 1977 531'.33 76-52418
ISBN 0-08-021294-8

This is a translation of the Polish book Zagadnienia Falowe w Teorii Plastycznosci *published by* Państwowe Wydawnictwo Naukowe, Warsaw.

Printed in Great Britain by William Clowes & Sons, Limited London, Beccles and Colchester

CONTENTS

PUBLISHER'S NOTE

In order to make this volume available as economically and as rapidly as possible much of the mathematics has been reproduced directly from the original Polish edition. Due to technical problems, however, it was not possible to match the typography of the new updated material with the original, and certain characters within the new text matter will appear different from those in the reproduced matter. The variations comprise

ν which appears as υ in the original,
ϵ which appears as ε in the original,
g which appears as g in the original;

no distinction is intended and it is hoped that they will in no way distract the reader.

PREFACE

The dynamics of inelastic continua is nowadays a domain of intense study. Society, aiming at its own welfare and safety but aware of limited natural resources, requires reliable answers regarding the mechanical behaviour of solids and the performance of structures under transient agencies. Impacts, blasts, collisions, thermal shocks, and irradiations constitute current hazards of modern technology. Assessments of damage and failure due to the propagation of stress waves, evaluations of permanent deformation and energy absorption under impact, methods of estimation of pressure transmitted through protective structures, calculations of structural reliability under transient random charges, are of importance for the development of technology and for a safe utilization of its products.

The last decade is associated with significant advances in the mechanical sciences, in particular as regards the inelastic behaviour of materials and structures. Mathematical models of complex material response were formulated within the framework of non-linear continuum mechanics and justified in often sophisticated experiments. Mechanisms of inelastic deformation were explored to a large extent and couplings between mechanical and other fields were studied. Methods of solving boundary value problems for the differential equations governing the dynamics of inelastic solids were developed and various problems of a direct technological interest regarding structural strength and the propagation of waves in elastic–plastic and rate sensitive materials were solved. Advanced specialized monographs are thus needed to facilitate an access to the accumulated knowledge and to its utilization in engineering service to society.

The present book largely contributes towards meeting such a demand in the domain of continuum dynamics. Dr. Nowacki has succeeded in producing both an original and a comprehensive presentation of principles, methods, and solutions regarding the propagation of waves in elastic–plastic and viscoplastic solids.

This monograph is comprehensive since it exposes wave propagation problems for the considered types of material response starting with discussions of constitutive relations, justifies the hypotheses introduced in specialized theories and the simplifications made in the analysis of particular problems, presents both analytical and numerical methods of solving problems, and gives a large number of solutions to specific problems of wave propagation in inelastic solids.

The book is original both in its outline and in its contents. It includes a number of contributions Dr. Nowacki has made to the study of wave propagation in elastic–plastic and viscoplastic continua. There is a competent exposition of the mathematical questions relating to hyperbolic differential equations, a presentation of analytical and numerical techniques resulting in effective solutions of problems concerning plane, cylindrical and spherical stress waves and thermal stress waves, and discussions of two-dimensional waves, the whole being supplemented by solved problems. All this makes the book useful both in study and in design.

By its contents and its competent and lucid presentation of problems of inelastic waves the book constitutes a significant and useful contribution to the domain of continuum dynamics. It has no equivalent counterpart in the existing literature and it will doubtless be appreciated by both specialists and students because of its originality and pertinence.

Warsaw, 13 *April*, 1976 A. SAWCZUK

INTRODUCTION

This book is devoted to wave problems in the theory of plasticity. Such problems arise in cases in which intensive dynamic loads acting on the elements of a structure are big enough to produce plastic deformation of the elements. To date a considerable number of contributions and monographs have been published in this field. The first papers referring to these problems appeared in the forties and the theory rapidly developed in the sixties. Many problems, particularly the one-dimensional ones, have been investigated fully. A few monographs exist presenting the theory of impact in continuous and discrete systems in a general way. Here we should quote, first of all, the monograph by Kolsky [71], published in 1953, next the book by Goldsmith [46] which appeared in 1960, and those by Rakhmatulin and Demianov [125] and Cristescu [34], published in 1961 and 1967 respectively. In these books the reader can find the general theory of impact in continuous plastic bodies. Since the publication of these books a number of problems have been solved which are significant from the standpoint of engineering practice. Numerous papers on the dynamics of inelastic structures have appeared. In connection with the rapid development of computational techniques a great number of methods have been devised for the numerical integration of quasi-linear and semi-linear partial differential equations. These investigations have been accompanied by the development of the pertinent analytical methods of solving initial and boundary value problems for inelastic bodies. They mainly refer to the problems of stress wave propagation in the cases of complicated stress states, the waves generated by multi-parameter loading, three-dimensional waves, and, finally, to thermal stress waves.

This book is confined to those wave problems which are such that they can be described by a system of hyperbolic partial differential equations of the first order, which are quasi-linear or semi-linear. A survey of the literature is given in the field of stress wave propagation in elastic/viscoplastic media and in particular the papers of Polish authors are discussed.

The book is arranged in the following way. The first chapter has been devoted to the fundamental equations of the dynamics of inelastic media. The dynamical properties of materials (metals and soils) are discussed offering an account of the most representative theories of plasticity and viscoplasticity. Chapter II considers the basic definitions of discontinuity surfaces and the conditions which have to be satisfied across these surfaces. Simultaneously, certain mathematical fundamentals have been given, referring to systems of differential equations, quasi-linear and semi-linear, of the first order. Also initial and boundary value problems for hyperbolic equations have been formulated. The remaining chapters have been devoted to methods of solving stress wave propagation problems, namely, one-dimensional plane waves, spherical and cylindrical waves, longitudinal–transverse waves, waves in beams and plates, and plane two-dimensional stress waves. Wave propagation problems for elastic–plastic and elastic/viscoplastic media have been treated in detail, as well as the most important problem of shock waves in metals and soils. The last chapter deals with thermal wave propagation problems, relating to waves generated by thermal shocks applied to the boundary of the media under consideration.

Wave propagation problems in strings and membranes are not discussed in this book. The reader is referred for an exhaustive treatment to the monographs by Rakhmatulin and Demianov [125] and by Cristescu [34]. Since the time these books appeared the subject has not been changed except for some insignificant modifications. The theory of impact in inelastic solids, in which wave phenomena are not taken into account, have also been neglected in this book. These problems have been discussed in a number of monographs, for example in [118] (chapter V) and in [47] (chapter XV, where the theory of impact in discrete mechanical systems is discussed).

I am much indebted and thankful to my colleague Dr. Bogdan Raniecki from the Institute of Fundamental Technical Research of the Polish Academy of Sciences for his valuable suggestions and numerous discussions during the preparation of the book.

CHAPTER I

FUNDAMENTAL EQUATIONS OF THE DYNAMICS OF INELASTIC MEDIA

1. Dynamical properties of materials

The dynamical properties of materials will be briefly presented based on the results of experiments. The reader can find more details of the experiments in the monographs by Rakhmatulin and Demianov [125], Kolsky [71], Cristescu [34], and Perzyna [114] and [118].

The investigation of the behaviour of materials under dynamic conditions has taken place in several different directions. The determination of the dynamic characteristic (constitutive relationship) constitutes the primary direction of the investigations. Also, much research has been done in order to find the plastic strain distribution along the specimens tested and to investigate the influence of transverse motion during the propagation of longitudinal waves in a test piece which produces the effect of wave dispersion. The object of numerous experimental investigations has been to determine the influence of temperature and that of radiation on the behaviour of metals under dynamic conditions.

In the case of static processes, the mechanical properties of material are described by the stress–strain diagram. In the case of the dynamic loadings, however, the stress–strain relation is influenced by the strain rate. The influence of a number of other phenomena is also observed, e.g. temperature, radiation, etc.

The fundamental properties of metals and soils will be presented in this section.

1.1. DYNAMICAL PROPERTIES OF METALS

The experimental investigations of Clark and Duwez [25], Manjoine [84], Hauser, Simmons and Dorn [51], Campbell and Ferguson [22], Marsh and Campbell [88], Lindholm [76]–[78], and of many other authors reveal that much higher stress is necessary to reach the yield limit in an impact loading compared with that for a slow one. In the case of a number of practically important materials (e.g. high-carbon steels) a relationship explicitly independent of the strain rate [125] $\sigma = \sigma(\epsilon)$, obtained during dynamic loading of a specimen, can be used to characterize the dynamic behaviour of the materials. This relationship essentially differs from the static characteristic. The dynamic and static characteristics for a specimen made of mild steel are presented in Fig. 1. Curve 1 denotes the static case, while 2 indicates the dynamic one. Curve 2 has been drawn using the method due to Rakhmatulin [125], based on the measurement of plastic deformation of the initial cross-section of the specimen under the effect of the impact as a function of the impact speed. The yield limit of the material increases for dynamic loading of the specimen. The character of the curve is determined by the type of material. It has been found in many experimental investigations that those metals with a distinct yield limit are particularly sensitive to strain rate. A good example of this behaviour of exhibiting high sensitivity to strain rate is that of mild steel and pure iron. The influence of strain rate on the change of yield limit for mild steel was the subject of investigations in [25], [84], [76]–[78], [88], [24], and in other papers. For example, Clark and Duwez [25] found that for mild steel

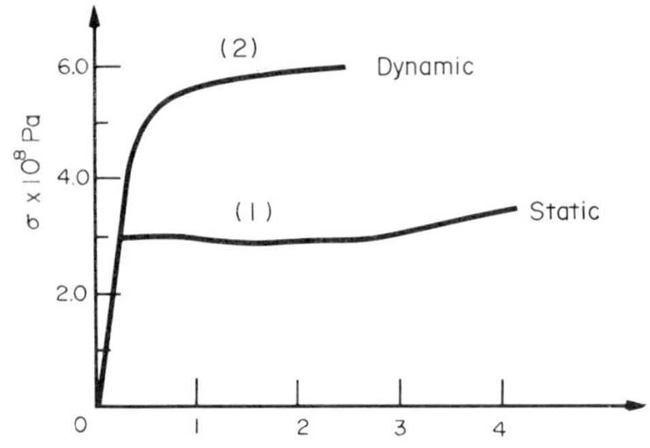

Fig. 1

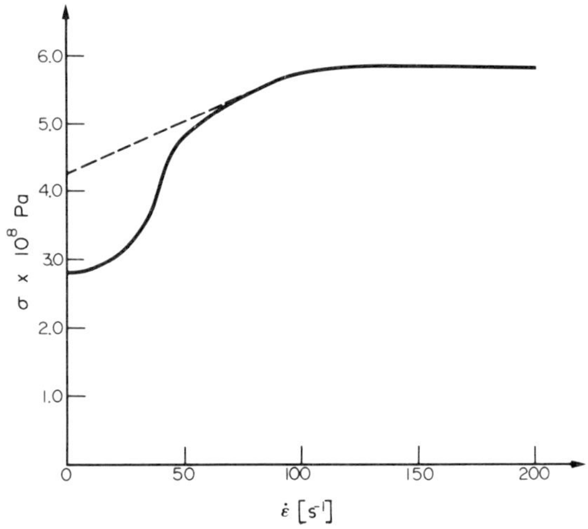

Fig. 2

(0.22% carbon) the yield limit increases with the increase of strain rate (Fig. 2) from about 2.71×10^8 Pa (the static yield limit) to a value of about 5.76×10^8 Pa for a strain rate of $\dot{\epsilon} \approx 200\ s^{-1}$. The yield limit increases up to the moment when it joins the curve representing the change of the conventional strength limit (dashed line in Fig. 2). The conventional strength limit increases in the range of strain rates from 0 to $\approx 200\ s^{-1}$.

The results of numerous experiments have shown that the yield limit for mild steel in dynamic loading situations can reach a value which is 2–3 times greater than that for static loading processes. It has also been noted that in a dynamic loading process the strain-hardening effect decreases in comparison with that for static loading.

Most of the investigations connected with the determination of dynamic material characteristics were performed under conditions of uniaxial tension or compression or for pure shear. Only a few experiments have been performed so far investigating the material characteristics in complex loading conditions. The investigations of Lindholm [76]–[78] belong to the pioneer ones in this field. The results of these experiments refer to aluminium and steel specimens under uniaxial tension and shear. The second stress tensor invariant against the second strain tensor invariant is shown in Fig. 3 for various strain rates [76]–[78]. The experiments were performed for a range of strain rates from $10^3\ s^{-1}$ to $4 \times 10^{-3}\ s^{-1}$. It is clear from the diagram that for higher strain rates the greater values of the stress intensity $\sqrt{J_2}$ correspond to a definite value of the strain intensity $\sqrt{I_2}$ = const. Lindholm [77] has performed identical experiments for steel specimens, obtaining similar results.

There already exists an extensive literature concerning the influence of strain rate as well as of temperature on the stress–strain relations. Each of these effects is considered separately. It has been found experimentally that at low temperatures the lower yield limit of metals does not depend on the strain rate, while at elevated temperatures even very small changes of strain rate cause a significant decrease or increase in the yield limit [22], [24].

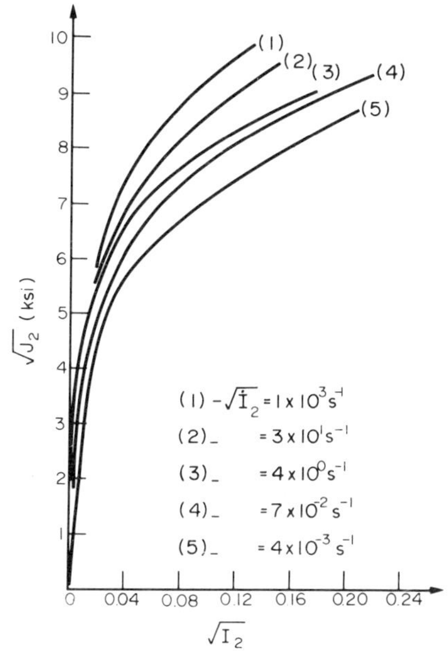

Fig. 3

The effect of radiation on the plastic properties of metals has also been investigated. As a rule the radiation of specimens leads to an increase of both yield limit and strength limit. The influence of strain rate is different for radiated metal than for unradiated. A radiated metal is considerably more sensitive to a change of strain rate [119].

1.2. DYNAMICAL PROPERTIES OF SOILS

In the case of quasi-static loading, soil settlement is produced by fluid seeping into and filling the cavities in the soil skeleton. Then the entire loading is transmitted by the soil skeleton and the soil deformation is entirely connected with the displacement of the soil skeleton grains. In the case of dynamic loading the soil behaves like a more homogeneous medium; the pressure in the fluid and air filling the soil skeleton is close to the pressure acting on the soil, particularly when the soil is very wet.

Compressibility is an important property of soils. For small loads of the order $10^4 - 5 \times 10^4$ Pa, the soil can be treated as an incompressible medium. The compressibility of soils differs significantly from that of metals, water, air, etc. When the stresses compressing a specimen increase the density of soil can increase significantly and the associated dilatational strain is irreversible due to the displacements of the soil grains and their crushing. During unloading this irreversible character of the processes mentioned is demonstrated by very slight changes of the density. It is noteworthy that in the case of dynamic loading the displacement of the soil grains due to their inertia lags behind the increase of pressure. This can be observed at the instant when the pressure attains its maximum value when the loading begins to develop. Though the pressure decreases a diminishing of the specimen volume still takes place during the first stage of unloading. This property for sand of different initial water content is shown in Fig. 4. The irreversibility of the dilatational compression decreases with the increase of water content and is caused by the elasticity of water filling the pores of the skeleton.

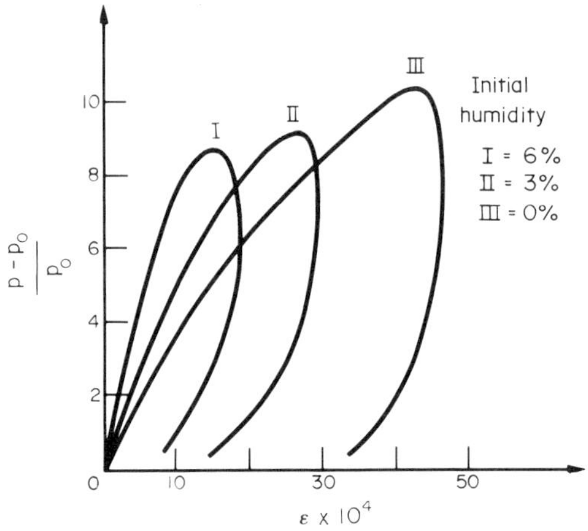

Fig. 4

Recent investigations have revealed that soil possesses rheological properties and is sensitive to change in the displacement rate. This sensitivity varies with the kind of soil. The papers of Grigorian [48], of Rakhmatulin, Sagomonian, and Alekseyev [126], and of other authors are devoted to the dynamic characteristics of soils. The influence of displacement-rate effects on soil behaviour is discussed in [126].

2. Basic theories of plasticity

The physical relationships of the basic theories of plasticity will now be presented including the strain theory of plasticity, the bilinear theory, and the theory of plastic flow. The rate effect on the stress–strain relations is not taken into account in these theories; nevertheless, they are frequently applied to the dynamic problems of plasticity due to the fairly good approximations that the theories provide for a definite class of cases.

2.1. STRAIN THEORY OF PLASTICITY

The constitutive equations of the theory of small elastic-plastic strains (equations of Nádai–Hencky–Iliushin) constitute a generalization of the physical relationships of the theory of elasticity and can be treated as a certain kind of extrapolation beyond the elastic state. The following three postulates are assumed in this theory, the first two being exactly the postulates of the theory of elasticity, namely:

(1) The principal directions of the stress tensor coincide with the principal directions of the strain tensor.
(2) Mean normal stress is proportional to the mean strain; the proportionality coefficient is the same in both elastic and plastic states.

(3) The stress intensity is a function of only the strain intensity, and should be determined for each material experimentally, i.e.

(2.1)
$$\sigma_i = 2m\left(\varepsilon_i\right)\varepsilon_i,$$

where $m(\epsilon_i)$ is a function of strain intensity only, and σ_i and ϵ_i denote the stress and strain intensities, respectively, and are defined by the formulae

$$\sigma_i = \left(\frac{3}{2} s_{ij} s_{ij}\right)^{1/2}, \qquad \varepsilon_i = \left(\frac{3}{2} e_{ij} e_{ij}\right)^{1/2}.$$

The system of equations for the strain theory takes the form [151]:

(2.2)
$$\tilde{s}_{ij} = \tilde{e}_{ij}, \qquad \sigma_i = 2m\left(\varepsilon_i\right)\varepsilon_i, \qquad \sigma_{kk} = 3K\varepsilon_{kk},$$

$$\tilde{s}_{ij} = \frac{s_{ij}}{\sigma_i}; \quad \tilde{e}_{ij} = \frac{e_{ij}}{\varepsilon_i}.$$

where $\tilde{s}_{ij}$ and $\tilde{e}_{ij}$ are the normalized deviatoric stress and strain tensors $K = \dfrac{1}{3}\left(2\mu + 3\lambda\right)$ and λ and μ denote the Lamé constants. After some algebra we obtain from (2.2) the constitutive equations

(2.3)
$$\sigma_{ij} = 2m\left(\varepsilon_i\right)\varepsilon_{ij} + \frac{1}{3}\left[3K - 2m\left(\varepsilon_i\right)\right]\varepsilon_{kk}\delta_{ij}.$$

Function $m(\epsilon_i)$ is determined on the basis of experimental data from relation (2.1). The relationship $\sigma_i = f(\epsilon_i)$ does not differ much from the relationship $\sigma = f(\epsilon)$ (Fig. 5) obtained for a uniaxial stress state. Function $f(\epsilon_i)$ can be obtained from function $f(\epsilon)$ by a change of scale of the coordinates (Fig. 5). The character of the diagram remains the same. In a uniaxial state of stress we obtain $\sigma_i = \sigma$, $\epsilon_i = (1 + \bar{\nu})\epsilon$, where $\bar{\nu}$ denotes the coefficient of transverse shrinkage.

The states described by (2.3) correspond to a loading process, i.e. to a process in which the stress intensity is an increasing function of time, i.e. for $d\epsilon_i/dt > 0$. If σ_i after reaching a certain value, e.g. $\sigma_i = \sigma_i^*$ (Fig. 5) at a point B, starts decreasing then unloading occurs. Then $d\epsilon_i/dt < 0$ and the unloading follows the straight line BC, parallel to Hooke's straight line OA.

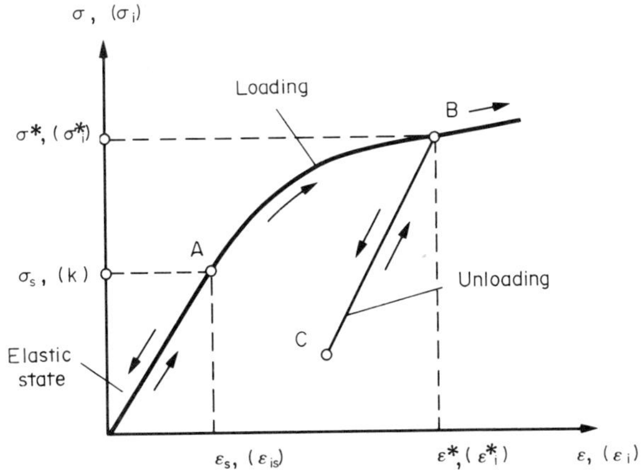

Fig. 5   Fig. 6

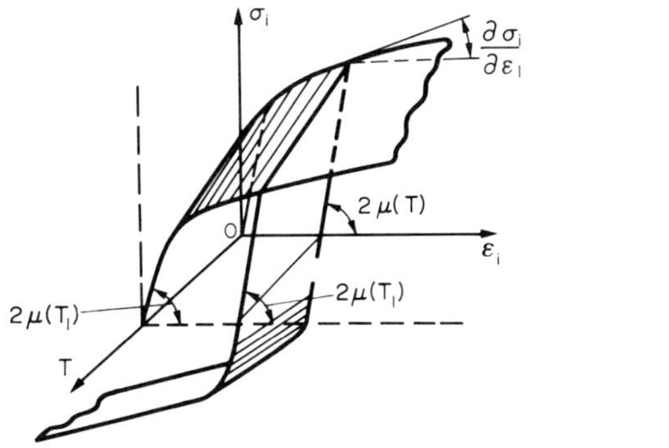

The physical equations for unloading take the form

$$s_{ij}^{*}-s_{ij}=2\mu\,(e_{ij}^{*}-e_{ij}),\qquad \sigma_{kk}^{*}-\sigma_{kk}=3K\,(\varepsilon_{kk}^{*}-\varepsilon_{kk}); \tag{2.4}$$

the asterisks denote the state of stress and strain corresponding to the initial unloading.

In the theory of small elastic–plastic strains, when the temperature field is taken into account, it is assumed that the stress intensity is a function of the strain intensity and temperature T:

$$\sigma_i = 2m\,(\varepsilon_i, T)\,\varepsilon_i, \tag{2.5}$$

and

$$\sigma_{kk} = 3K\varepsilon_{kk}-9K\alpha T, \tag{2.6}$$

where α denotes the coefficient of thermal dilatational expansion. The function $m(\epsilon_i, T)$ is determined from simple tension experiments on cylindrical specimens at different temperatures. The function $m(\epsilon_i, T)$ represents a certain surface in the $(\sigma_i, \epsilon_i, T)$ space (Fig. 6) [150]. Within the elastic strain range $\left(\sigma_i \leqslant \sigma_s(T)\dfrac{1}{\sqrt{3}}\right.$, where $\sigma_s(T)$ denotes temperature dependent elastic $\left.\text{limit}\right)$ the surface becomes a plane, the equation of which takes the form

$$\sigma_i = 2\mu\,(T)\,\varepsilon_i, \tag{2.7}$$

where $\mu(T)$ denotes the temperature dependent shear modulus.

An active loading process at an arbitrary point in the medium is represented by a curve on the surface (2.5) in the $(\sigma_i, \epsilon_i, T)$ space. The unloading deformation process in this space is represented by a curve in a plane parallel to the plane (2.7):

$$\sigma_i = 2\mu\,(T)\,(\varepsilon_i-\varepsilon_i^{*}). \tag{2.8}$$

Such a geometrical interpretation of the deformation process leads us to ascertain that active loading at each element of a body (when the mechanical characteristics are dependent on temperature) takes place for the condition $d\epsilon_i/dt > 0$ while unloading occurs for the condition $d\epsilon_i/dt \leqslant 0$.

The constitutive equations of the strain theory for small strains therefore take the following form:

for active plastic loading:

$$\sigma_{ij} = 2m\,(\varepsilon_i, T)\,\varepsilon_{ij}+\frac{1}{3}\,[3K-2m\,(\varepsilon_i, T)]\;\varepsilon_{kk}\,\delta_{ij}-9K\alpha T\delta_{ij} \tag{2.9}$$

and in an unloading zone:

$$\sigma_{ij}^{*}-\sigma_{ij} = 2\mu\,(T)\,(\bar{e}_{ij}^{*}-e_{ij})+K\,(T)\frac{\mu\,(T_1)}{\mu\,(T)}\,(\varepsilon_{kk}^{*}-\alpha T)\,\delta_{ij}-K\,(T)\,(\varepsilon_{kk}-\alpha T)\delta_{ij} \tag{2.10}$$

where the asterisk denotes the corresponding values at a moment when the unloading process begins (i.e. when $d\epsilon_i/dt = 0$), $\bar{e}^*_{ij} = e^{*p}_{ij} + \frac{\mu(T_1)}{\mu(T)} e^{*e}_{ij}$, indices e and p denote the elastic and plastic components of the deviatoric strain tensor $e_{ij} = e^e_{ij} + e^p_{ij}$.

If the material constants do not depend on temperature the physical equations of the strain theory with the temperature field taken into account take the form:

in a zone of active plastic loading ($d\epsilon_i/dt > 0$):

$$\sigma_{ij} = 2m(\varepsilon_i)\,\varepsilon_{ij} + \frac{1}{3}[3K - 2m(\varepsilon_i)]\,\varepsilon_{kk}\,\delta_{ij} - 9K\alpha T\delta_{ij} \tag{2.11}$$

while in a zone of unloading ($d\epsilon_i/dt \leqslant 0$):

$$s^*_{ij} - s_{ij} = 2\mu(e^*_{ij} - e_{ij}), \qquad \varepsilon_{kk} = \frac{\sigma_{kk}}{3K} + 3\alpha T. \tag{2.12}$$

2.2. BILINEAR THEORY

The constitutive equations for a complex stress state [1] take into account the material plastic compressibility in such a way that the stress–strain relations for the plastic body are bilinear. For the active plastic loading $\left(dJ_2/dt > 0,\ \text{where } J_2 = \frac{1}{2} s_{ij}\, s_{ij} \text{ denotes the second invariant of the deviatoric tensor}\right)$ these relations take the following form:

$$\begin{aligned} e_{ij} &= \frac{1}{2\mu_1} s_{ij} + \left(\frac{1}{2\mu_2} - \frac{1}{2\mu_1}\right)(s_{ij} - s^0_{ij}), \\ \varepsilon_{ii} &= \frac{1}{3K_1}\sigma_{ii} + \left(\frac{1}{3K_2} - \frac{1}{3K_1}\right)(\sigma_{ii} - \sigma^0_{ii}), \end{aligned} \tag{2.13}$$

where σ^0_{ij} is the initial stress tensor corresponding to the transition point of material into plastic state, μ_1 and K_1 are the shear and bulk moduli, respectively, and μ_2 and K_2 denote the material constants in the plastic state:

$$\mu_2 = \frac{E_2}{2(1+\nu_2)}, \qquad 3K_2 = \frac{E_2}{1 - 2\nu_2}. \tag{2.14}$$

In the case of uniaxial tension the coefficient ν_2 is defined as the ratio of the transverse deformation in the plastic zone to the longitudinal deformation also in the plastic zone:

$$\nu_2 = -\frac{\varepsilon^p_2}{\varepsilon^p_1}. \tag{2.15}$$

Thus ν_2 and μ_2 in (2.14) play a role of Poisson's ratio and shear modulus respectively in the plastic range.

For a uniaxial state of stress the equations of bilinear theory become those of the strain theory. The application of the bilinear theory in problems involving complex stress states has the advantage, when compared with the other theories of plasticity, that the equations can be integrated in both elastic and plastic states (since both the deviatoric and spherical parts of the stress and strain tensors are linearly related in both elastic and plastic zones). This virtue of the theory enables us to construct effectively solutions to many boundary value problems; on the other hand, the theory contains certain simplifications of a physical nature.

2.3. THEORY OF PLASTIC FLOW

The physical equations for the theory of plastic flow will be presented both for the case of an ideally plastic material and for a material with strain hardening. We shall present the yield conditions and the conditions characterizing the state of medium whether we have a loading process, an unloading process, or a neutral state.

Ideally plastic material. In a complex state of stress the onset of deformation in the plastic material is determined by the yield condition which in stress space is a smooth, convex surface called the yield surface. At each point of the surface there exists a unique tangent plane.

The yield condition is given by

$$F(\sigma_{ij}) = 0, \tag{2.16}$$

where F is an even function with respect to σ_{ij} such that the condition $F < 0$ determines the elastic state and $F = 0$ the plastic state (Fig. 7). It is assumed that there exists function $F_1 = F_1(\epsilon_{ij})$ playing the role of a potential for the rate of plastic strain. Usually function F_1 is identified with the function F constituting the yield condition. Thus

$$\dot{\varepsilon}^p_{ij} = \lambda \frac{\partial F}{\partial \sigma_{ij}}, \tag{2.17}$$

where $\dot{\varepsilon}^p_{ij}$ denotes the plastic part of the strain rate and λ is a parameter ($\lambda > 0$).

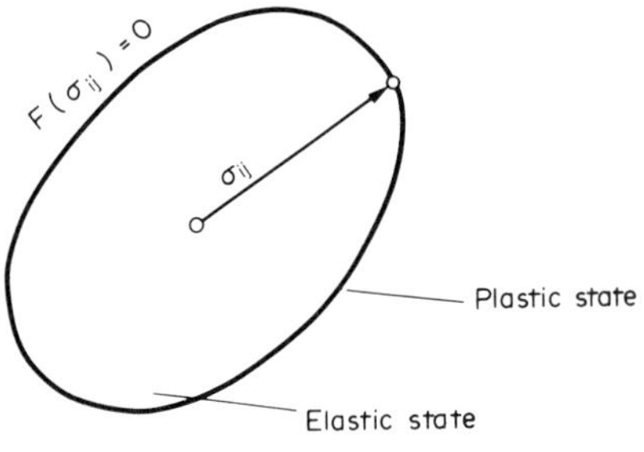

Fig. 7

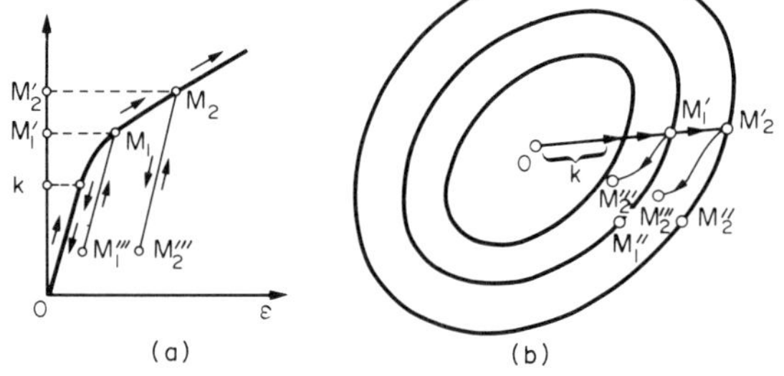

Fig. 8

If $F = 0$ then the material is in a plastic state; the existence of plastic strain increments is determined by whether the given state corresponds to loading or unloading:

the loading state of an ideally plastic material takes place if

$$\frac{\partial F}{\partial \sigma_{ij}} \dot{\sigma}_{ij} = 0, \tag{2.18}$$

while the state of unloading if

(2.19) $$\frac{\partial F}{\partial \sigma_{ij}} \dot{\sigma}_{ij} < 0.$$

The physical relationships for the ideally plastic media are obtained by adding to (2.17) the elastic strain rates $\left(\dot{e}^{e}_{ij} = \frac{1}{2\mu} \dot{s}_{ij}, \dot{\varepsilon}_{kk} = \frac{1}{3K} \dot{\sigma}_{ii}\right)$ and by taking into account the fact that the plastic strain has no effect on the volume change $\epsilon^{p}_{kk} = 0$:

(2.20) $$\dot{e}_{ij} = \frac{\dot{s}_{ij}}{2\mu} + \lambda \frac{\partial F}{\partial \sigma_{ij}}, \qquad \dot{\varepsilon}_{kk} = \frac{1}{3K} \dot{\sigma}_{kk}.$$

Assuming, in the special case, the Huber–Mises yield condition

(2.21) $$F(\sigma_{ij}) = \frac{1}{2} s_{ij} s_{ij} - k^2 = 0,$$

where k denotes the yield limit for pure shear, one obtains

(2.22) $$\dot{e}_{ij} = \frac{\dot{s}_{ij}}{2\mu} + \lambda s_{ij}, \qquad \dot{\varepsilon}_{kk} = \frac{1}{3K} \dot{\sigma}_{kk},$$

called the Prandtl–Reuss equations. Here the parameter λ is determined as $\lambda = \dot{W}^{p}/2k^2$, where $\dot{W}^{p}$ denotes the power of plastic strain of a body element of unit volume.

Material hardening. Suppose that for a uniaxial stress state the $\sigma - \epsilon$ diagram is assumed as given in Fig. 8a; thus the hardening phenomenon is taken into account. Generalizing the property of hardening during loading to the case of a complex stress state, the notion of the existence of consecutive yield surfaces is introduced in the nine-dimensional stress space (Fig. 8b). The consecutive yield surfaces correspond to points M_i' on the axis $\epsilon = 0$. The surfaces can be symmetric and their form can be the same as the initial yield surface. In this case, during the loading process, isotropic expansion of the yield surface takes place, i.e. isotropic hardening occurs. Unloading processes correspond to changes of stress state along the path $M_i'M_i'''$, while neutral states correspond to the changes along the path $M_i'M_i''$ (without any additional increase of plastic strains).

In the general case the yield condition, for a material exhibiting hardening, takes the form

(2.23) $$F(\sigma_{ij}, \varepsilon^{p}_{ij}, \varkappa) = 0,$$

where $\varkappa$ denotes the hardening parameter, defined by the plastic work,

(2.24) $$\varkappa = W^{p} = \int_{0}^{t} \sigma_{ij}\, d\varepsilon^{p}_{ij}.$$

Thus the parameter $\varkappa$ does not affect the form of the yield surface during the plastic straining process. The increase in its value causes gradually the expansion of the yield surface. For a material which has not yet experienced plastic strain, we obviously obtain $\varkappa = 0$.

The process of active loading from a plastic state to another plastic state is accompanied by a plastic strain, then

$$\frac{\partial F}{\partial \sigma_{ij}} \dot{\sigma}_{ij} > 0, \qquad F = 0. \tag{2.25}$$

For unloading no increase of plastic strain occurs, therefore $\dot{\varkappa} = 0$, and the unloading condition takes the form

$$\frac{\partial F}{\partial \sigma_{ij}} \dot{\sigma}_{ij} < 0, \qquad F = 0. \tag{2.26}$$

The neutral state is characterized by the condition

$$\frac{\partial F}{\partial \sigma_{ij}} \dot{\sigma}_{ij} = 0, \qquad F = 0. \tag{2.27}$$

All these conditions hold only for stable materials.

In what follows we shall use the yield condition for a material exhibiting the property of hardening in the following form,

$$F(\sigma_{ij}, \varepsilon_{ij}^{p}) = 0. \tag{2.23'}$$

The physical relationship takes the form

$$\dot{\varepsilon}_{ij}^{p} = \lambda \frac{\partial F}{\partial \sigma_{ij}} \tag{2.28}$$

(it results from the assumption that the direction of strain rate vector $\dot{\epsilon}_{ij}^{p}$ is orthogonal to the yield surface and independent of the direction of stress rate tensor $\dot{\sigma}_{ij}$), where λ is a non-negative function defined as

$$\lambda = -\frac{\dfrac{\partial F}{\partial \sigma_{ij}} \dot{\sigma}_{ij}}{\dfrac{\partial F}{\partial \varepsilon_{kl}^{p}} \dfrac{\partial F}{\partial \sigma_{kl}}} \tag{2.29}$$

Since, during the loading process, the numerator of expression (2.29) is positive, it follows, by virtue of (2.25), that $(\partial F/\partial \epsilon_{ij}^{p})(\partial F/\partial \epsilon_{ij}) < 0$.

Generally, the physical equations describing an isotropically hardening material can be written down in the form [151]:

$$\dot{\varepsilon}_{ij} = A_{ijkl} \dot{\sigma}_{kl} \qquad \text{or} \qquad \dot{\sigma}_{ij} = B_{ijkl} \dot{\varepsilon}_{kl}, \tag{2.30}$$

where

$$A_{ijkl} = \begin{cases} H_{ijkl} + h \dfrac{\partial F}{\partial \sigma_{ij}} \dfrac{\partial F}{\partial \sigma_{kl}}, & \text{if} \quad F = 0 \text{ and } \dfrac{\partial F}{\partial \sigma_{kl}} \dot{\sigma}_{kl} > 0, \\ H_{ijkl}, \quad \text{if} & F < 0 \text{ or } F = 0, \text{ but then } \dfrac{\partial F}{\partial \sigma_{kl}} \dot{\sigma}_{kl} \leqslant 0, \end{cases}$$

$$(2.31)\qquad B_{ikjl} = \begin{cases} H^{-1}_{ijkl} - \dfrac{h\dfrac{E}{1+\nu}\dfrac{\partial F}{\partial \sigma_{ij}}\dfrac{\partial F}{\partial \sigma_{kl}}}{\dfrac{1+\nu}{E}+h\left(\dfrac{\partial F}{\partial \sigma_{mn}}\dfrac{\partial F}{\partial \sigma_{mn}}\right)}, & \text{if} \quad F=0 \text{ and } \dfrac{\partial F}{\partial \sigma_{kl}}\dot{\varepsilon}_{kl} > 0, \\ H^{-1}_{ijkl}, \quad \text{if} \quad F<0 \text{ or } F=0, \text{ but then } \dfrac{\partial F}{\partial \sigma_{kl}}\dot{\varepsilon}_{kl} \leqslant 0. \end{cases}$$

At the same time

$$(2.32)\qquad H_{ijkl} = \frac{1+\nu}{E}\delta_{ik}\delta_{jl} - \frac{\nu}{E}\delta_{ij}\delta_{kl}.$$

Function h, occurring in (2.31), is a scalar function that can be dependent on the current stress state and the loading history $h = -1\Big/\left(\dfrac{\partial F}{\partial \varepsilon_{ij}}\dfrac{\partial F}{\partial \sigma_{ij}}\right)$.

3. Strain rate sensitive plastic media

3.1. ELASTIC/VISCOPLASTIC MEDIUM

Numerous experimental investigations of the dynamic properties of plastic materials reveal that many materials behave differently under static and dynamic loading. It has been shown that the main reason is due to strain rate sensitivity of the materials.

According to Naghdi and Murch [91] we distinguish between elastic–viscoplastic and elastic/viscoplastic media. An elastic–viscoplastic material is one that exhibits distinct viscous properties in both elastic and plastic regimes. By the term elastic/viscoplastic medium we mean a material that possesses viscous properties only in plastic regimes, and is ideally elastic. The notion of an elastic/viscoplastic medium constitutes an idealization that enables us to simplify the discussion, first of all when discussing the problem of determining the yield criterion. Both media are discussed in detail by Perzyna in his monograph [114]. We confine ourselves to a presentation of the constitutive equations for elastic/viscoplastic media only in the case of small strains. Investigations in this direction were commenced by Hohenemser and Prager [52]. Following this the idea was developed in papers for one-dimensional problems by Sokolovskii [147] and Malvern [83], who proved that, based on the assumptions of Hohenemser and Prager, certain dynamic properties of strain rate sensitive media can be described. The theory of Hohenemser and Prager was generalized by Perzyna [115], [116].

We present the constitutive equations derived by Perzyna [115] for strain rate sensitive media.

The initial yield condition (since the elastic region does not exhibit viscous properties) does not differ from known criteria from the classical theory of plasticity (compare p. 8).

It is assumed that the yield function takes the form

$$(3.1)\qquad F(\sigma_{ij}, \varepsilon^p_{ij}) = \frac{f(\sigma_{ij}, \varepsilon^p_{ij})}{\varkappa} - 1,$$

where $f(\sigma_{ij}, \epsilon_{ij}^p)$ depends on the stress state σ_{ij} and on the plastic strain state ϵ_{ij}^p (it is assumed that the strain rate can be decomposed into elastic and inelastic parts $\dot{\epsilon}_{ij} = \dot{\epsilon}_{ij}^e + \dot{\epsilon}_{ij}^p$, where $\dot{\epsilon}_{ij}^p$ represents the coupling of the viscous and plastic effects). $\varkappa$ denotes the hardening parameter defined by

$$\varkappa = \varkappa(W^p) = \varkappa\left(\int_0^t \sigma_{kl}\, d\varepsilon_{kl}^p\right); \tag{3.2}$$

W^p denotes the plastic strain energy of the material.

It is assumed that the yield surface $F = 0$, considered in nine-dimensional stress space, is regular and convex.

Perzyna [116], taking into account the above assumptions, has proposed, for strain rate sensitive materials, the following constitutive equations:

$$\dot{\varepsilon}_{ij} = \frac{1}{2\mu}\dot{s}_{ij} + \frac{1-2\nu}{E}\dot{\sigma}_{kk}\,\delta_{ij} + \gamma\langle\Phi(F)\rangle\frac{\partial f}{\partial\sigma_{ij}}, \tag{3.3}$$

where ϵ_{ij}, s_{ij} are the components of the strain tensor and of the deviatoric stress tensor respectively. E denotes the Young modulus, ν is the Poisson ratio, γ is the coefficient of material viscosity, Φ is generally a non-linear function of the argument F, and the symbol $\langle\Phi(F)\rangle$ is defined in the following way:

$$\langle\Phi(F)\rangle = \begin{cases} 0 & \text{for} \quad F \leqslant 0, \\ \Phi(F) & \text{for} \quad F > 0. \end{cases} \tag{3.4}$$

Function $\Phi(F)$ is determined from the results of experimental investigations, concerned with the dynamic properties of the material. A pertinent choice of the function permits a description of the strain rate effect on the yield limit.

The constitutive equations (3.3) can be also represented in another form, namely

$$\dot{e}_{ij} = \frac{1}{2\mu}\dot{s}_{ij} + \gamma\langle\Phi(F)\rangle\frac{\partial f}{\partial\sigma_{ij}}, \qquad \dot{\varepsilon}_{ii} = \frac{1}{3K}\dot{\sigma}_{ii}, \tag{3.5}$$

where e_{ij} are the deviatoric strain tensor components and K is the bulk modulus.

It is apparent from the above equations that the inelastic strain rate is a function of the stress state, i.e. of the difference between the real state and the state corresponding to the static yield condition. The function of the stress state difference determines, in accordance with Maxwell's viscosity law, the inelastic strain rate. On the other hand, the elastic strain tensor components are independent of the strain rate. Material hardening is also taken into account in the constitutive equations (3.3) or (3.5). Both isotropic as well as anisotropic hardening can be described by the function F.

For the sake of a more detailed analysis of constitutive equations (3.3) or (3.5) we consider the inelastic part of the strain rate

$$\dot{\varepsilon}_{ij}^p = \gamma\Phi(F)\frac{\partial f}{\partial\sigma_{ij}}. \tag{3.6}$$

After some simple algebra we obtain, from (3.6), the relationship representing the dynamic yield condition for an elastic/viscoplastic material exhibiting isotropic or anisotropic hardening,

(3.7) $$f(\sigma_{ij}, \varepsilon_{ij}^p) = \varkappa(W^p)\left\{1 + \Phi^{-1}\left[\frac{(I_2^p)^{1/2}}{\gamma}\left(\frac{1}{2}\frac{\partial f}{\partial \sigma_{kl}}\frac{\partial f}{\partial \sigma_{kl}}\right)^{-1/2}\right]\right\},$$

where Φ^{-1} denotes the inverse function corresponding to Φ, I_2^p is the second invariant of the inelastic strain rate tensor, $I_2^p = \frac{1}{2}\dot{\varepsilon}_{ij}^p \dot{\varepsilon}_{ij}^p$.

Equation (3.7) determines the change of the current yield surface during a dynamic process involving inelastic straining. The change of the current yield surface is due to the isotropic or anisotropic hardening of the material as well as due to rheological effects which exert their influence through the strain rate.

A strain rate sensitive plastic material exhibiting isotropic hardening is a particular case of an elastic/viscoplastic material. Let us assume that the function F has the following form,

(3.8) $$F = \frac{f(\sigma_{ij})}{\varkappa} - 1,$$

i.e. function f depends solely on the stress state. Confining the discussion to the Huber–Mises yield condition, i.e. assuming that

(3.9) $$f(\sigma_{ij}) = (J_2)^{1/2},$$

where J_2 denotes the second invariant of the stress deviator, we obtain from (3.5) the equations

(3.10) $$\dot{e}_{ij} = \frac{1}{2\mu}\dot{s}_{ij} + \frac{\gamma}{2}\left\langle \Phi\left(\frac{\sqrt{J_2}}{\varkappa} - 1\right)\right\rangle \frac{s_{ij}}{\sqrt{J_2}}, \qquad \dot{\varepsilon}_{ii} = \frac{1}{3K}\dot{\sigma}_{ii}.$$

The dynamic yield condition assumes the form

(3.11) $$\sqrt{J_2} = \varkappa(W^p)\left[1 + \Phi^{-1}\left(\frac{\sqrt{I_2^p}}{\gamma}\right)\right].$$

In the case of a uniaxial stress state, (3.10) take the form of known relations proposed by Malvern [83]:

(3.12) $$\dot{\varepsilon} = \frac{\dot{\sigma}}{E} + \gamma^* \left\langle \Phi\left[\frac{\sigma}{f_1(\varepsilon^p)} - 1\right]\right\rangle,$$

where ϵ denotes strain, σ is the stress, the viscosity constant is defined as $\gamma^* = \gamma/\sqrt{3}$, and $f_1(\epsilon^p)$ is the static material characteristic for simple tension. The relationship (3.12) is presented in Fig. 9a. For $\dot{\epsilon}$ = const the curves are parallel to the static curve.

It should be pointed out that for very high rates of stress and strain changes the non-linear term in (3.12) $\gamma^* \left\langle \Phi\left(\frac{\sigma}{f(\varepsilon^p)} - 1\right)\right\rangle$ is negligible with respect to the terms $\dot{\epsilon}$ and $\dot{\sigma}/E$. Then the medium behaves like a linearly elastic body. On the other hand, for small rates of stress and strain changes it results from (3.12) that $\sigma = f_1(\epsilon^p)$, i.e. the motion of the medium is described by the static characteristic of the material.

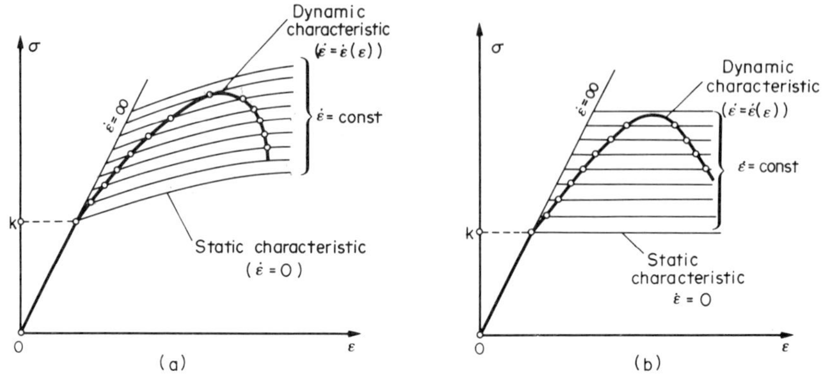

Fig. 9

Kaliski [55] has proposed a certain modification of the constitutive equations (3.10) for isotropic hardening. The equations take the following form:

$$\dot{e}_{ij} = \frac{1}{2\mu}\dot{s}_{ij} + \gamma \left\langle \Phi\left(\frac{\sigma_i}{f_1(\varepsilon_i)} - 1\right)\right\rangle \frac{s_{ij}}{\sigma_i}, \qquad \dot{\varepsilon}_{ii} = \frac{1}{3K}\dot{\sigma}_{ii}. \tag{3.13}$$

Function $f_1(\epsilon_i)$ represents the static stress intensity against the strain intensity for a material with hardening (Fig. 10). Here a hypothesis was assumed that the relationship $\sigma_i = f_1(\epsilon_i)$ for a complex stress state is identical to equation $\tau = f_1(\gamma)$ for pure shear.

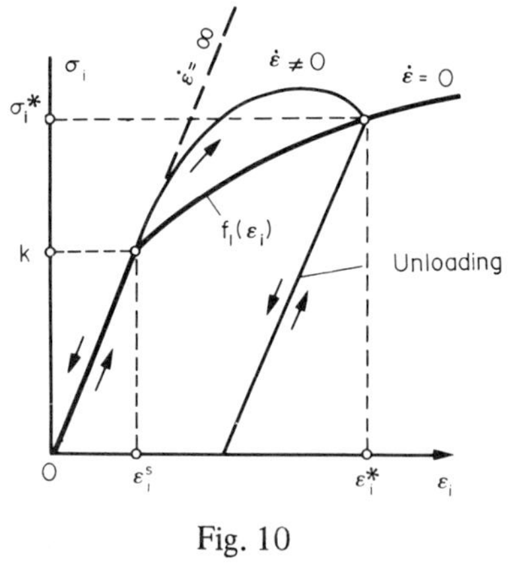

Fig. 10

The difference between (3.10) and (3.13) consists in the definition of parameter $\varkappa$. Constitutive equations (3.10) lead, in the limiting case of $\gamma = \infty$, to the theory of plastic flow as it was described in [114]. In the case of a uniaxial stress state (3.13) become the relations (3.12) of Malvern. In the case when the loading process is slow in time, (3.13) become the strain theory equations.

If the argument of function Φ in (3.13) is assumed to take the form $F = (\sigma_i/f_1(\epsilon_i)-1)$ (like in the strain theory), then the mathematical analysis is considerably simplified as compared with using $\varkappa$ defined by (3.2).

Function $\Phi(F)$ occurring in (3.10), (3.12), and (3.13) will be determined at p. 19.

The constitutive equations for elastic/viscoplastic media, presented up to this point, were given in a cartesian coordinate system. Now we formulate constitutive equations (3.3) in a curvilinear coordinate system in terms of the contravariant components of the stress tensor and the strain rate tensor, using a natural basis.

Equations (3.3) take the following form:

$$d^{ij} = \frac{1}{2\mu}\dot{s}^{ij} + \frac{1-2\nu}{E}\dot{\sigma}^k_k g^{ij} + \gamma\langle\Phi(F)\rangle\frac{\partial f}{\partial\sigma_{ij}}, \tag{3.14}$$

where

$$s^{ij} = \sigma^{ij} - \frac{1}{3}\sigma^k_k g^{ij}, \quad \sigma^k_k = g_{km}\sigma^{mk}, \tag{3.15}$$

d^{ij} denotes the strain rate tensor, defined in the following manner:

$$d^k_l = \frac{1}{2}\left(\frac{Dv^k}{\partial q^l} + \frac{Dv^l}{\partial q^k}\right), \tag{3.16}$$

where $v^i = \dfrac{Dq^i}{dt}$ denotes the contravariant components of the velocity vector, $\dfrac{Dv^i}{\partial q^j}$ denotes the contravariant derivative of the velocity vector, defined as follows:

$$\frac{Dv^k}{\partial q^l} = \frac{\partial v^k}{\partial q^l} + \Gamma^k_{jl}v^j, \tag{3.17}$$

q^i are the coordinates of a point in the curvilinear coordinate system, g^{ij} are the components of the metric tensor and Γ^k_{ij} denotes the Christoffel symbols of the second kind (compare p.28).

In the case of small strains, the strain tensor takes the following form in orthogonal curvilinear coordinates:

$$\begin{aligned} \epsilon_{ij} &= \frac{1}{2}\left(g_{\underline{ii}}\frac{\partial u^i}{\partial q^j} + g_{\underline{jj}}\frac{\partial u^j}{\partial q^i}\right), \text{ for } i \neq j, \\ \epsilon_{ii} &= \frac{\partial u^i}{\partial q^i} + \frac{1}{2}\frac{\partial g_{\underline{ii}}}{\partial q^k}u^k, \text{ for } i = j,\ i \text{ not summed,} \end{aligned} \tag{3.18}$$

where $\mathbf{u}$ is the displacement vector. The covariant components of the strain tensor are defined by the equation

$$\epsilon^{ij} = g^{im}g^{jn}\epsilon_{mn}. \tag{3.19}$$

The physical components of the stress tensor $\sigma^{(ij)}$, the physical components of the strain tensor $\epsilon^{(ij)}$, and the physical components of the velocity vector $v^{(i)}$ are defined by the equations

$$\sigma^{(ij)} = \sigma^{\underline{ij}} h_{\underline{i}} h_{\underline{j}}, \quad \epsilon^{(ij)} = \epsilon^{\underline{ij}} h_{\underline{i}} h_{\underline{j}}, \quad v^{(i)} = v^{\underline{i}} h_{\underline{i}}, \tag{3.20}$$

where $h_i = \sqrt{g_{\underline{ii}}}$.

3.2. ELASTIC/VISCO-IDEALLY-PLASTIC MEDIA

The constitutive equations describing elastic/visco-ideally-plastic media can be obtained if we assume that function F is independent of the strains, i.e.

$$F = \frac{f(J_2, J_3)}{c} - 1, \tag{3.21}$$

where J_3 is the third invariant of the deviatoric stress tensor, while c denotes a material constant.

In this case, the constitutive equations take the form

$$\dot{e}_{ij} = \frac{1}{2\mu} \dot{s}_{ij} + \gamma \left\langle \Phi\left[\frac{f(J_2, J_3)}{c} - 1\right]\right\rangle \frac{\partial f}{\partial \sigma_{ij}}, \quad \dot{\varepsilon}_{ii} = \frac{1}{3K} \dot{\sigma}_{ii}. \tag{3.22}$$

The dynamic yield condition for the function in the form (3.21) reads as follows:

$$f(J_2, J_3) = c\left\{1 + \Phi^{-1}\left[\frac{\sqrt{I_2^p}}{\gamma}\left(\frac{1}{2}\frac{\partial f}{\partial \sigma_{pq}}\frac{\partial f}{\partial \sigma_{pq}}\right)^{-1/2}\right]\right\}. \tag{3.23}$$

Equations (3.22) were proposed by Perzyna [115].

If we assume that the function F takes the form

$$F = \frac{\sqrt{J_2}}{k} - 1, \tag{3.24}$$

where k denotes the yield limit for pure shear, then the constitutive equations (3.22) take the following form:

$$\dot{e}_{ij} = \frac{1}{2\mu} \dot{s}_{ij} + \frac{\gamma}{2}\left\langle \Phi\left[\frac{\sqrt{J_2}}{k} - 1\right]\right\rangle \frac{s_{ij}}{\sqrt{J_2}}, \quad \dot{\varepsilon}_{ii} = \frac{1}{3K} \dot{\sigma}_{ii}. \tag{3.25}$$

Then the dynamic yield condition takes the form

$$\sqrt{J_2} = k\left[1 + \Phi^{-1}\left(\frac{\sqrt{I_2^p}}{2\gamma}\right)\right]. \tag{3.26}$$

The relationship between $\sqrt{J_2}$ and $\sqrt{I_2^p}$, given by (3.26), is plotted in Fig. 11 for an arbitrary function F.

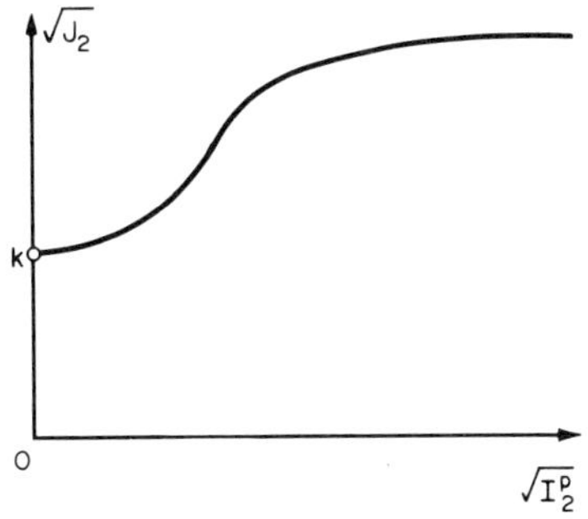

Fig. 11

For an elastic/viscoplastic medium, on account of the fact that J_2 is a function of strain rate, the viscoplastic strain occurs, provided $J_2 > k^2$, independently of the sign of the time derivative $\dot{J}_2$. The sign of the derivative $\dot{J}_2$ is essential for the determination of the medium state (loading, unloading, neutral state) in the strain theory of plasticity (compare section 2.1).

In the particular case when the function $\Phi(F)$ is assumed to be linear, i.e. $\Phi(F) = F$, we obtain from (3.25) the equations introduced by Freudenthal [41]. If it is also assumed that the elastic strains are small compared with the inelastic strains, one obtains the constitutive equations of Hohenemser and Prager [52].

The constitutive equations describing an elastic/visco-ideally-plastic medium (3.22) in terms of the contravariant components (in a curvilinear system of coordinates) and the natural basis take the following form:

$$d^{ij} = \frac{1}{2\mu}\dot{s}^{ij} + \frac{1-2\nu}{3E}\dot{\sigma}_k^k g^{ij} + \gamma \langle \Phi\left(\frac{\sqrt{J_2}}{k} - 1\right)\rangle \frac{s^{ij}}{\sqrt{J_2'}}, \tag{3.27}$$

where

$$J_2 = \frac{1}{2}s^{ij}s_{ij} = \frac{1}{2}s^{ij}s^{kl}g_{ki}g_{lj}, \qquad \dot{\sigma}_k^k = g_{km}\dot{\sigma}^{mk}.$$

In the case of a uniaxial stress state (3.25) are reduced to the form

$$\dot{\varepsilon} = \frac{\dot{\sigma}}{E} + \gamma^* \langle \Phi\left(\frac{|\sigma|}{\sigma_s} - 1\right)\rangle \text{sign}\ \sigma, \tag{3.28}$$

where σ_s denotes the yield limit for simple tension. This model was introduced by Sokolovskii [147]. Relationship (3.28) is presented in Fig. 9b. For $\dot{\epsilon}$ = const (3.28) represents straight lines parallel to the static characteristic.

In order to explain the essence of the above model, we mention here the following example [147]. Let us assume that σ increases linearly with time for $0 \leqslant t \leqslant t_0$, reaching the value p_m, and then decreases linearly to zero in time $t_0 \leqslant t \leqslant t_1$, i.e.

$$\sigma = \begin{cases} p_m \dfrac{t}{t_0} & \text{for} \quad 0 \leqslant t \leqslant t_0, \\ p_m \left(1 - \dfrac{t - t_0}{t_1 - t_0}\right) & \text{for} \quad t_0 \leqslant t \leqslant t_1, \end{cases}$$

where $p_m > \sigma_s > 0$.

Eliminating time from (3.28) and integrating, we obtain for the "loading" process

$$\varepsilon = \begin{cases} \dfrac{\sigma}{E} & \text{for} \quad 0 \leqslant \sigma \leqslant \sigma_s \\ \dfrac{\sigma}{E} + \dfrac{\gamma^* \sigma_s t_0}{2p_m}\left(\dfrac{\sigma}{\sigma_s} - 1\right)^2 & \text{for} \quad \sigma_s \leqslant \sigma \leqslant p_m, \end{cases}$$

while for the "unloading" process we have

$$\varepsilon - \varepsilon_r = \begin{cases} \dfrac{\sigma}{E} - \dfrac{\gamma^*(t_1 - t)\sigma_s}{2p_m}\left(\dfrac{\sigma}{\sigma_s} - 1\right)^2 & \text{for} \quad \sigma_s \leqslant \sigma \leqslant p_m, \\ \dfrac{\sigma}{E} & \text{for} \quad 0 \leqslant \sigma \leqslant \sigma_s, \end{cases}$$

where

$$\varepsilon_r = \frac{\gamma^* \sigma_s t_1}{2p_m}\left(\frac{p_m}{\sigma_s} - 1\right)^2.$$

These relations are presented in Fig. 12. Line *OAB* corresponds to the "loading" process, line *BCD* to the "unloading" process. It should be emphasized that the same physical law (3.28) governs the process of "loading" and that of "unloading". Nevertheless, the irreversibility of the process "loading–unloading" is conserved, as shown in Fig. 12.

The function $\Phi(F)$ appearing in constitutive equations (3.10), (3.12), (3.25), and (3.28) is to be determined from experimental data. It is assumed, when determining function $\Phi(F)$, that the influence of the elastic part of the strain is negligible as compared with the viscoplastic part of the strain. Moreover, the hypothesis is made that the curve $\sqrt{J_2} - \sqrt{I_2^p}$ for the complex stress state is exactly the same as the curve $\sigma - \dot{\varepsilon}$ for the uniaxial stress state. The validity of the hypothesis in a particular case of a complex stress state was confirmed in the experimental investigations of Lindholm [76]–[78] for aluminium and steel specimen under simultaneous tension and torsion. In Fig. 13 we see the relationship $\sqrt{J_2} - \sqrt{I_2}$ determined on the basis of Lindholm's [77] experimental investigations. In the experiments thin cylindrical specimens were employed, subjected to instant simultaneous tension and torsion. The circles in Fig. 13 designate the experimental points corresponding to the case of simple tension, the triangles – pure torsion; the squares, however, denote the points corresponding to the computed values of stress and strain intensities for the complex stress state (simultaneous torsion and tension). The curves $\sqrt{J_2}$ against $\sqrt{I_2}$ in the cases of simple tension, pure torsion, and the complex stress state are identical. One could presume that the similarity hypothesis of curve $\sqrt{J_2} - \sqrt{\dot{I}_2^p}$ to the curve $\sigma - \dot{\epsilon}$ for the uniaxial stress state is generally true for any complex stress state.

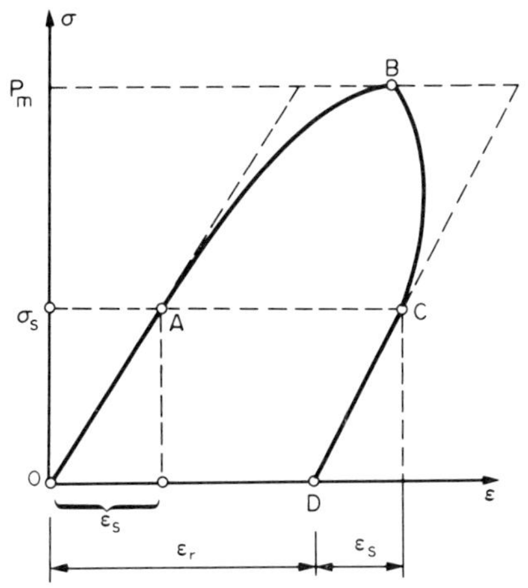

Fig. 12

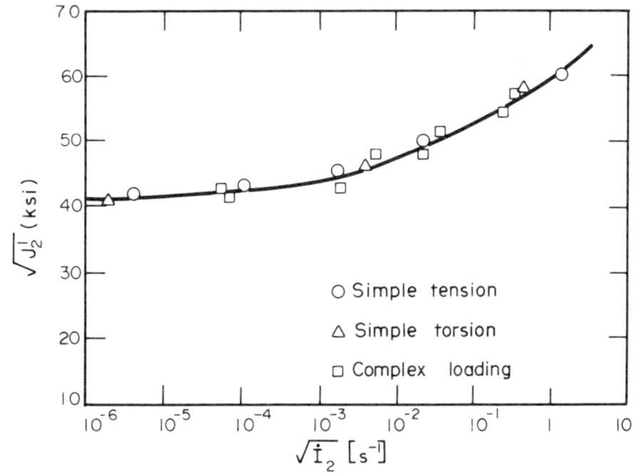

Fig. 13

Perzyna [115], on the basis of experimental investigations of metals, has determined the function $\Phi(F)$ assuming various forms for it, for example

$$\Phi(F) = \sum_{\alpha=1}^{N} A_\alpha[\exp F^\alpha - 1], \qquad \Phi(F) = \sum_{\alpha=1}^{N} B_\alpha F^\alpha. \tag{3.29}$$

The coefficients in (3.29) were determined from the experimental data obtained by Clark and Duwez [25] who found the change of yield limit as a function of strain rate for mild steel within the range of strain rate of $0 \leqslant \dot{\epsilon} \leqslant 200\ \text{s}^{-1}$, i.e. within the range where this change is important. The experimental curve is presented in Fig. 14 by the continuous line. The dashed lines denote the approximation to the experimental curve by functions (3.29) in the case $N = 5$. The values of the constants A_α, B_α $(\alpha = 1, \ldots, 5)$ in the case of function F for a uniaxial stress state $\left(F = \dfrac{|\sigma|}{\sigma_s} - 1\right)$ are the following ones:

$A_1 = 217.56$, $A_2 = -654.11$, $A_3 = 874.52$, $A_4 = -484.15$, $A_5 = 93.56$
$B_1 = 337.53$, $B_2 = -1470.56$, $B_3 = 3271.71$, $B_4 = -3339.98$, $B_5 = 1280.06$.

By means of the similarity hypothesis concerning the $\sqrt{J_2} - \sqrt{\dot{I}_2^p}$ and $\sigma - \dot{\epsilon}$ curves the above constants can be used in a complex stress state [77], [114].

In the case of a medium governed by (3.13), function $\Phi(F)$ is determined in a slightly different way. We shall now show how to deal with the function $\Phi(F)$ in the case of a uniaxial stress state. In order to construct the entire range of the function $\gamma^*\Phi\left(\dfrac{\sigma}{f_1(\varepsilon)} - 1\right)$ we have to know the complete relationship $\sigma = \sigma(\epsilon, \dot{\epsilon})$. The following method of determining the relaxation function $\Phi(F)$ was proposed in [57]. If the surfaces $\sigma = \sigma(\epsilon_s, \dot{\epsilon})$ and $\sigma = \sigma(\epsilon_w, \dot{\epsilon})$ are known from the experiments (ϵ_w denotes the strain at the strength limit and ϵ_s the strain at the elastic limit), then the function $\sigma = \sigma(\epsilon, \dot{\epsilon})$ can be determined by interpolation between these

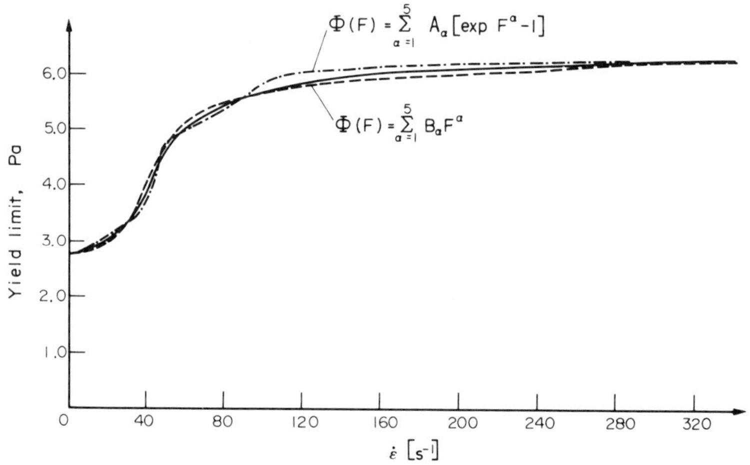

Fig. 14

surfaces. To simplify the numerical calculations both the surfaces $\sigma(\epsilon_s, \dot{\epsilon})$ and $\sigma(\epsilon_w, \dot{\epsilon})$ can be approximated by the planes. Then the values of the interpolated function $\sigma = \sigma(\epsilon, \dot{\epsilon})$ are located on the planes connecting the straight-line segments for $\epsilon = \epsilon_s$ and $\epsilon = \epsilon_w$. The function

$\gamma^* \Phi\left[\dfrac{\sigma}{f_1(\varepsilon)} - 1\right]$ has the dimension of the strain rate. For example, for the nth approximation

segment we obtain (in the case of a linearly hardening material)

$$\gamma^*\Phi(F) = \left[\left(1 - \frac{\varepsilon - \varepsilon_s}{\varepsilon_w - \varepsilon_s}\right)\frac{\sigma - \sigma_{sn1}}{\sigma_{sn2} - \sigma_{sn1}} + \frac{\varepsilon - \varepsilon_s}{\varepsilon_w - \varepsilon_s}\,\frac{\sigma - \sigma_{wn1}}{\sigma_{wn2} - \sigma_{wn1}}\right](\dot{\varepsilon}_{n2} - \dot{\varepsilon}_{n1}), \tag{3.30}$$

where the indices $n2$ and $n1$ denote the parameter values on the left-hand boundary and on the right-hand boundary of the approximation segment, respectively. The calculation of the functions $\gamma^*\Phi(F)$ written down in the above form can be made arbitrarily accurate by taking into account an appropriate number of extrapolation planes.

In order to determine function $\Phi\left(\dfrac{\sigma_i}{f_1(\varepsilon_i)} - 1\right)$ the hypothesis is made [57] that the relationships between the intensities of stress σ_i, strain rates $\dot{\epsilon}_i$, and strains ϵ_i are identical to those between σ, $\dot{\epsilon}$, ϵ in the uniaxial stress state.

3.3. ELASTIC/VISCOPLASTIC MEDIUM WITH TEMPERATURE DEPENDENT PROPERTIES

The constitutive equations describing the behaviour of strain rate sensitive materials are discussed in [111]. It is assumed there that in the constitutive equations for an elastic/viscoplastic medium (3.10) the yield limit, viscosity coefficient, γ, and relaxation function, $\Phi(F)$, are temperature dependent. The equations take the following rorms:

$$\dot{e}_{ij} = \frac{1}{2\mu}\dot{s}_{ij} + \gamma(T)\langle\Phi\left[\frac{f(\sigma_{ij}, \varepsilon^p_{ij})}{\varkappa(T)} - 1\right]\rangle\frac{\partial f}{\partial\sigma_{ij}},$$

(3.31)

$$\dot{\varepsilon}_{ii} = \frac{1}{3K}\dot{\sigma}_{ii} + 3\alpha K\dot{T}.$$

An analysis of (3.31) was carried out in [117] on the basis of experimental results. Since the available experimental data referred to conditions of simple tension or compression (3.31) were also considered for the case of a uniaxial stress state. The discussion was confined only to the case of an elastic/visco-ideally-plastic medium; moreover, only the inelastic part of strain rate was taken into account. It was shown in the paper that the best agreement with the experimental results is obtained if it is assumed that only the yield limit and viscosity coefficient depend on temperature while, simultaneously, the relaxation function $\Phi(F)$ is temperature independent.

The identical discussion could be carried out in the case of the constitutive equations for an elastic/viscoplastic medium, when material hardening is taken into account, by the introduction into the argument of the function $\Phi(F)$ of a hardening parameter dependent on temperature $\varkappa(T)$, or, in the case of (3.12), by assuming that the static curve for material hardening depends on temperature $f_1(\epsilon^p, T)$.

4. Soil dynamics equations

We shall present the equations describing the behaviour of soils in the case of dynamic loading. First of all we shall present the soil dynamics equations introduced by Grigorian [48]. Based on this model a number of boundary problems have been solved, both quasi-static and dynamic ones, connected with the processes of stress wave propagation in soils produced by various types of detonations.

We also present the constitutive equations for soils exhibiting strain rate sensitivity. These equations were proposed by Olszak and Perzyna [113].

Concluding the chapter we shall present a certain method of describing the mechanical properties of soil, based on the strain theory using the so-called rigid unloading model, i.e. under the assumption that the strain intensity does not change during the unloading process. In the case of one-dimensional problems it leads to the assumption that during unloading the body behaves like an ideally rigid one.

4.1. GRIGORIAN'S EQUATIONS

The soil dynamics equations presented at this point were derived by Grigorian [48].

If an element of the medium undergoes an irreversible change of volume p then the pressure acting on it ϱ is related to the density by the equation

(4.1) $$p = f_1(\varrho).$$

The irreversible change of density (volume) occurs only during loading (see section 1.2). Equation (4.1) holds only when $dp/dt > 0$. In the case of unloading the following relationship holds:

(4.2) $$p = f_2(\varrho, p^*),$$

where p^* denotes the maximum pressure (Fig. 15) acting on the considered element of the medium during the preceding irreversible change of volume. On the basis of (4.1) one can introduce the density value ϱ^* corresponding to the pressure p^*.

(4.3) $$p^* = f_1(\varrho^*).$$

The values p^* or ϱ^* can be treated as the parameters characterizing the final (irreversible) voluminal deformation.

In the process of unloading from point (p^*, ϱ^*) on the curve given by (4.3), there exists a reversible change of volume; the point describing the process is displaced along the curve given by (4.2). This point can reach the position p^0, ϱ^0 corresponding to a state of the considered element in which it is not able to undergo any greater hydrostatic tensile stress. The set of such states is illustrated by curve

(4.4) $$p^0 = \varphi(\varrho^0).$$

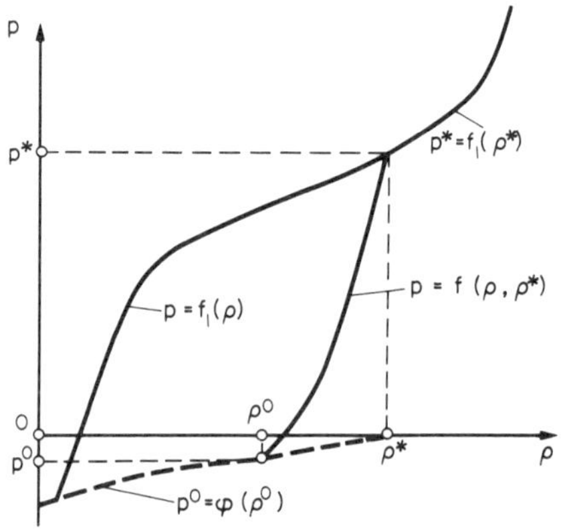

Fig. 15

Equations (4.1)–(4.3) can be rewritten in the following forms:

(4.5) $$\begin{aligned} p &= f(\varrho, \varrho^*)\, H(\varrho-\varrho^0)\, H(\varrho^*-\varrho), \\ p^* &= f(\varrho^*, \varrho^*) \equiv f_1(\varrho^*), \\ p^0 &= f(\varrho^0, \varrho^*) = \varphi(\varrho^0); \end{aligned}$$

H denotes the Heaviside function. It results from the condition

(4.6) $$\frac{d\varrho^*}{dt} = \frac{d\varrho}{dt}\, H(\varrho-\varrho^*)\, H\left(\frac{d\varrho}{dt}\right)$$

that the irreversible strain occurs only with the increase of ϱ^*.

It is assumed, in accordance with the fundamentals of the theory of plastic flow, that for conditions in which the shear strains cannot be exclusively elastic, a part of the infinitesimal strain is produced plastically (irreversibly) and is proportional to the deviatoric stress tensor.

The yield condition is assumed to take the form of the second invariant of stress deviator being dependent on the pressure p:

$$J_2 = \frac{1}{2} s_{ij} s_{ij} = F(p), \tag{4.7}$$

where $s_{ij} = \sigma_{ij} - p\delta_{ij}$ and F is a non-decreasing function of its argument.

The yield condition for this model (4.7) differs from those models used in the theory of plasticity for metals, where it is assumed that J_2 is either constant in the course of plastic deformation (ideal plasticity) or depends on the characteristic of plastic strain (hardening). Relationship (4.7) is a condition of the ideally plastic type whose yield limit depends on the first stress tensor invariant, i.e. pressure p. It is a condition of the Mises–Schleicher type.

In order to describe the components of elastic shear strain the Jaumann time derivative of the deviatoric stress tensor is employed. Thus we obtain the following relationships between the components of the deviatoric stress tensor and those of the deviatoric strain rate tensor:

$$\mu\left(\dot{e}_{ij} - \frac{1}{3}\dot{e}_{kk}\delta_{ij}\right) = \frac{\tilde{d}s_{ij}}{dt} + \lambda s_{ij}, \tag{4.8}$$

where symbol $\tilde{d}s_{ij}/dt$ denotes Jaumann's stress tensor derivative defined by the equation

$$\frac{\tilde{d}s_{ij}}{dt} = \frac{ds_{ij}}{dt} - s_{ik}\omega_{jk} - s_{jk}\omega_{ik}, \tag{4.9}$$

where

$$2\omega_{ij} = \frac{\partial v_i}{\partial x_j} - \frac{\partial v_j}{\partial x_i} \quad \text{and} \quad \dot{e}_{ij} = \frac{\partial v_i}{\partial x_j} + \frac{\partial v_j}{\partial x_i}.$$

Parameter λ should be positive in the course of plastic straining. In the case of small displacements and strains, (4.8) transforms into the ordinary Hooke's law.

Parameter λ can be determined by means of the yield condition (4.7). Multiplying (4.8) by s_{ij} and summing we obtain

$$s_{ij}\mu\left(\dot{e}_{ij} - \frac{1}{3}\dot{e}_{kk}\delta_{ij}\right) = \mu s_{ij}\dot{e}_{ij} = s_{ij}\frac{\tilde{d}s_{ij}}{dt} + \lambda s_{ij}s_{ij} = \mu\dot{W} \tag{4.10}$$

where $\dot{W}$ denotes the rate of change of the distortion strain energy.

Making use of the relation

$$s_{ij}\frac{\tilde{d}s_{ij}}{dt} = s_{ij}\frac{ds_{ij}}{dt} = \frac{1}{2}\frac{d(s_{ij}s_{ij})}{dt} = \frac{dJ_2}{dt} \tag{4.11}$$

and taking into account the yield condition (4.7) we obtain from (4.10)

$$\lambda = \frac{\mu\dot{W} - F'(p)\dfrac{dp}{dt}}{2F(p)}. \tag{4.12}$$

Formula (4.12) holds if $J_2 = F(p)$ and if $\lambda > 0$, i.e. if $\mu\dot{W} - F'(p)\dfrac{dp}{dt} > 0$. Otherwise $\lambda \equiv 0$. The parameter λ can also be expressed by a single analytical formula, namely

$$\lambda = \frac{\mu\dot{W} - F'(p)\dfrac{dp}{dt}}{2F(p)} H[J_2 - F(p)]\, H\left[\mu\dot{W} - F'(p)\frac{dp}{dt}\right]. \tag{4.13}$$

Thus formulae (4.8), (4.9), and (4.13) provide a complete description of the shear strains.

Experimental results obtained in the case of sandy soils have confirmed the pertinence of the yield condition of the type (4.7), and have established that for such soils in real conditions (for $p < 15 \times 10^5$ Pa) the function $F(p)$ takes the following form:

$$F(p) = (\alpha p + \beta)^2, \tag{4.14}$$

where α and β are constants. The description of the experiments, by means of which function $F(p)$ and functions (4.1) and (4.3) can be determined, is given in [126].

4.2. ELASTIC/VISCO-IDEALLY-PLASTIC SOILS

Olszak and Perzyna [113] proposed constitutive equations for soils with rheological effects taken into account, namely the sensitivity of soils on the strain rate change.

A particular case of the static yield function was assumed in [113] to describe the behaviour of the elastic/visco-ideally-plastic soil in the dynamic case:

$$F = \frac{f(\sigma_{ij})}{k} - 1, \quad f(\sigma_{ij}) = \alpha J_1' + \sqrt{J_2}, \tag{4.15}$$

where J_1' is the first invariant of the stress tensor, α is the parameter characterizing the soil dilatation rate, and k denotes a constant representing the yield limit of soil.

Constitutive equations (2.3) in the case of an elastic/visco-ideally-plastic medium, provided F is assumed in the form given by (4.15), take the following form:

$$\dot{\varepsilon}_{ij} = \frac{1}{2\mu}\dot{s}_{ij} + \frac{1-2\nu}{E}\dot{\sigma}_{kk}\delta_{ij} + \gamma\left\langle \Phi\left[\frac{\alpha J_1 + \sqrt{J_2}}{k} - 1\right]\right\rangle\left(\alpha\delta_{ij} + \frac{s_{ij}}{2\sqrt{J_2}}\right). \tag{4.16}$$

The dynamic yield condition for the soil is given in the form

$$\alpha J_1' + \sqrt{J_2} = k\left\{1 + \Phi^{-1}\left[\frac{\sqrt{I_2^p}}{\gamma}\left(\frac{3}{2}\alpha^2 + \frac{1}{4}\right)^{-1/2}\right]\right\}. \tag{4.17}$$

The rate of voluminal dilatation of the soil, from (4.16), is expressed as follows:

$$\dot{\varepsilon}_{ii} = \frac{1-2\nu}{E}\dot{\sigma}_{ii} + 3\alpha\gamma\left\langle \Phi\left[\frac{\alpha J_1' + \sqrt{J_2}}{k} - 1\right]\right\rangle. \tag{4.18}$$

It is apparent from this equation that the inelastic strains are accompanied by volume change if $\alpha \neq 0$. This property is called the soil dilatation.

It can be shown in the case when $\gamma \to \infty$ that the equations of the flow theory are obtainable from (4.16). If $\gamma \to \infty$ we obtain from (4.17) and similarly from (4.15) that

$$f(\sigma_{ij}) = \alpha J_1' + \sqrt{J_2} = k, \quad \text{or} \quad F = 0; \tag{4.19}$$

thus, it results from the definition of function $\Phi(F)$ that in this limiting case the quantity $\gamma\Phi(F)$ is an undetermined parameter, i.e.

$$\dot{\varepsilon}_{ij}^{p} = \lambda \frac{\partial f}{\partial \sigma_{ij}} = \lambda \left(\alpha\delta_{ij} + \frac{s_{ij}}{2\sqrt{J_2}}\right) = \frac{\sqrt{I_2^p}}{\sqrt{3\alpha^2/2 + 1/4}} \left[\alpha\delta_{ij} + \frac{s_{ij}}{2\sqrt{J_2}}\right]. \tag{4.20}$$

The above equations comprise the constitutive equations for the case of elastic-ideally-plastic soils.

4.3. RIGID UNLOADING

For certain materials, in particular for soils under high pressure, unloading in the pressure–voluminal strain diagram is practically perpendicular to the strain axis. The relationships $(p - p_0)/p_0$ as a function of ϵ, for sands of various degrees of wetness, are presented in Fig. 16. If the initial moisture content of the sand decreases, then the process of unloading occurs with smaller and smaller strain change. A small strain change for unloading is even more apparent in the case of high pressures of the order of several hundred atmospheres. The characteristics for a mixture of sand and clay with an initial moisture content of about 15% are presented in Fig. 16. We see that in the unloading process the change in the strain of the medium is very small. For practical purposes we can assume that the strains do not change during the unloading process, consequently

$$\varepsilon = \varepsilon^* = \text{const}, \tag{4.21}$$

where ϵ^* denotes the maximum value of strain in the process (Fig. 17). The medium governed during unloading by (4.21) behaves like a rigid body: its strain does not change with time. Accordingly, the motion of the medium during unloading is described by ordinary differential equations.

We can generally assume on the basis of experimental results that the characteristic $\sigma = \sigma(\epsilon)$ together with rigid unloading constitutes a good approximation to the dynamic properties of dry sandy soils.

A medium with the above properties has been called a plastic gas (Rakhmatulin and Demianov [125]), i.e. a gas that is incompressible during unloading.

Rigid unloading has been generalized also to the case of complex stress states. Assuming, for example, the medium in the loading region to be described by the equations of the Hencky–Iliushin strain theory, we then adopt the condition of constant strain intensity during unloading, i.e.

$$\varepsilon_i = \varepsilon_i^* = \text{const}. \tag{4.22}$$

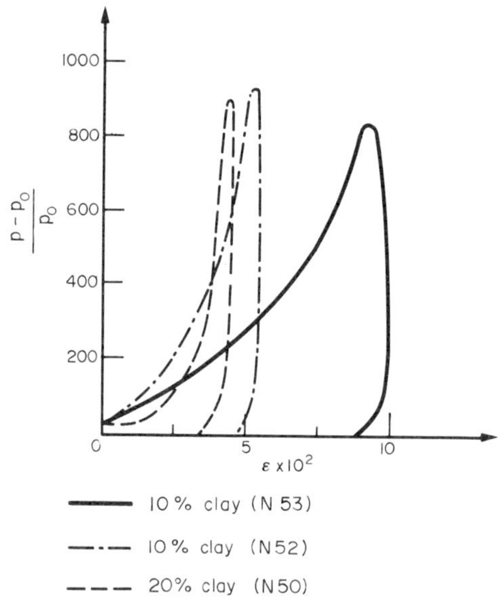

Fig. 16

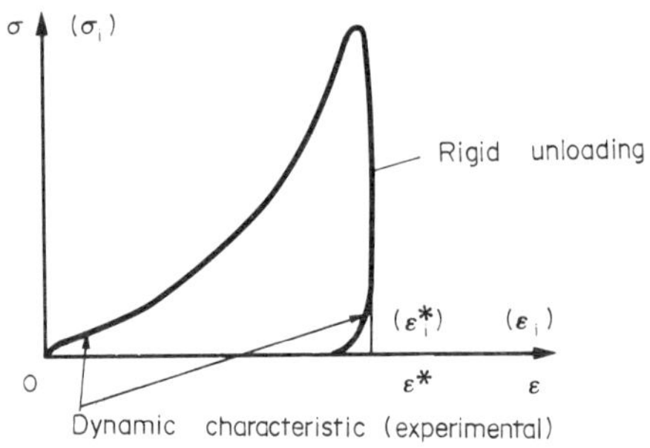

Fig. 17

This condition does not mean, in a complex stress state, that the medium is "rigid" as it was in the case of uniaxial state; it permits the time change of the strain tensor components during the unloading process.

By means of the assumption of the rigid-unloading model, the partial differential equations describing the motion of the medium in one-dimensional problems reduce, in the unloading process, to ordinary differential equations or to Volterra integral equations of the second kind. A number of practically important problems of the wave propagation in soils have been solved on the basis of a model with rigid unloading. The solutions of these problems will be discussed in detail on p. 73. Also the estimation of the error arising on account of rigid unloading instead of the real characteristic will be given (Fig. 17).

5. Equations of motion and continuity

The equation of motion can be derived from the principle of the momentum conservation of a body of volume V bounded by a surface S,

$$\frac{d}{dt}\int_V \varrho v_i \, dV = \int_V X_i \, dV + \int_S t_i \, dS, \tag{5.1}$$

where ϱ is the density of the medium, $v_i = du_i/dt$ are the components of the velocity vector of an element of the continuous medium, X_i are the components of the body force vector,

$$t_i = \sigma_{ji} n_j \tag{5.2}$$

denotes the components of the stress vector applied to surface S, n_j determines the direction of the normal to the surface at the point of application of vector t_i.

Making use of the relation

$$\int_S t_i\, dS = \int_S \sigma_{ji}\, n_j\, dS = \int_V \sigma_{ji,j}\, dV, \tag{5.3}$$

we can reduce (5.1), for an element of constant mass ϱdV in rectangular coordinates, to the following form:

$$\varrho \frac{dv_i}{dt} = \sigma_{ji,j} + X_i \tag{5.4}$$

or, making use of the definition of material derivative, we obtain

$$\varrho \frac{\partial v_i}{\partial t} + \varrho v_{i,j} v_j = \sigma_{ji,j} + X_i. \tag{5.4'}$$

It can be proved that in the case of small deformation gradients the equations of motion take the following form:

$$\varrho \frac{\partial v_i}{\partial t} = \sigma_{ji,j} + X_i. \tag{5.5}$$

Now we present the equations of motion in an arbitrary system of curvilinear coordinates.

Let us designate, in space E_3, a point M whose cartesian rectangular coordinates will be denoted by x^1, x^2, x^3, and curvilinear coordinates by q^1, q^2, q^3. Let us denote the unit vectors along the axes x^1, x^2, x^3 by $\mathbf{i}_1, \mathbf{i}_2, \mathbf{i}_3$. They constitute a constant orthonormal basis. Simultaneously, let us consider at point M the vectors

$$\mathbf{g}_i = \frac{\partial M}{\partial q^i} = \frac{\partial x^k}{\partial q^i}\, \mathbf{i}^k. \tag{5.6}$$

These vectors are tangential to the curvilinear coordinates. They constitute a basis (non-orthogonal) called the natural basis at point M. The basis varies from point to point.

The components of the fundamental metric tensor g_{ij} are determined as follows:

$$g_{ij} = \mathbf{g}_i \cdot \mathbf{g}_j = \frac{\partial x^k}{\partial q^i} \frac{\partial x^k}{\partial q^j}. \tag{5.7}$$

The determinant g of the matrix g_{ij} takes the form

$$g = \left| g_{ik} \right| = \left| \frac{\partial x^r}{\partial q^s} \right| \left| \frac{\partial x^r}{\partial q^s} \right|. \tag{5.8}$$

The Christoffel symbols of the first kind Γ_{ijk} are expressed by the metric tensor g_{ij} in the following manner:

$$\Gamma_{ijk} = \frac{1}{2} \left(\frac{\partial g_{ij}}{\partial q^k} + \frac{\partial g_{ik}}{\partial q^j} - \frac{\partial g_{jk}}{\partial q^i} \right). \tag{5.9}$$

The Christoffel symbols of the second kind Γ^i_{jk} are determined from the formula

$$\Gamma^i_{jk} = g^{ih}\,\Gamma_{hjk}, \tag{5.10}$$

with the symmetry condition with respect to the lower indices $\Gamma^k_{lm} = \Gamma^k_{ml}$.

In the case of curvilinear orthogonal coordinates the Christoffel symbols are as follows:

$$\begin{aligned}
&\Gamma^i_{jk} = 0, \text{ if all three indices } i, j, k \text{ are distinct},\\
&\Gamma^i_{ij} = \frac{1}{2g_{\underline{ii}}}\,\frac{\partial g_{\underline{ii}}}{\partial q^j},\\
&\Gamma^i_{jj} = -\frac{1}{2g_{\underline{ii}}}\,\frac{\partial g_{\underline{jj}}}{\partial q^i} \quad \text{provided } i \neq j
\end{aligned} \tag{5.11}$$

(the underlined indices are not to be summed).

The equations of motion (5.5) assume, in the curvilinear coordinates, the following forms:

$$\frac{D\sigma_i^j}{\partial q^j} = \varrho a_i - X_i \tag{5.12}$$

or

$$\frac{D\sigma^{ij}}{\partial q^j} = \varrho a^i - X^i, \tag{5.13}$$

where $\mathbf{a}$ is the acceleration vector and the symbol $\dfrac{D}{(\partial q^j)}$ denotes the covariant derivative.

The covariant components of the acceleration vector a_i have the following form:

$$a_i = \frac{d}{dt}(g_{ij}v^j) - \frac{1}{2}\frac{\partial g_{hk}}{\partial q^i} v^h v^k, \tag{5.14}$$

while the contravariant components of the acceleration vector are expressed as follows:

$$a^i = g^{ih}a_h = \frac{\partial v^i}{\partial t} + \frac{\partial v^i}{\partial q^k} v^k + g^{ih}\left(\frac{\partial g_{hj}}{\partial q^r} v^r v^j - \frac{1}{2}\frac{\partial g_{pq}}{\partial q^h} v^p v^q\right), \tag{5.15}$$

where v^i denote the contravariant components of the velocity vector $v^i = \left(\dfrac{dq^i}{dt}\right)$.

Equation (5.15) can be expressed in a simpler form using the definitions of the Christoffel symbols (5.9) and (5.10), viz.,

$$a^i = \frac{\partial v^i}{\partial t} + \frac{\partial v^i}{\partial q^k} v^k + \Gamma^i_{hk} v^h v^k. \tag{5.16}$$

The equations of motion, (5.12) and (5.13), in an arbitrary system of curvilinear coordinates have the following forms, respectively:

$$\frac{D\sigma_i^j}{\partial q^j} = \varrho \left[\frac{d}{dt} (g_{ij}v^j) - \frac{1}{2} \frac{\partial g_{hk}}{\partial q^i} v^h v^k \right] - X_i, \tag{5.17}$$

$$\frac{D\sigma^{ij}}{\partial q^j} = \varrho \left[\frac{\partial v^i}{\partial t} + \frac{\partial v^i}{\partial q^k} v^k + \Gamma^i_{hk} v^h v^k \right] - X^i. \tag{5.18}$$

In the case of orthogonal curvilinear coordinates equations (5.17) and (5.18), by means of expressions (5.11), can be expressed in the following form:

$$\frac{1}{\sqrt{g}} \frac{\partial}{\partial q^j} \left[\sqrt{g}\, \sigma^{ij} g_{\underline{ii}} \right] - \frac{1}{2} \left[\sigma^{jj} \frac{\partial g_{jj}}{\partial q^i} \right] = g_{\underline{ii}} \left(\varrho a^i - X^i \right), \tag{5.19}$$

where a^i is given by formula (5.16).

Now we shall derive the continuity equations.

From the condition of mass conservation which asserts that, for an arbitrary motion of continuous medium, the mass passing through an area element dS surrounding the volume V is in equilibrium with the change of mass within that volume we have

$$\int_S \rho v_i n_i dS + \frac{\partial}{\partial t} \int_V \rho dV = 0, \tag{5.20}$$

where n_i denote the direction cosines of the normal to the area element dS. From this we obtain the continuity equation

$$\frac{d\rho}{dt} + \rho v_{i,i} = 0. \tag{5.21}$$

In the case of curvilinear coordinates the continuity equation takes the following form:

$$\sqrt{g} \frac{\partial \varrho}{\partial t} + \frac{\partial}{\partial q^i} \left(\varrho \sqrt{g}\, v^i \right) = 0. \tag{5.22}$$

CHAPTER II

WAVES, CONTINUITY CONDITIONS, MATHEMATICAL FUNDAMENTALS

6. Discontinuity fronts, definitions

In a medium of volume V, the surface separating two regions of volumes V_1 and V_2 will be denoted by $S^*(t)$ (Fig. 18). By $\mathbf{u}(x^i, t)$ we denote a continuous function in volume V, describing the displacement of material points of the medium. It is assumed that certain derivatives of the function $\mathbf{u}(x^i, t)$ with respect to coordinates x^i and t experience sudden changes (jumps) when passing across $S^*(t)$. Such a surface is called a discontinuity surface of nth order if all derivatives with respect to x^i or t up to and including $n-1$ are continuous, and if certain derivatives with respect to x^i or t of the nth order are discontinuous across that surface.

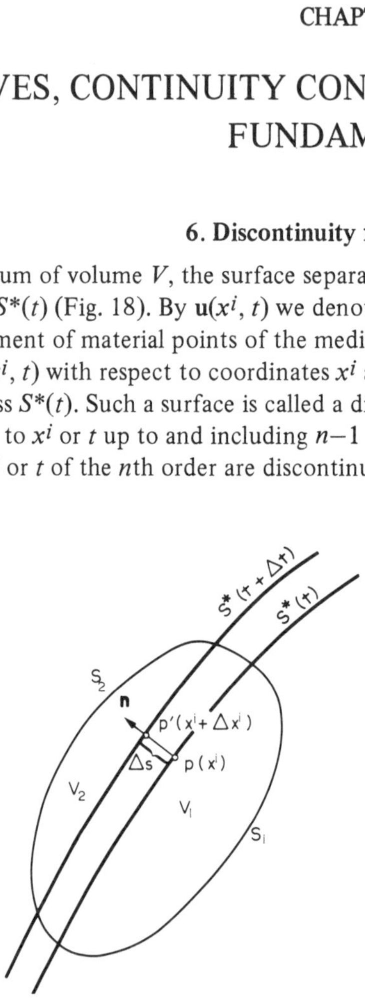

Fig. 18

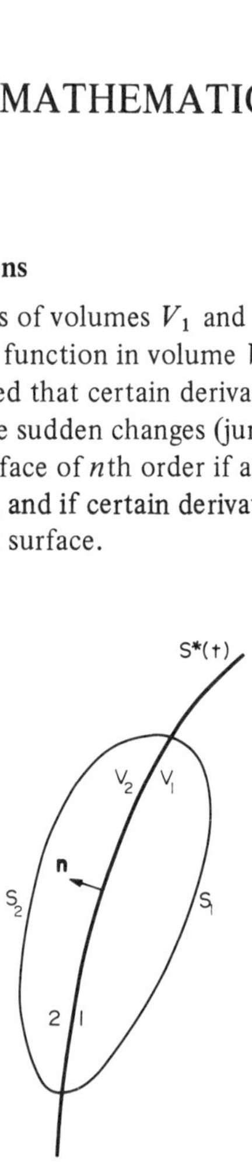

Fig. 19

If surface $S^*(t)$ moves through the medium we say that the discontinuity propagates; surface $S^*(t)$ is then called a *wave.*

If the discontinuity is of the first order then such a wave is called a *wave of strong discontinuity*, if the discontinuity is of the order $n \geqslant 2$ then it is called an *ordinary wave.* A discontinuity of zero order denotes a discontinuity in the medium itself. Thus if the stress tensor field σ_{ij} and the field of material particle velocity, v_i, have discontinuities across surface $S^*(t)$, then the surface constitutes a *strong discontinuity wave.* If the stress tensor field, σ_{ij}, and the particle velocity, v_i, are continuous functions across $S^*(t)$, whereas any of their first derivative is discontinuous then such a wave is called a *weak discontinuity wave* or *acceleration wave.*

Denoting by $S^*(t)$ a position of the wave surface in coordinates x^i at time t, by $S^*(t + \Delta t)$ its position at time $t + \Delta t$, and by Δs the length of the normal to $S^*(t)$ at point $P(x^i)$ between the surfaces $S^*(t)$ and $S^*(t + \Delta t)$ (Fig. 18), we define the *velocity of wave propagation* as follows:

$$G = \lim_{\Delta t \to 0} \frac{\Delta s}{\Delta t}. \tag{6.1}$$

Let us denote by $S^*(t)$ a surface in a medium separating two regions designated 1 and 2, respectively (Fig. 19). It is assumed that the normal to the surface is directed from region 1 to 2. The equation of the surface at instant t takes the form $F(x^i, t) = 0$.

We shall now define the following kinds of waves: a wave of plastic loading, a plastic wave, an elastic wave, an unloading wave, and a shock wave.

1. A surface $S^*(t)$ (Fig. 19), separating two regions in an elastic–plastic or in an elastic/viscoplastic medium, moving in the direction from region 1 to 2, where material particles in region 2 are either in an elastic or an unloading state, while in region 1 they are in a plastic state, is called a *wave of plastic loading.* If the equation of surface $S^*(t)$ at the instant t is $F(x^i, t) = 0$, then the velocity vector of the plastic loading wave is equal to

$$\mathbf{G}_1 = -\frac{\partial F}{\partial t}\mathbf{n} = -\frac{\partial F}{\partial t}\frac{\operatorname{grad} F}{|\operatorname{grad} F|}. \tag{6.2}$$

2. If material particles are in plastic state in both regions 1 and 2, then the moving surface $S^*(t)$, separating these regions, is called a *plastic wave.* If, on the other hand, in each region there is an elastic state (or a state of elastic unloading) then surface $S^*(t)$ is called an *elastic wave.*

3. Let us consider another case. Assume that the material particles in region 2 are undergoing a process of plastic loading whereas in region 1 the material particles are taking part in an unloading process (the particles in this region were in the past plastically deformed). If surface $S^*(t)$ moves towards region 2 then it constitutes a wave that is called an *unloading wave.* The velocity vector of the unloading wave at the points of the unloading wave surface takes the form

$$\mathbf{G}_{\mathrm{unl}} = -\frac{\partial F}{\partial t}\mathbf{n} = -\frac{\partial F}{\partial t}\frac{\operatorname{grad} F}{|\operatorname{grad} F|}, \tag{6.3}$$

where, as before, the vector of the normal to surface $S^*(t)$ is directed from region 1 to 2.

In the case of an elastic/viscoplastic medium the definition of an unloading wave is exactly the same as in the case of an elastic–plastic medium; however, the conditions which determine the wave front are different; this will be shown below.

4. A *shock wave* constitutes a particular case of a strong discontinuity wave. It is characterized by discontinuities in both the stress tensor and the velocity (that is the discontinuities of the first derivatives of displacement) across its front, even though the stress tensor changes continuously in time at the place from which the wave starts to propagate. In other words the generation of a shock wave does not depend on a discontinuity of the boundary conditions but its existence is caused by the physical behaviour of the medium (i.e. its

material characteristic) and the deformation of the medium.† In the case of small strains shock waves may exist in elastic–plastic and non-linearly elastic media. A shock wave cannot exist in those elastic/viscoplastic media whose constitutive equations were given in Chapter I.

Frequently in the American literature the term "shock wave" is used instead of the wave of strong discontinuity. We shall clearly distinguish between both these notions.

The definition of Riemann waves given in Chapter III (p. 47) while the definition of simple waves in the case of a state of plane strain is given in Chapter V (p. 145).

7. Dynamic continuity conditions across discontinuity fronts

We shall now determine the dynamic continuity conditions which hold across discontinuity fronts. Let us denote by $S^*(t)$ the surface separating region V into two parts V_1 and V_2 (Fig. 19) and by S_1 and S_2 those parts of surface S which bound volumes V_1 and V_2 respectively. The normal velocity of surface $S^*(t)$ in the direction of volume V_2 is denoted by G. Since the change of volume V is connected only with the motion of material particles, then the normal velocity v_n of the points on the surface S is equal to $v_i n^i$. In the particular case, however, when the strain in region V is determined by the motion of material particles only, then $v_i n^i = G$.

The following equation holds (Hadamard [50] and Mandel [85]):

$$\frac{d}{dt}\int_V f(x,t)\,dV = \int_V \frac{\partial f}{\partial t}\,dV + \int_S Gf(x,t)\,dS, \tag{7.1}$$

whenever $f(x, t)$ is a continuous and differentiable function of the coordinates x^i and of time t, and G is the velocity of the surface S in the direction of its external normal with respect to a fixed coordinate system.

By means of (7.1) the following relations [152] can be written for regions 1 and 2 (Fig. 19):

$$\begin{aligned} \frac{d}{dt}\int_{V_1} f(x,t)\,dV &= \int_{V_1} \frac{\partial f}{\partial t}\,dV + \int_{S_1} f v_n\,dS + \int_{S^*} fG\,dS, \\ \frac{d}{dt}\int_{V_2} f(x,t)\,dV &= \int_{V_2} \frac{\partial f}{\partial t}\,dV + \int_{S_2} f v_n\,dS - \int_{S^*} fG\,dS. \end{aligned} \tag{7.2}$$

The region of integration S^* is the part of the surface S contained within volume V.

Let us consider the case of a strong discontinuity wave, i.e. the case when the function $f(x, t)$ possesses a discontinuity across the surface $S^*(t)$. Denoting by the indices 1 and 2 the values of the function $f(x, t)$ on the sides of $S^*(t)$ in regions 1 and 2 respectively, we obtain from formulae (7.2) the following relation:

$$\frac{d}{dt}\int_V f(x,t)\,dV = \int_V \frac{\partial f}{\partial t}\,dV + \int_S f v_n\,dS + \int_{S^*} (f_1 - f_2)\,G\,dS. \tag{7.3}$$

†This applies to strains only. In the case of small strains the existence of a shock wave depends only on the material characteristic.

Assuming that the function $f(x, t)$ is the density of the medium $\varrho(x, t)$, we obtain

$$f_1(x,t) = \varrho_1(x,t), \quad f_2(x,t) = \varrho_2(x,t), \tag{7.4}$$

and making use of the condition of mass conservation (5.6), where it is assumed that the mass does not change during the deformation, we have

$$\frac{d}{dt}\int_V \varrho \, dV = 0, \tag{7.5}$$

and we obtain from (7.3) the following expression:

$$\int_V \frac{\partial \varrho}{\partial t} dV + \int_{S_1} \varrho v_n \, dS + \int_{S_2} \varrho v_n \, dS + \int_{S^*} (\varrho_1 - \varrho_2) G \, dS = 0. \tag{7.6}$$

Next we pass to the limit in (7.6) in which $V \to 0$ for a fixed instant t in such a way that in the limit the volume passes into part S_0^* of the surface S^*. Then we obtain

$$0 - \int_{S_0^*} \varrho_1 v_{1n} \, dS + \int_{S_0^*} \varrho_2 v_{2n} \, dS + \int_{S_0^*} (\varrho_1 - \varrho_2) G \, dS = 0, \tag{7.7}$$

where v_{1n} and v_{2n} denote the velocities of particles normal to surface S^* from the sides in regions 1 and 2 respectively.

Since the region of integration S_0^* is arbitrary, we obtain from (7.7) the following condition:

$$\varrho_1(v_{1n} - G) = \varrho_2(v_{2n} - G). \tag{7.8}$$

In the case when surface $S^*(t)$ moves in the opposite direction, i.e. in the direction of region 1, condition (7.8) assumes the form

$$\varrho_1(v_{1n} + G) = \varrho_2(v_{2n} + G). \tag{7.8'}$$

Now we assume in formula (7.3) that

$$f_1(x,t) = \varrho_1 v_{1i}, \quad f_2(x,t) = \varrho_2 v_{2i}, \quad (i = 1,2,3), \tag{7.9}$$

hence we obtain the expression

(7.10)

$$\frac{d}{dt}\int_V \varrho v_i \, dV = \int_V \frac{\partial}{\partial t}(\varrho v_i) \, dV + \int_{S_1} \varrho v_n v_i \, dS + \int_{S_2} \varrho v_n v_i \, dS + \int_{S^*} (\varrho_1 v_{1i} + \varrho_2 v_{2i}) G \, dS.$$

If use is made of the principle of conservation of momentum (5.1), in the case in which there are no forces (i.e. when $X_i \equiv 0$), or if

$$\frac{d}{dt}\int_V \varrho v_i \, dV = \int_S t_i \, dS, \tag{7.11}$$

then the left-hand side of (7.10) can be replaced by the integral of the stress vector $t_i = \sigma_{ji} n^j$ over surface S. Thus (7.10) takes the form

$$\int_S \sigma_{ji} n^j dS = \int_V \frac{\partial}{\partial t}(\varrho v_i) \, dV + \int_{S_1} \varrho v_n v_i \, dS + \int_{S_2} \varrho v_n v_i \, dS + \int_{S^*} (\varrho_1 v_{1i} - \varrho_2 v_{2i}) G \, dS. \tag{7.12}$$

Performing the identical passage to the limit with $V \to 0$, as was done before, we obtain

$$\int_{S_0^*} (\sigma_{2ij} - \sigma_{1ij}) n^j \, dS = -\int_{S_0^*} \varrho_1 v_{1i} v_{1n} \, dS + \int_{S_0^*} \varrho_2 v_{2i} v_{2n} dS + \int_{S_0^*} (\varrho_1 v_{1i} - \varrho_2 v_{2i}) G \, dS, \tag{7.13}$$

where $\sigma_{1\,ij}$, $\sigma_{2\,ij}$ are the components of the stress tensor on the sides 1 and 2 of the surface S^* respectively. On account of the arbitrariness of region S_0^*, relation (7.13) takes the form

$$(\sigma_{2ij} - \sigma_{1ij}) n^j = -\varrho_1 v_{1i}(v_{1n} - G) + \varrho_2 v_{2i}(v_{2n} - G). \tag{7.14}$$

Next, making use of relation (7.8), we obtain

$$(\sigma_{2ij} - \sigma_{1ij}) n^j = \varrho_1 (v_{1n} - G)(v_{2i} - v_{1i}). \tag{7.15}$$

Introducing the following notation $[\sigma_{ij}]$, $[v_i]$ for the magnitudes of jumps of stress and velocity across surface $S^*(t)$:

$$[\sigma_{ij}] = \sigma_{2ij} - \sigma_{1ij}, \quad [v_i] = v_{2i} - v_{1i}, \tag{7.16}$$

and we obtain from (7.15)

$$[\sigma_{ij}] n^j = \varrho_1 (v_{1n} - G)[v_i]. \tag{7.17}$$

This condition is called the *condition of dynamic continuity*. If surface $S^*(t)$ moves in the direction of region 1, not as before, then the condition of dynamic continuity takes the form

$$[\sigma_{ij}] n^j = \varrho_1 (v_{1n} + G)[v_i]. \tag{7.17)}$$

On account of the fact that velocity v_{1n} is small compared with the wave front velocity, the condition of dynamic continuity reduces to the following form:

$$[\sigma_{ij}] n^j = -\varrho_1 G [v_i]. \tag{7.18}$$

8. Conditions of kinematic continuity across discontinuity fronts

Now we pass to the definition of the conditions of kinematic continuity. Again we consider two positions of the discontinuity surface: $S^*(t)$ and $S^*(t + \Delta t)$ (Fig. 18). We assume that function $u(x^i, t)$ is a continuous, differentiable function of the variables x^i and t on both sides of the moving surface $S^*(t)$. The indices 1 and 2 denote the values of function $u(x^i, t)$ at the sides 1 and 2 of the surface $S^*(t)$ respectively. By $\Delta(u_2 - u_1)$ we denote the difference in the value of $(u_2 - u_1)$ at points P' and P (Fig. 18), i.e.

$$\frac{\Delta(u_2-u_1)}{\Delta t} = \frac{\Delta[u]}{\Delta t} = \frac{(u_2'-u_1')-(u_2-u_1)}{\Delta t} = \frac{u_2'-u_2}{\Delta t} - \frac{u_1'-u_1}{\Delta t}, \tag{8.1}$$

where "primes" denote the values at point P'.

Passing, in (8.1), to the limit as $\Delta t \to 0$, we obtain

$$\frac{\delta[u]}{\delta t} = \frac{\delta u_2}{\delta t} - \frac{\delta u_1}{\delta t} = \left[\frac{\delta u}{\delta t}\right], \tag{8.2}$$

where symbol δ denotes the derivatives with respect to time of the quantities $[u]$, u_2, and u_1.

On each side of surface $S^*(t)$ the following two approximate equations hold:

$$\begin{aligned} \frac{u_1'-u_1}{\Delta t} &= \left(\frac{\partial u}{\partial x^i}\right)_1 \frac{\Delta x^i}{\Delta t} + \left(\frac{\partial u}{\partial t}\right)_1, \\ \frac{u_2'-u_2}{\Delta t} &= \left(\frac{\partial u}{\partial x^i}\right)_2 \frac{\Delta x^i}{\Delta t} + \left(\frac{\partial u}{\partial t}\right)_2. \end{aligned} \tag{8.3}$$

Passing to the limit $\Delta t \to 0$, and taking into account the relation

$$\frac{\delta x^i}{\delta t} = \lim_{\Delta t \to 0} \frac{\Delta x^i}{\Delta t} = \lim_{\Delta t \to 0} \left(\frac{\Delta s}{\Delta t}\right) n^i = G n^i,$$

in the derivation of which use was made of the definition of G, the normal velocity of propagation of the surface $S^*(t)$, we obtain

$$\begin{aligned} \frac{\delta u_1}{\delta t} &= \left(\frac{\partial u}{\partial x^i}\right)_1 G n^i + \left(\frac{\partial u}{\partial t}\right)_1, \\ \frac{\delta u_2}{\delta t} &= \left(\frac{\partial u}{\partial x^i}\right)_2 G n^i + \left(\frac{\partial u}{\partial t}\right)_2. \end{aligned} \tag{8.4}$$

Subtracting the equations and utilizing (8.2) we obtain the condition

$$\left[\frac{\partial u}{\partial t}\right] = -G n^i \left[\frac{\partial u}{\partial x^i}\right] + \frac{\delta[u]}{\delta t}, \tag{8.5}$$

which is called the *condition of kinematic continuity* of the first order for function $u(x^i, t)$.

When function u is continuous passing across surface $S^*(t)$, then $[u] = 0$, and the condition (8.5) reduces to the form

$$\left[\frac{\partial u}{\partial t}\right] = -Gn^i\left[\frac{\partial u}{\partial x^i}\right]. \tag{8.6}$$

If the wave front $S^*(t)$ travels in the opposite direction to the sense of normal n, i.e. in the direction $2 \to 1$, then the condition of kinematic continuity (8.6) becomes

$$\left[\frac{\partial u}{\partial t}\right] = Gn^i\left[\frac{\partial u}{\partial x^i}\right]. \tag{8.7}$$

In the case when function $u(x^i, t)$ denotes the displacement of points of a medium, then we obtain from (8.6)

$$[v_j] = -Gn^i[u_{j,i}] = -Gn^i[u_{(j,i)}+u_{[j,i]}], \quad \text{i.e.} \quad [v_j] = -Gn^i[\varepsilon_{ij}+\omega_{ij}], \tag{8.8}$$

where

$$v_j = \frac{\partial u_j}{\partial t}, \qquad \varepsilon_{ij} = u_{(j,i)}, \qquad \omega_{ij} = u_{[j,i]}.$$

If the motion is irrotational, then $\omega_{ij} = 0$ and therefore the condition of kinematic continuity takes the form

$$[v_j] = -Gn^i[\varepsilon_{ij}]. \tag{8.9}$$

9. Mathematical methods applied in problems of plastic wave propagation

Numerous wave problems concerning elastic–plastic or elastic/viscoplastic media reduce to solving initial or boundary value problems for a system of first-order partial differential equations of the hyperbolic type. Since, in most of the problems discussed in this book, the motion of the medium is described by a single spatial variable and time, we shall confine ourselves to systems of equations with two independent variables only [33].

9.1. SYSTEMS OF QUASI-LINEAR PARTIAL DIFFERENTIAL EQUATIONS OF THE FIRST ORDER WITH TWO INDEPENDENT VARIABLES x AND t

The general form for partial differential equations of the first order with respect to functions $u_1, \ldots, u_n$ takes the following form:

$$F_i(u_{1,x}, \ldots, u_{n,x}, u_{1,t}, \ldots, u_{n,t}, u_1, \ldots, u_n, x, t) = 0 \qquad i = 1, 2, \ldots, k. \tag{9.1}$$

The system is determinate provided $n = k$. The system of (9.1) can be represented in another form if we introduce the vectors

$$u = \{u_1, \ldots, u_n\}, \qquad u_{,x} = \{u_{1,x}, \ldots u_{n,x}\}, \qquad u_{,t} = \{u_{1,t}, \ldots, u_{n,t}\}, \tag{9.2}$$

namely

$$F_i(u_{,x}, u_{,t}, u, x, t) = 0, \qquad i = 1, 2, \ldots, n. \tag{9.3}$$

Functions $u_i = u_i(x, t)$ possessing continuous first derivatives and satisfying the equations of system (9.3) are called *solutions of the system of equations.*

In the case when functions F_i are linear functions with respect to quantities $\mathbf{u}_{,x}$ and $\mathbf{u}_{,t}$, the system of differential equations (9.3) is called a *system of quasi-linear equations.* If functions F_i are linear with respect to the set of variables $\mathbf{u}$, $\mathbf{u}_{,x}$, $\mathbf{u}_{,t}$, then system (9.3) is called a *system of linear equations.*

A system of quasi-linear partial differential equations of the first order can be written down in the form

$$(9.4)\qquad L_j[\mathbf{u}] = \sum_{i=1}^{n} a_{ij} u_{i,x} + b_{ij} u_{i,t} + c_j = 0, \qquad j = 1, 2, \ldots, n,$$

where coefficients a_{ij}, b_{ij}, and c_j are functions of x, t, and $\mathbf{u}$ only. In the case when coefficients a_{ij} and b_{ij} are independent of $\mathbf{u}$, the system of (9.4) is called *a semi-linear system.* However, if also c_j is linearly dependent on $\mathbf{u}$ then such a system is called a *linear system.* If coefficients c_j are equal to zero then the system is called a *homogeneous system.*

The system of (9.4) can be represented also in the matrix form. We introduce vectors determined by (9.2), vector $\mathbf{c} = \{c_1, \ldots c_n\}$ and matrices $\mathbf{A}$ and $\mathbf{B}$ whose elements are the coefficients a_{ij} and b_{ij} respectively.

$$(9.5)\qquad \mathbf{A} = \begin{bmatrix} a_{11} & \ldots & a_{1n} \\ \cdot & \cdot\ \cdot\ \cdot & \cdot \\ a_{n1} & \ldots & a_{nn} \end{bmatrix}, \qquad \mathbf{B} = \begin{bmatrix} b_{11} & \ldots & b_{1n} \\ \cdot & \cdot\ \cdot\ \cdot & \cdot \\ b_{n1} & \ldots & b_{nn} \end{bmatrix}.$$

Next, assuming the notation $a_i b_i$ for $\sum_i a_i b_i$ (in order to avoid writing the summation sign in cases which do not lead to any ambiguity), we obtain

$$(9.6)\qquad L[\mathbf{u}] = \mathbf{A}\, u_{,x} + \mathbf{B} u_{,t} + \mathbf{c} = 0.$$

9.2. CHARACTERISTIC CURVES AND CHARACTERISTIC EQUATIONS

The system of (9.4) can be solved with respect to derivatives $u_{i,x}$, at a certain point of the phase plane (x_0, t_0), if and only if the matrix of coefficients a_{ij} is non-singular at point x_0, t_0. If matrix a_{ij} is singular at point (x_0, t_0), then the line $x = x_0$ is called a characteristic line at point (x_0, t_0), otherwise the line is called a non-characteristic one [33].

The solution of Cauchy's problem in which the initial values for u_i are given on an arbitrary analytical curve C, expressed in the form $\xi(x, t) = 0$ (belonging to the family of curves $\xi(x, t) =$ const), can be reduced to the case for which the initial set is an axis of coordinates by introducing new independent variables ξ and η instead of the variables x and t. The system of (9.4) in terms of coordinates ξ and η will take the following form:

$$(9.7)\qquad L_j[\mathbf{u}] = \sum_{i=1}^{n} [(a_{ij}\xi_{,x} + b_{ij}\xi_{,t}) u_{i,\xi} + (a_{ij}\eta_{,x} + b_{ij}\eta_{,t}) u_{i,\eta}] + c_j = 0.$$

Curve C is called a characteristic at point (x_0, t_0) if straight line $\xi = \xi_0$ is a characteristic for the transformed system (9.7) at the corresponding point (ξ_0, η_0).

The characteristics of the system of (9.4) will now be determined. It is assumed that in the region under consideration one of matrices (9.5), e.g. $\mathbf{B}$, is non-singular, i.e. that $|b_{ij}| \neq 0$. Moreover, it is assumed that the coefficients of the system of equations are differentiable.

We consider (9.4) and we pose Cauchy's problem, i.e. we determine on an arbitrary curve C, described by the equation $\varphi(x, t) = 0$ (where $\varphi_{,x}^2+\varphi_{,t}^2 \neq 0$), on which the initial values of the vector $\mathbf{u}$ are prescribed, the first derivatives $\mathbf{u}_{,x}$ in such a way that (9.4) $L[\mathbf{u}] = 0$ holds for the obtained region.

First of all, one should observe that the inner derivative $\mathbf{u}_{,t}\,\varphi_{,x}-\mathbf{u}_{,x}\,\varphi_{,t}$ is known on curve C. On curve C the following relation between $\mathbf{u}_{,t}$ and $\mathbf{u}_{,x}$ holds:

$$\mathbf{u}_{,t} = -a\mathbf{u}_{,x} + \frac{d\mathbf{u}}{dt}, \tag{9.8}$$

where $a = -\varphi_{,t}/\varphi_{,x}$.

Substituting (9.8) into (9.4) we obtain on curve C

$$L_j[\mathbf{u}] = (a_{ij}-ab_{ij})\,u_{i,x}+b_{ij}\,u_{i,t}+c_j = 0, \quad j = 1, 2, \ldots, n \tag{9.9}$$

i.e. a system of equations in the n derivatives.

The condition

$$Q = |\mathbf{A}-a\mathbf{B}| \neq 0, \tag{9.10}$$

where Q is called the *characteristic determinant* of the system of (9.4), constitutes a necessary and sufficient condition for the unique determination of all the first derivatives on C.

If $Q \neq 0$ on curves $\varphi(x, t) = \text{const}$, then the curves are *non-characteristic.* Such an arbitrary curve can be combined with the "band" for which (9.4) is valid.

If $a(x, t)$ is a real solution of the algebraic equation $Q = 0$ which is of nth degree in a, then the curves C determined by the ordinary differential equation

$$\frac{dx}{dt} = -\frac{\varphi_{,t}}{\varphi_{,x}} = a \tag{9.11}$$

constitute the *characteristic curves.*

If the equation $Q = 0$ does not possess real roots, then all the curves are non-characteristic; local continuation of the initial data is then always possible. In this case the system of equations is called an *elliptic system.* In the case when equation $Q = 0$ possesses n real roots and all the roots are distinct, such a system is said to be *completely hyperbolic.*

If a is a real root of the equation $Q = 0$, then, on curve C, the following system of equations can be solved with respect to vector $\mathbf{l}$ with components $\mathbf{l} = \{l_1, \ldots, l_n\}$:

$$\mathbf{l}\,(\mathbf{A}-a\mathbf{B}) = 0. \tag{9.12}$$

In this case the linear combination of (9.4) $l_jL_j[\mathbf{u}]$ can be written down in the form

$$l_j\,L_j[\mathbf{u}] = l_j\,b_{ij}(u_{i,t}+au_{i,x})+l_j\,c_j = 0 \tag{9.13}$$

or in the matrix form

$$\mathbf{l}L[\mathbf{u}] = \mathbf{lB}\,(\mathbf{u}_{,t}+a\mathbf{u}_{,x})+\mathbf{lc} = 0. \tag{9.14}$$

If we denote the derivative of function $u_i(x, t)$ with respect to t in the direction $a = dx/dt$ by

$$\frac{du_i}{dt} = u_{i,t} + au_{i,x} \tag{9.15}$$

then it is easy to see that (9.13) or (9.14) contains a linear combination of the derivatives du_i/dt.

The equation

$$a = \frac{dx}{dt} \tag{9.16}$$

denotes for all functions $u_i(x, t)$ a general direction of differentiation in (9.14) which is a characteristic direction for the system of (9.9).

Thus we have the following relations along the characteristic directions (9.16) of the system of (9.9):

$$\mathbf{l}L[\mathbf{u}] = \mathbf{lB}\left(\frac{d\mathbf{u}}{dt}\right) + \mathbf{lc} = 0. \tag{9.17}$$

These equations describe the *relations on the characteristics.* Vector $\mathbf{l} = \{l_1, \ldots, l_n\}$ is determined from (9.12). Thus, in the case of a system of equations of the hyperbolic type, i.e. when there exist n families of characteristic curves, the system of (9.4) can be replaced by the equivalent system of (9.17) in which each equation contains differentiation only in one characteristic direction only.

The discussion presented here can easily be generalized to the case of a system of equations of the first order of n independent variables (compare [33]).

Let us now consider the case of a system of two partial differential equations of the first order with independent variables x and t. Thus we consider the following system of equations:

$$L[\mathbf{u}] = \mathbf{A}\mathbf{u}_{,x} + \mathbf{B}\mathbf{u}_{,t} + \mathbf{c} = 0, \tag{9.18}$$

where

$$\mathbf{u} = \{u_1, u_2\}, \qquad \mathbf{c} = \{c_1, c_2\},$$

$$\mathbf{A} = \begin{bmatrix} a_{11} & a_{21} \\ a_{12} & a_{22} \end{bmatrix}, \qquad \mathbf{B} = \begin{bmatrix} b_{11} & b_{21} \\ b_{12} & b_{22} \end{bmatrix}. \tag{9.19}$$

The characteristic directions, in accordance with condition (9.10), are determined from the condition $|\mathbf{A} - a\mathbf{B}| = 0$. We obtain an algebraic equation of the second degree for a, the roots of which take the following form:

$$a_{1,2} = \frac{a_{12}b_{21} + a_{21}b_{12} - b_{11}a_{22} - a_{11}b_{22} + \sqrt{\Delta}}{2(b_{11}b_{22} - b_{12}b_{21})} \tag{9.20}$$

where

$$\Delta = (a_{12}b_{21} + a_{21}b_{12} - b_{11}a_{22} - a_{11}b_{22}) - 4(b_{11}b_{22} - b_{12}b_{21})(a_{11}a_{22} - a_{12}a_{21}).$$

The system of (9.18) is of:
hyperbolic type, if $\Delta > 0$,
parabolic type, if $\Delta = 0$,
elliptic type, if $\Delta < 0$.

The differential equations of the characteristics take the following form:

$$dx = a_{1,2}\, dt. \tag{9.21}$$

From (9.17) we obtain the relations along the characteristic directions $a_{1,2}$ in the form

$$(l_1 b_{11} + l_2 b_{12})\, du_1 + (l_1 b_{21} + l_2 b_{22})\, du_2 + (l_1 c_1 + l_2 c_2)\, dt = 0, \tag{9.22}$$

where vector $\mathbf{l} = \{l_1, l_2\}$ is determined from (9.12).

9.3. INITIAL AND BOUNDARY VALUE PROBLEMS FOR QUASI-LINEAR SYSTEMS OF TWO HYPERBOLIC DIFFERENTIAL EQUATIONS OF THE FIRST ORDER WITH TWO INDEPENDENT VARIABLES

As a rule we cannot find exact solutions of the systems of partial differential equations describing wave problems in elastic–plastic and elastic/viscoplastic bodies. This being the case, approximate methods are then applied [32], [79], [80], [134], [137]. These methods are based on the approximation of derivatives in the differential equations by difference quotients and then on solving the systems of difference equations instead of the systems of differential equations. The difference methods used should ensure convergence to the exact solution of differential equations. The stability of these methods constitutes a necessary condition for this convergence.

We shall now present the finite difference method applied along the characteristics. Let us consider a system of two homogeneous quasi-linear equations of the first order (9.18) (with $c = 0$):

$$\mathbf{A}\mathbf{u}_{,x} + \mathbf{B}\mathbf{u}_{,t} = 0. \tag{9.23}$$

We assume that the system of equations is hyperbolic, i.e. that it possesses two distinct characteristic directions $a_{1,2}$ determined by (9.20). In the case when (9.18) constitutes a system of semi-linear equations, i.e. when $\mathbf{u} = \mathbf{u}\,(x, t)$, $\mathbf{A} = \mathbf{A}\,(x, t)$, $\mathbf{B} = \mathbf{B}\,(x, t)$, and $\mathbf{c} = \mathbf{c}\,(x, t, \mathbf{u})$, the procedure is similar.

The solution of both initial and boundary value problems will be obtained by an approximate method consisting of replacing the differential equations of the characteristics in the hodograph plane, and in the phase plane by equations in finite difference form. It should be observed that in particular cases the exact analytical solution of initial and boundary value problems can be constructed by means of the method of characteristics.

Now we present the solution of the *Cauchy problem* for (9.23) in region D of phase plane (x, t) (Fig. 20), bounded by the non-characteristic curve $C: t = \varphi\,(x)$, spatially oriented, on which the boundary conditions† $u_1(x, \varphi\,(x))$ and $u_2(x, \varphi\,(x))$ are prescribed, and by characteristics led out from points A_0 and B_0. Two families of characteristics – positive and negative – can be led out from curve C, covering the entire region D by a net of characteristics (Fig. 20). Let us replace the differential equations of the characteristics (9.21) by equations

†The initial conditions are prescribed on a spatially orientated curve; the boundary conditions, however, are prescribed on a time-oriented curve. In the (x, t) plane curve C with equation $t = \varphi\,(x)$ is time-oriented at point (x_0, t_0) with respect to (9.23) provided the condition $|\varphi'(x_0)| > a$ or $\varphi'(x_0) = \mp\infty$ holds. On the other hand, curve C is spatially oriented if $|\varphi'(x_0)| < a$.

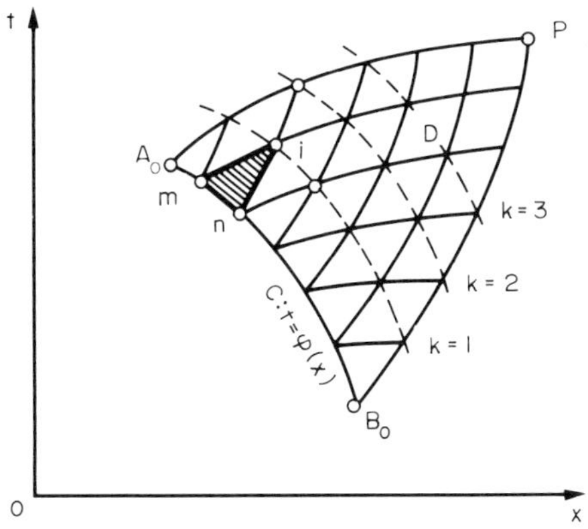

Fig. 20

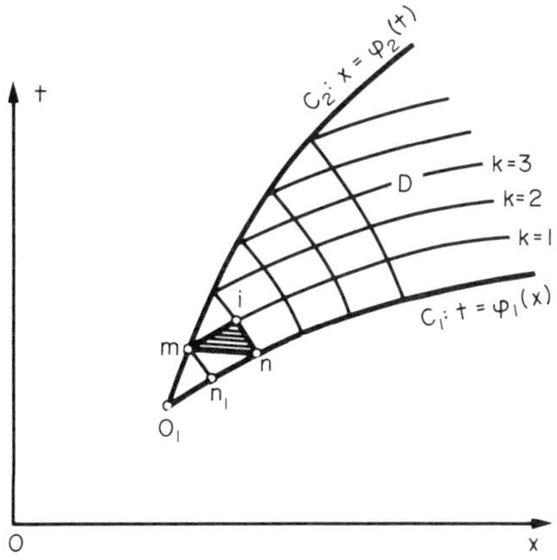

Fig. 21

in finite difference form:

$$x - x_m = a_1(u_{1m}, u_{2m}, \ldots)(t - t_m), \qquad (a_1 > a_2) \tag{9.24}$$
$$x - x_n = a_2(u_{1n}, u_{2n}, \ldots)(t - t_n)$$

and (9.22) (with $c_1 = c_2 = 0$) also by equations in terms of finite differences:

$$u_1 - u_{1m} = -\eta(u_{1m}, u_{2m}, \ldots)(u_2 - u_{2m}), \tag{9.25}$$
$$u_1 - u_{1n} = +\eta(u_{1n}, u_{2n}, \ldots)(u_2 - u_{2n}),$$

where

$$\eta(u_1, u_2, x, t) = \frac{l_1 b_{21} + l_2 b_{22}}{l_1 b_{11} + l_2 b_{12}}.$$

If the values x_m, x_n, t_m, t_n and the values of functions $u_{1m}, u_{2m}, u_{1n}, u_{2n}$ are known, i.e. the values of functions u_1, u_2 at points m and n (Fig. 20), then on solving the above systems of equations we find the values x, t, u_1, u_2 at point i corresponding to the point of intersection of the characteristics departing from points m and n. Thus, commencing the construction of the solution in each row ($k = 1, 2, \ldots$), from curve C we obtain the solution in the entire region D.

Now we shall discuss the *Picard problem* for a homogeneous system of quasi-linear equations (9.23) in region D (Fig. 21), bounded in the phase plane (x, t) by the characteristic curve C_1, $t = \varphi_1(x)$, with the prescribed values of functions $(x, \varphi_1(x))$, $u_2(x, \varphi_1(x))$, and by curve C_2, $x = \varphi_2(t)$, which is time-oriented, on which function $u_2(\varphi_2(t), t)$ is prescribed. The solution is constructed based on (9.24) and (9.25). Since the values $x_{n1}, t_{n1}, u_{1n_1}, u_{2n_1}$, and u_{2m} are known, we obtain from $(9.24)_2$ and $(9.25)_2$

$$x - x_{n_1} = a_2(u_{1n_1}, u_{2n_1}, \ldots)(t - t_{n_1}), \tag{9.26}$$
$$u_1 - u_{1n_1} = \eta(u_{1n_1}, u_{2n_1}, \ldots)(u_2 - u_{2n_1}),$$

the position of point m on curve C_2 (from the first of equations (9.26)), and the value of the function u_1 at this point. Next, knowing the values $x_m, x_n, t_m, t_n, u_{1n}, u_{2n}, u_{1m}$ and u_{2m}, as in the Cauchy problem, making use of (9.24) and (9.25), we find the solution for point i. Then we pass to the row $k = 2$, repeating the whole procedure; next we pass to the row $k = 3$, and so on.

If curve C_2, $t = \varphi_2(x)$ (Fig. 21), is also a characteristic one on which the values of functions $u_1(x, \varphi_2(x))$ and $u_2(x, \varphi_2(x))$, are known, then the *Darboux problem* for (9.23) can be solved in region D, bounded by characteristics C_1 and C_2 and by characteristics emanating from point P. On the other hand, if curve C_1, $x = \varphi_1(t)$,and curve C_2, $x = \varphi_2(t)$ (Fig. 21). are time-oriented non-characteristic curves, while the values of functions $u_1(\varphi_1(t), t)$ and $u_2(\varphi_1(t), t)$ are known on curve C_1, and the values of function $u_1(\varphi_2(t), t)$ are given on curve C_2, then the solution of the *Goursat problem* for (9.23) can be found in region D. In such a case region D is bounded by the characteristic curves departing from point P and by non-characteristic curves C_1 and C_2. The solutions of the Darboux and Goursat problems are constructed on the basis of (9.24) and (9.25).

On account of the fact that, generally speaking, in the case of the system of quasi-linear equations (9.23) the equations of the characteristics depend on the solution $\mathbf{u}$, the method of solution by means of finite differences along the characteristics is quite complicated.

The iterative method of Courant [33] is frequently used in the solving of a system of two first order quasi-linear or semi-linear equations of the hyperbolic type with two independent variables

$$\mathbf{A}\mathbf{u}_{,x} + \mathbf{B}\mathbf{u}_{,t} + \mathbf{c} = 0. \tag{9.27}$$

The method consists in reducing the system of partial differential equations to one of ordinary differential equations. In the equations only differentiation in the characteristic directions occurs. Next, the method of successive approximations is applied to solve the ordinary differential equations. A review of the approximate methods of solution of wave problems in inelastic media when the motion is described by a system of first-order hyperbolic, partial differential equations (almost-linear and quasi-linear) can be found in [32], [79], [120], [134], and [137]. In [120] the reader can find an evaluative review of five different methods of solving boundary value problems for inelastic bodies and of their effectiveness in solving these problems. It has been shown in the numerical calculations performed for the problem of a shear, cylindrical wave propagating in an elastic/viscoplastic medium that the results obtained by means of the iterative Courant method, by the method of direct integration and by the method of finite differences along the characteristics are convergent. The maximum difference between the results obtained by the application of the above methods did not exceed 3% (in the considered time interval).

In the case of a system of two linear partial differential equations of the first order with constant coefficients with two independent functions u_1 and u_2, the system can be reduced to a single partial differential equation with constant coefficients for functions u_1 or u_2. The solutions of initial value problems can then be found in an analytical form [33], [72], [79]. Also in these cases the Laplace integral transform can be applied (compare, for example, p. 101). However, this method is useless in some cases, namely when, simultaneously with the solution of a given system of equations in a region, an unknown boundary of the region has to be determined. This occurs, for example, when an unloading wave in an elastic–plastic medium is to be found (for an assumed sectionally linear material characteristic).

CHAPTER III

ONE-DIMENSIONAL PLANE WAVES

10. Propagation of a plane, longitudinal wave of loading in a semi-infinite, homogeneous, elastic–plastic bar

We shall consider the process of the propagation of plastic strain in a semi-infinite elastic–plastic bar produced by a dynamic loading $p(t)$ applied at the bar end and non-decreasing in time (i.e. $dp/dt > 0$). The problem will be discussed in a Lagrangian coordinate system, the x-axis will coincide with the bar axis, and the origin of the coordinate system, $x = 0$, is assumed to be at the end of the bar. It is also assumed that the bar, in the course of deformation, does not buckle and that the influence of the transverse deformation of the bar on the process of propagation of the longitudinal waves is negligibly small. We shall only consider small strains and we shall assume that the bar density during the deformation process does not change. The only non-vanishing stress tensor component is $\sigma_{xx} = \sigma$, the non-zero strain tensor components are $\epsilon_{xx} = \epsilon$ and $\epsilon_{yy} = -\nu\epsilon$.

If we neglect body forces, the equation of motion (5.5), in this case, reduces to the form

$$\varrho \frac{\partial v}{\partial t} = \frac{\partial \sigma}{\partial x} \tag{10.1}$$

Assuming the constitutive equations of the strain theory of plasticity, we have for a one-dimensional stress state

$$\sigma = \sigma(\varepsilon). \tag{10.2}$$

For the time being we assume (at this point in the book) that the function $\sigma(\epsilon)$ is a monotonically increasing function of ϵ (Fig. 22) and that for each ϵ the derivative $d\sigma/d\epsilon$ is a monotonically decreasing function (i.e. $d^2\sigma/d\epsilon^2 < 0$). If $d\sigma/d\epsilon$ is a monotonically increasing function, i.e. if

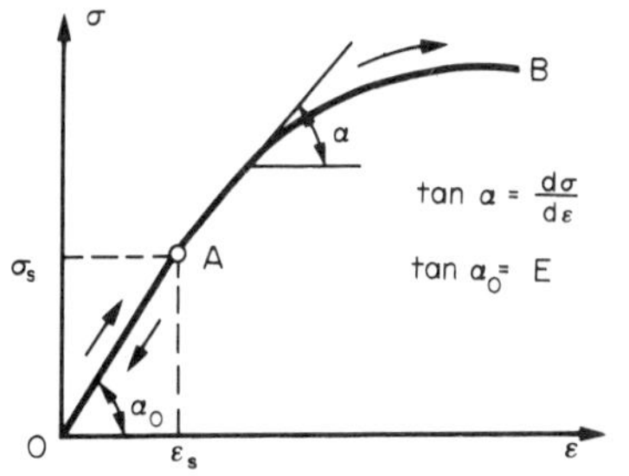

Fig. 22

$d^2\sigma/d\epsilon^2 > 0$ then shock waves appear in the bar; this case will be discussed at p. 67. For stresses $\sigma < \sigma_S$ (where σ_S is the elastic limit) the $\sigma - \epsilon$ relationship, in accordance with Hooke's law, is linear, i.e.

$$\sigma = E\varepsilon \quad \text{for} \quad \sigma \leqslant \sigma_s, \tag{10.3}$$

where E is Young's modulus.

From the continuity equation (5.7), for the case of small strains, we obtain the relationship

$$\frac{\partial \varepsilon}{\partial t} = \frac{\partial v}{\partial x}. \tag{10.4}$$

Taking into account the unique relation $\sigma = \sigma(\epsilon)$ on the loading curve and introducing the symbol,

$$a^2(\sigma) = \frac{1}{\varrho}\frac{d\sigma}{d\varepsilon}, \tag{10.5}$$

where $d\sigma/d\epsilon$ denotes the tangent to the $\sigma - \epsilon$ curve in Fig. 22, we have

$$\frac{\partial \varepsilon}{\partial t} = \frac{d\varepsilon}{d\sigma}\frac{\partial \sigma}{\partial t} = \frac{1}{\varrho a^2(\sigma)}\frac{\partial \sigma}{\partial t}. \tag{10.6}$$

Making use of (10.4) we obtain from (10.6) the following system of two partial differential equations of the first order,

$$\varrho\frac{\partial v}{\partial t} = \frac{\partial \sigma}{\partial x}, \qquad \frac{\partial v}{\partial x} = \frac{1}{\varrho a^2(\sigma)}\frac{\partial \sigma}{\partial t}, \tag{10.7}$$

with two unknown functions $v(x, t)$ and $\sigma(x, t)$.

The system of (10.7) is equivalent to an equation of the second order in terms of displacement $u(x, t)$. Taking into account that $v = \partial u/\partial t$, $\epsilon = \partial u/\partial x$, we obtain

$$\frac{\partial^2 u}{\partial t^2} - a^2(\varepsilon)\frac{\partial^2 u}{\partial x^2} = 0. \tag{10.8}$$

$a(\epsilon)$, in this equation, denotes the speed of longitudinal wave propagation in the bar.

On account of the fact that the speed of wave propagation is, in general, a function of stress, the system of (10.7) constitutes a system of first-order hyperbolic quasi-linear partial differential equations. We shall determine the characteristics for this system and the relations on characteristics. The matrices **A** and **B** of (9.18) for the system of (10.7) have the following form:

$$\mathbf{A} = \begin{bmatrix} 0 & -\dfrac{1}{\varrho} \\ 1 & 0 \end{bmatrix}, \qquad \mathbf{B} = \begin{bmatrix} 1 & 0 \\ 0 & -\dfrac{1}{\varrho a^2(\sigma)} \end{bmatrix}.$$

On the basis of the condition (compare Chapter I):

$$|\mathbf{A}-a\mathbf{B}| = -a^2 \frac{1}{\varrho a^2(\sigma)} + \frac{1}{\varrho} = 0,$$

we obtain the following characteristic directions:

$$a = \mp a(\sigma). \tag{10.9}$$

Since $a = dx/dt$, then the characteristics of the system of (10.7) are obtained by integration of the differential equations for the characteristics

$$dx = \mp a(\sigma)\, dt. \tag{10.10}$$

In general these equations cannot be integrated over the plane (x, t) before solving the problem, since quantity a is a function of stress $\sigma(x, t)$.

The relations (9.17), yielding the equation

$$l_1 \frac{du_1}{dt} - l_2 \frac{1}{\varrho a^2(\sigma)} \frac{du_2}{dt} = 0,$$

have to be satisfied on the characteristics.

From (9.12) we obtain the following relation:

$$l_2 = \mp a(\sigma)\, l_1,$$

hence the corresponding relations

$$dv \pm \frac{1}{\varrho a(\sigma)} d\sigma = 0 \tag{10.11}$$

are valid along the characteristics $dx = \mp a(\sigma)\, dt$.

These relations are called the differential equations of characteristics in the (σ, v) or hodograph plane. On integrating we obtain

$$v = \mp \frac{1}{\varrho} \int_0^\sigma \frac{d\sigma_1}{a(\sigma_1)} + C_{1,2} \quad \text{for} \quad dx = \mp a(\sigma)\, dt. \tag{10.12}$$

Thus the system of first-order quasi-linear partial differential equations (10.7), by the use of the methods of characteristics, has been replaced by the equivalent system of ordinary differential equations (10.11) along the characteristics.

Now we consider the simplest case of the propagation of the loading waves in a homogeneous, semi-infinite bar, the initial state of which was undisturbed.

We determine the solution of (10.8) with the following prescribed initial conditions (Cauchy conditions):

$$u(x, 0) = \left.\frac{\partial u(x,t)}{\partial t}\right|_{t=0} = 0 \quad \text{for} \quad t = 0,\ x > 0, \tag{10.13}$$

and the boundary conditions

(10.14) $$\sigma(0,t) = -p(t) \quad (p(t) > 0) \quad \text{for} \quad x = 0, \; t \geqslant 0,$$

where $p'(t) \geqslant 0$, in order to ensure that it is a loading process.

The initial conditions (10.13) mean that the bar at the initial instant was in the undeformed state and at rest. The validity of the initial conditions (10.13) is connected with the solution of Cauchy's problem for region I (Fig. 23) bounded by the x-axis and the positive characteristic OA. Drawing from an arbitrary point $P(x, t)$ of region I the positive and the negative characteristics until they intersect the x-axis, we obtain from (10.12)

(10.15) $$v(x,t) = -\frac{1}{\varrho}\int_0^{\sigma} \frac{d\sigma_1}{a(\sigma_1)}, \quad v(x,t) = +\frac{1}{\varrho}\int_0^{\sigma} \frac{d\sigma_1}{a(\sigma_1)},$$

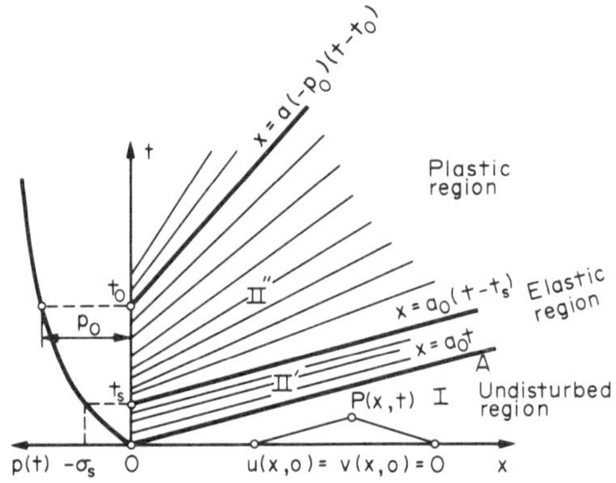

Fig. 23

since $C_1 = C_2 = 0$ from the initial conditions (10.13). It follows from (10.15) that in region I we have $u(x, t) = v(x, t) \equiv 0$, and $a = a(0) = a_0$, where $a_0 = \sqrt{E/\varrho}$ denotes the speed of propagation of the longitudinal elastic waves in the bar. Thus region I is bounded at the top by the characteristic OA which has the equation $x = a_0 t$. Now we determine the solution behind the wave front $x = a_0 t$, where the interaction of the boundary condition is already apparent. Along the negative characteristics for $t > x/a_0$, we have the relation

$$v = -\frac{1}{\varrho}\int_0^{\sigma} \frac{d\sigma_1}{a(\sigma_1)} + C_2.$$

Since these characteristics commence at the characteristic $x = a_0 t$, where $v = \sigma = 0$, we have $C_2 = 0$. Next, making use of the relations on the positive characteristics we obtain for region II: $v = \text{const}$, $\sigma = \text{const}$, and also $\epsilon = \text{const}$ along the positive characteristics. Thus, along each positive characteristic in region II, the stress σ, the particle velocity v, and the strain ϵ possess the same values as at the origin of the bar at the time instant corresponding to the given characteristic. On account of the fact that the speed of wave propagation is a function of stress (or of strain), which is constant along each characteristic, the characteristics are straight lines of the form:

$$x = a(\sigma)(t - t_0).$$

If, for example, a compressive stress at the end of the bar reaches value $-p_0$ at time t_0, then the stress at point x from the end reaches value $-p_0$ after time

$$t = t_0 + \frac{x}{a(-p_0)}.$$

The stress on the surface of wave $x = a(-p_0)(t - t_0)$ is constant and equal to $-p_0$. Such waves are called *Riemann waves.*

The stress at the end of the bar increases in time (Fig. 23), reaching at instant $t = t_s$ value $\sigma = \sigma_s$ corresponding to the elastic limit. We divide region II on the phase plane into two subregions II′ and II″. Region II′ is governed by elastic state; the Riemann waves propagate in it with constant speeds equal to a_0, since according to the relation $\sigma = \sigma(\epsilon)$ (Fig. 22) for $\sigma < \sigma_s$ we have $a_0 = \sqrt{E/\varrho}$. In region II″ the velocity of the Riemann waves varies depending on the value of the pressure at the end of the bar ($x = 0$), i.e. on the value of the pressure at instant $t = t_0$ when the wave is originated. On account of the assumed function $\sigma = \sigma(\epsilon).(d^2\sigma/d\epsilon^2 < 0)$ the velocity of successive Riemann waves is smaller and smaller. As the disturbance is carried away from the end of the bar, the initial stress region (at $x = 0$) is "washed out" in time. If the values of stress, velocity, and strain along the characteristics are known (σ = const, v = const, ϵ = const along $x = a(\sigma)t$ + const), then the stress, velocity, and strain can easily be determined at an arbitrary cross-section of the bar x at any instant t.

We shall now discuss certain particular cases of loading wave propagation in a semi-infinite homogeneous elastic–plastic bar.

First we consider the case of wave propagation in an initially undisturbed bar for which the σ–ϵ relation is of the form as in Fig. 22, and that $\sigma_s \neq 0$. The pressure $p(t)$ monotonically increasing in time $0 \leqslant t < t^*$ up to value p_0 ($p_0 > |\sigma_s|$), then constant, equal to p_0 for $t > t^*$ (Fig. 24) is prescribed at the end $x = 0$. Region I is undisturbed on account of the homogeneous

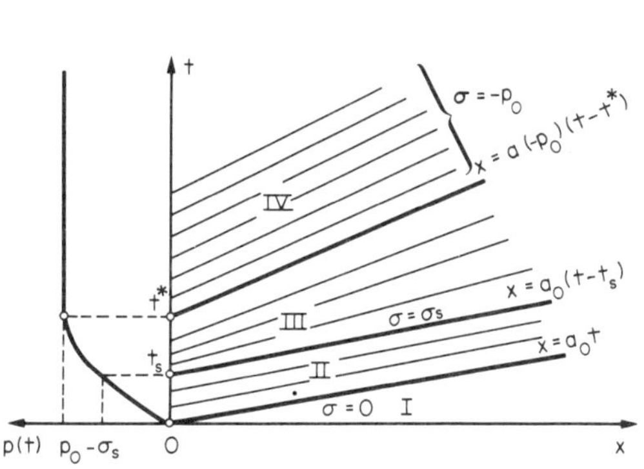

Fig. 24

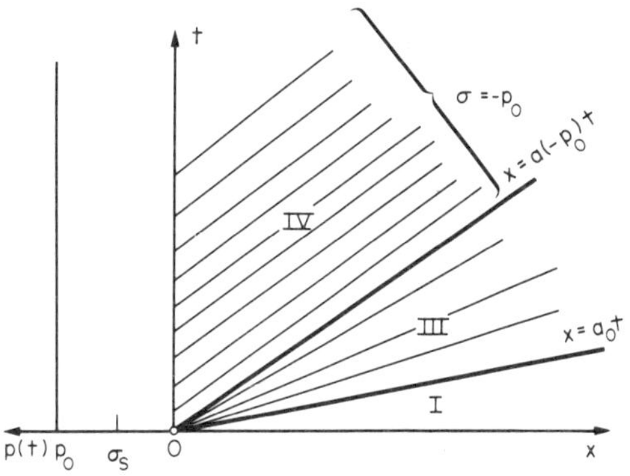

Fig. 25

initial conditions. In region II the elastic waves propagate; in this region the stress is equal to $\sigma = -p(t - x/a_0)$. In region III, bounded by characteristic $x = a_0(t - t_s)$, plastic deformation occurs (Reimann waves propagate there). Finally, region IV is also a region of plastic strain. Due to the assumed boundary condition the stress there is constant and equal to $\sigma = -p_0$.

Passing to the limit with $t^* \to 0$ we obtain the case of the sudden loading of the bar by a constant value of pressure $p_0(p_0 > |\sigma_s|)$. The picture of the solution is shown in Fig. 25. The passage to the limit will be performed in two stages, namely, first the time t_s tends to zero, then region II vanishes; on the characteristic $x = a_0 t$ (Fig. 25) we obtain the value of stress $\sigma = \sigma_s$, therefore the wave is a wave of strong discontinuity. Next we let $t^* \to t_s = 0$ obtaining region III in Fig. 24 in the form of a triangle contained between characteristics $x = a_0 t$ and $x = a(-p_0)t$ (Fig. 25). Region IV obviously remains unchanged.

If it is assumed that the plastic limit is disregarded in the $\sigma - \epsilon$ relation (Fig. 22), i.e. that $\sigma_s \equiv 0$ (such a model can approximate, for example, a dry sandy soil), then in the case of the problem presented in Fig. 24 we have to pass to a limit with $\sigma_s \to 0$. Then regions II and III constitute a single region in which Riemann waves of variable velocity (decreasing with the increase of stress) will propagate. In the case of sudden loading (Fig. 25), if we pass to the limit $\sigma_s \to 0$ we obtain the solution as in Fig. 26. In this case there is no wave of strong discontinuity.

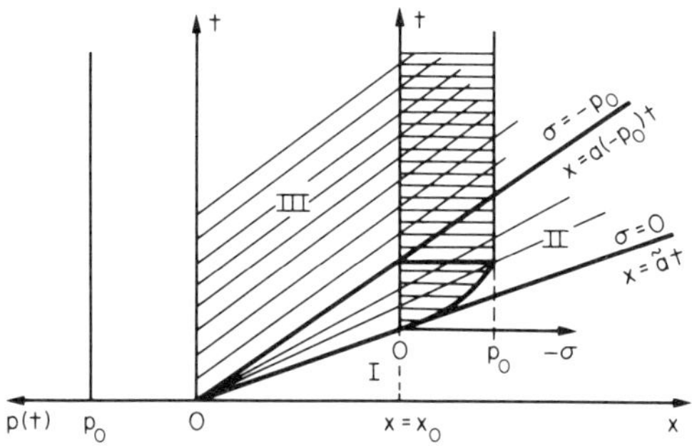

Fig. 26

On the fastest propagating wave $x = \tilde{a}t$, where $\tilde{a} = \sqrt{\left.\frac{1}{\varrho}\frac{d\sigma}{d\varepsilon}\right|_{\sigma=0}}$, the stress is equal to zero.

Behind it the waves of smaller velocity propagate, carrying on their fronts greater and greater stresses (region II). Finally, for $t \geqslant x/a(-p_0)$ the waves propagate a constant value of stress equal to p_0 (region III). In this case we can observe an interesting phenomenon, namely that the pressure, although suddenly applied at the end of the bar, gently increases from zero to the maximum value p_0 at an arbitrary cross-section x = const. Besides this, the gradient of the increase decreases with the distance from the end $x = 0$. Such a variation of stress in time at the cross-section $x = x_0$ is shown in Fig. 26.

Frequently a particular approximation of the real $\sigma - \epsilon$ diagram presented in Fig. 22 is assumed. In this, the curvilinear segment AB is approximated by a straight line (Fig. 27b). In the process of loading such a model a sudden change occurs in the wave propagation velocity as the load changes, namely for $\sigma \leqslant \sigma_s$ we have as before $a_0 = \sqrt{E/\varrho}$ while for $\sigma > \sigma_s$ we obtain $a(\sigma) = a_1 = \sqrt{E_1/\varrho}$ = const, where E_1 denotes the linear hardening modulus of the material. For steel we get $E_1 \approx (0.003 - 0.01)E$, for dry soils $E_1 \approx (0.05 - 0.1)E$.

The assumption that the $\sigma - \epsilon$ function is bilinear leads to a change in the picture of the wave solutions in the phase plane. Let us first consider the case when the passage from the elastic segment in the $\sigma - \epsilon$ diagram to the hardening segment is smooth along an arc AA_1 ($d\sigma/d\epsilon$ is a continuous function for any value of σ) (Fig. 27a). The solution for such a $\sigma - \epsilon$ relation and for a pressure monotonically increasing in time is presented in Fig. 28a. The difference as compared with the case shown in Fig. 23 consists in the fact that in region IV the disturbances

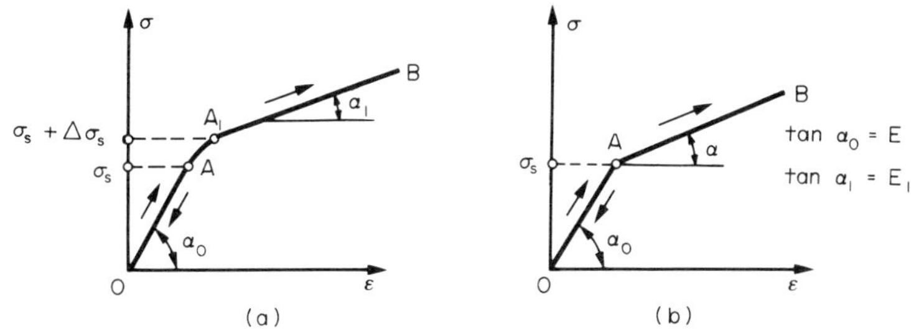

Fig. 27

propagate with constant velocity equal to a_1. If now point A^1 in the $\sigma - \epsilon$ daigram (Fig. 27a) tends to point A, i.e. if $\Delta\sigma_s \to 0$, then the stress in region III tends to a constant value $\sigma = \sigma_s$. This case (for a bilinear $\sigma - \epsilon$ relation as in Fig. 27b) is shown in Fig. 28b.

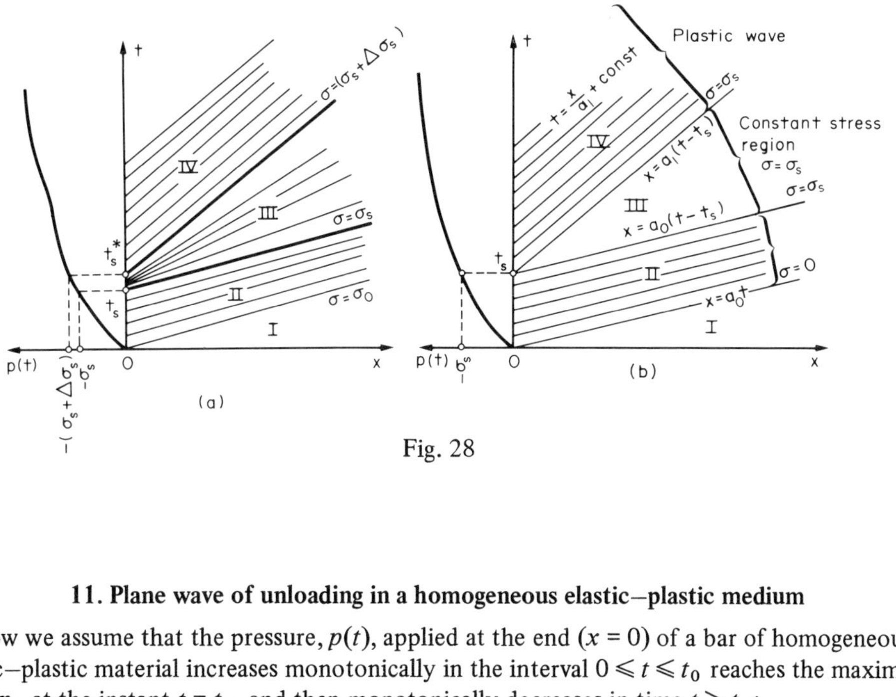

Fig. 28

11. Plane wave of unloading in a homogeneous elastic–plastic medium

Now we assume that the pressure, $p(t)$, applied at the end ($x = 0$) of a bar of homogeneous, elastic–plastic material increases monotonically in the interval $0 \leqslant t \leqslant t_0$ reaches the maximum value p_0 at the instant $t = t_0$, and then monotonically decreases in time $t > t_0$:

$$\sigma(0, t) = \begin{cases} -p_1(t), & p_1(t) > 0, \dfrac{dp_1}{dt} \geqslant 0 \quad \text{for} \quad 0 \leqslant t \leqslant t_0, \\ -p_2(t), & p_2(t) > 0, \dfrac{dp_2}{dt} \leqslant 0 \quad \text{for} \quad t \geqslant t_0, \end{cases} \tag{11.1}$$

and $p_1(t_0) = p_2(t_0) = p$.

Let us assume that the material characteristic $\sigma = \sigma(\epsilon)$ (Fig. 29) for the loading process consists of a straight line segment for $\sigma \leqslant \sigma_s$ (line OA) and a curve convex with respect to the

σ-axis for $\sigma \geqslant \sigma_s$ (curve AB) ($d^2\sigma/d\epsilon^2 < 0$); on the other hand, the unloading process (line BC) is elastic, i.e. stress σ and strain ϵ are linearly dependent,

$$\sigma - \sigma_0(x) = E(\varepsilon - \varepsilon_0(x)), \tag{11.2}$$

where $\sigma_0(x)$ and $\epsilon_0(x)$ denote the maximum stress and strain values, respectively, in the deformation process for a given cross-section of the bar. Simultaneously we assume that $d\sigma/d\epsilon$ is a continuous function of ϵ during the loading process.

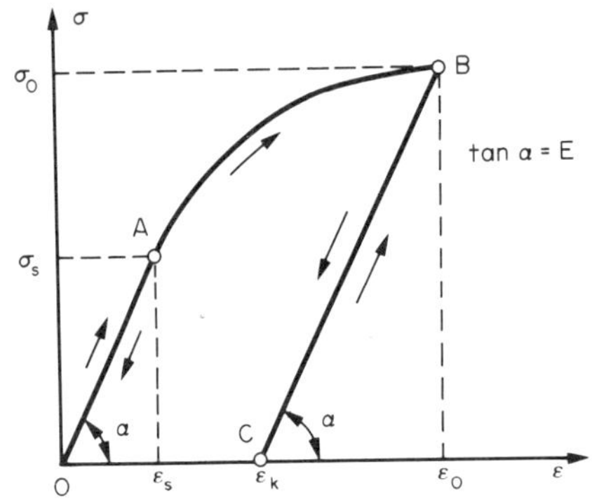

Fig. 29

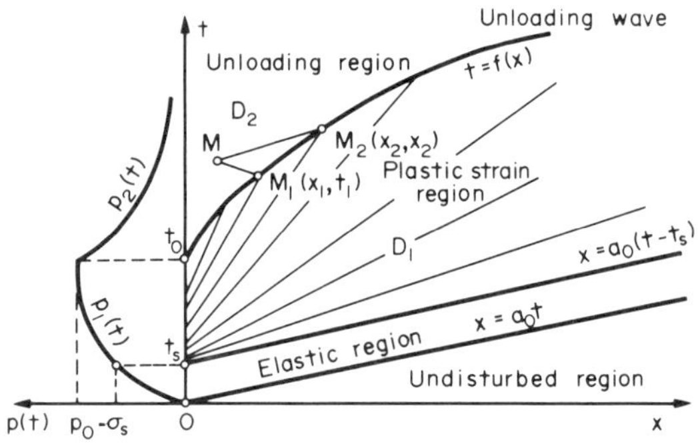

Fig. 30

At a given cross-section x of the bar (Fig. 30) the strain of an element increases continuously with time, in a manner which depends on the form of curve $\sigma = \sigma(\epsilon)$ and on the form of pressure $p_1(t)$, up to the value $\epsilon_0(x)$, which is smaller, however, than the maximum strain at the bar end $x = 0$. Next the strain of the element will decrease in a way which also depends on the character of the pressure variation at the end of bar $p_2(t)$ and on relation (11.2). Thus, in the phase plane, at a certain time t, from which the unloading process of the element begins, corresponding to the given x, there exists a curve, called the unloading wave, the equation of which is denoted in Fig. 30 as $t = f(x)$. Below the curve there is the process of plastic loading, above it the process of unloading. The problem of the determination of unloading waves was first presented by Rakhmatulin [124] in 1945.

The problem of the propagation of an unloading plane wave in a semi-infinite, homogeneous bar will be formulated in the following manner.

Determine functions $\sigma(x, t)$, $v(x, t)$, and $f(x)$ such that the curve described by the differentiable function $f(x)$ divides the phase plane (x, t) (for $x \geqslant 0, t > 0$) into two regions D_1 and D_2 (Fig. 30) in which functions $\sigma(x, t)$ and $v(x, t)$ are continuous in the region $x \geqslant 0$ and continuously differentiable in each of the regions D_1 and D_2 and satisfy the following equations respectively:

region D_1 (the region in which the material particles are in the process of plastic loading) is governed by the equations

$$\varrho \frac{\partial v}{\partial t} = \frac{\partial \sigma}{\partial x}, \quad \frac{\partial v}{\partial x} = \frac{1}{\varrho a^2(\sigma)} \frac{\partial \sigma}{\partial t} \quad \text{for} \quad \frac{d\varepsilon}{dt} \geqslant 0, \tag{11.3}$$

region D_2 (unloading region) is governed by (10.1), (10.4), and (11.2) which can be rewritten in the form

$$\frac{\partial v}{\partial t} = a_0^2 \frac{\partial \varepsilon}{\partial x} + \frac{1}{\varrho}\frac{d\sigma_0(x)}{dx} - a_0^2 \frac{d\varepsilon_0(x)}{dx} \qquad \text{for} \quad \frac{d\varepsilon}{dt} < 0. \tag{11.4}$$
$$\frac{\partial v}{\partial x} = \frac{\partial \varepsilon}{\partial t}$$

The shape of the unloading wave and function $\epsilon_0(x)$, unknown before the problem is solved, are to be found by means of the simultaneous solution of the systems of differential equations (11.3) and (11.4) which describe the deformation process, respectively, in the loading region and the elastic unloading region. Simultaneously, the boundary condition (11.1), as well as the continuity conditions of the velocity, stress, and strain, are to be satisfied on the unloading wave.

In the case of the boundary condition in the form (11.1) the unloading wave starts propagating from the end of the bar $x = 0$ at the instant $t = t_0$ (Fig. 30). The conditions prevailing at point $(0, t_0)$ correspond to the maximum value of stress at the bar end.

The solution in region D_1, i.e. up to the instant of the arrival of the unloading wave, is identical with that in the case considered in the preceding section, presented in Fig. 23 by regions II′ and II″. In region D_1, Riemann waves propagate, the equations of which are in the form $x = a(\sigma)t$. According to (10.12) and (10.5) the velocity of material particles in region D_1 1 can be expressed by the formula ($C_2 = 0$):

$$v = -\frac{1}{\varrho}\int_0^{\sigma} \frac{d\sigma_1}{a(\sigma_1)} = -\int_0^{\varepsilon} a(\varepsilon_1)\, d\varepsilon_1. \tag{11.5}$$

The following conditions should be satisfied on the unloading wave:

$$\varepsilon = \frac{\partial u}{\partial x} = \varepsilon_0(x) \qquad \text{for} \quad t = f(x) = \frac{x}{a[\varepsilon_0(x)]}, \tag{11.6}$$
$$v = \frac{\partial u}{\partial t} = v_0(x) = -\int_0^{\varepsilon_0} a(\varepsilon)\, d\varepsilon$$

where $\epsilon_0(x)$ and $v_0(x)$ are computed from the solution in region D_1 for $t = f(x)$.

Thus we have to solve the system of (11.4) with conditions of the Cauchy type (11.6) on the unloading wave and for the prescribed boundary condition (11.1). The system of (11.4) can be reduced to an equation of the second order in terms of displacement $u(x, t)$, namely:

$$\frac{\partial^2 u}{\partial t^2} - a_0^2 \frac{\partial^2 u}{\partial x^2} = F(x), \tag{11.7}$$

where

$$F(x) = \frac{1}{\varrho}\frac{d\sigma_0(x)}{dx} - a_0^2\frac{d\varepsilon_0(x)}{dx}. \tag{11.8}$$

The general solution of (11.7) takes the form

$$u(x,t) = \Phi_1(a_0 t - x) + \Phi_2(a_0 t + x) - \frac{1}{a_0^2}\int_0^x d\xi \int_0^\xi F(\bar{\xi})\, d\bar{\xi}. \tag{11.9}$$

On taking into account (11.8) we obtain

$$u(x,t) = \Phi_1(a_0 t - x) + \Phi_2(a_0 t + x) - \frac{1}{E}\int_0^x [\sigma_0(\xi) - E\varepsilon_0(\xi)]\, d\xi. \tag{11.10}$$

The validity of conditions (11.6) and of the boundary condition (11.1), which can be presented in the form

$$\left.\frac{\partial u}{\partial x}\right|_{x=0} = \varepsilon(0,t) = e(t), \tag{11.11}$$

where $\epsilon(0, t)$ is uniquely determined from the $\sigma - \epsilon$ diagram by $\sigma(0, t)$, enables us to determine uniquely functions Φ_1, Φ_2 and $\epsilon_0(x)$. For the given $\epsilon_0(x)$ we can determine $\sigma_0(x)$ from relation (11.2).

Let us consider a slightly simpler case. Namely, we assume that the pressure at the end of the bar $x = 0$ has been suddenly applied, its value at the initial instant is p_0, then it decreases monotonically (Fig. 31), i.e. it is assumed that the boundary condition (11.1) takes the form

$$\sigma(0,t) = -p(t), \quad p(t) > 0, \quad \frac{dp}{dt} \leqslant 0 \quad \text{for} \quad t \geqslant 0, \tag{11.1'}$$

or in the strains

$$\varepsilon(0,t) = e(t), \quad e(t) < 0, \quad \frac{de}{dt} \geqslant 0 \quad \text{for} \quad t \geqslant 0. \tag{11.1''}$$

In this case the unloading wave starts propagating from the origin of the coordinate system.

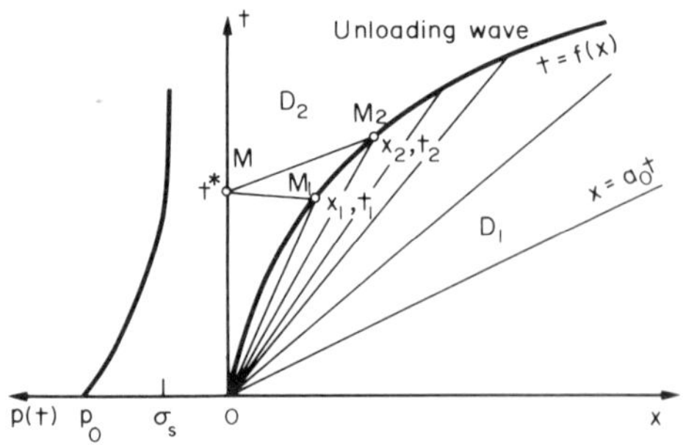

Fig. 31

Substituting (11.10) into the boundary condition (11.1″) we obtain the relation

$$-\Phi_1'(a_0 t)+\Phi_2'(a_0 t)-\frac{\sigma_0(0)}{E}+\varepsilon_0(0)=e(t), \tag{11.12}$$

on the basis of which we have

$$\Phi_2'(a_0 t+x)=\Phi_1'(a_0 t+x)+\frac{\sigma_0(0)}{E}-\varepsilon_0(0)+e\left(\frac{x}{a_0}+t\right). \tag{11.13}$$

Substituting next (11.10) into the initial conditions on the unloading wave $t=f(x)$ (11.6) we obtain

$$\begin{aligned} &\Phi_1'(a_0 f(x)-x)+\Phi_2'(a_0 f(x)+x)=\frac{v_0(x)}{a_0},\\ &-\Phi_1'(a_0 f(x)-x)+\Phi_2'(a_0 f(x)+x)-\frac{\sigma_0(x)}{E}=0. \end{aligned} \tag{11.14}$$

If we take into account (11.6) and the equations $x=a(\epsilon_0)f(x)$ and $\sigma_0(x)=\sigma_0[\epsilon_0(x)]$ and then eliminate, by means of (11.13), function Φ_2' we obtain from (11.14)

$$\begin{aligned} &\Phi_1'\left[x\left(1+\frac{1}{\alpha}\right)\right]+\Phi_1'\left[x\left(\frac{1}{\alpha}-1\right)\right]=-\int_0^{\varepsilon_0}\alpha\,d\varepsilon-\frac{\sigma_0(0)}{E}-\varepsilon_0(0)-e\left[\frac{x}{a_0}\left(\frac{1}{\alpha}+1\right)\right],\\ &\Phi_1'\left[x\left(1+\frac{1}{\alpha}\right)\right]-\Phi_1'\left[x\left(\frac{1}{\alpha}-1\right)\right]=\frac{\sigma_0[\varepsilon_0(x)]}{E}+\varepsilon_0(0)-\frac{\sigma_0(0)}{E}-e\left[\frac{x}{a_0}\left(\frac{1}{\alpha}+1\right)\right], \end{aligned} \tag{11.15}$$

whence

$$\begin{aligned} &\Phi_1'\left[x\left(1+\frac{1}{\alpha}\right)\right]=\varepsilon_0(0)-\frac{\sigma_0(0)}{E}-e\left[\frac{x}{a_0}\left(1+\frac{1}{\alpha}\right)\right]+\frac{1}{2}\left[\frac{\sigma_0(\varepsilon_0)}{E}-\int_0^{\varepsilon_0}\alpha\,d\varepsilon\right],\\ &\Phi_1'\left[x\left(\frac{1}{\alpha}-1\right)\right]=-\frac{1}{2}\left[\frac{\sigma_0(\varepsilon_0)}{E}+\int_0^{\varepsilon_0}\alpha\,d\varepsilon\right], \end{aligned} \tag{11.16}$$

where $\alpha=a(\epsilon)/a_0$.

From an arbitrary point M at the end of the bar $x=0$ (Fig. 31) we can draw the negative and the positive elastic characteristics which intersect with the unloading wave $t=f(x)$. The characteristics intersect the unloading wave at points $M_1(x_1, t_1)$ and $M_2(x_2, t_2)$ respectively. From the elastic characteristics MM_1 and MM_2 and the characteristics OM_1 and OM_2 in the loading region we obtain the following relations:

$$x_1\left(\frac{a_0}{a[\varepsilon_0(x)]}+1\right)=a_0 t^*,\qquad x_2\left(\frac{a_0}{a[\varepsilon_0(x)]}-1\right)=a_0 t^*, \tag{11.17}$$

where t^* denotes the time measured at the end of the bar $x=0$. Taking into account (11.17) we are able to eliminate $\Phi_1'(a_0 t^*)$ from the system of (11.16). We obtain the functional equation for $\epsilon_0(x)$

(11.18)

$$\varepsilon_0(0)-\frac{\sigma_0(0)}{E}-e(t^*)+\frac{1}{2}\int_0^{\varepsilon_0[x_1(t^*)]}\frac{a(\varepsilon)}{a_0}\left(\frac{a(\varepsilon)}{a_0}-1\right)d\varepsilon+\frac{1}{2}\int_0^{\varepsilon_0[x_2(t^*)]}\frac{a(\varepsilon)}{a_0}\left(\frac{a(\varepsilon)}{a_0}+1\right)d\varepsilon=0.$$

If we know the distribution of the permanent strains $\epsilon_k(x)$ from an experiment, in a semi-infinite elastic–plastic bar, and if the following relations are valid,

(11.19)
$$\varepsilon_k(x)=\varepsilon_0(x)-\frac{\sigma_0(x)}{E},$$

then for the given dynamic relation $\sigma=\sigma(\epsilon)$ we can determine, using (11.18), the nature of the strain changes at the end of the bar.

In [145] the problem has been considered of the propagation of an unloading wave in the case when the pressure is suddenly applied at the end and then monotonically decreases (the boundary condition (11.1′)). The existence and the uniqueness of the function describing the unloading wave have been proved. It has been proved that the reciprocal of the unloading wave speed is contained within the interval

(11.20)
$$\frac{1}{a[\varepsilon_0(x)]}\geqslant\frac{df}{dx}\geqslant\frac{1}{a_0},$$

and that as $t\to\infty$ the velocity of propagation of the unloading wave tends asymptotically to that of the propagation velocity of elastic waves.

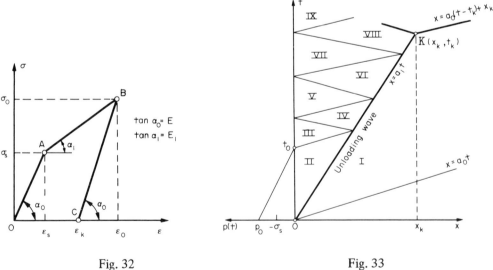

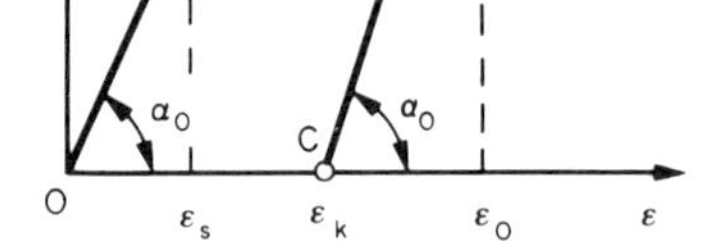

Fig. 32

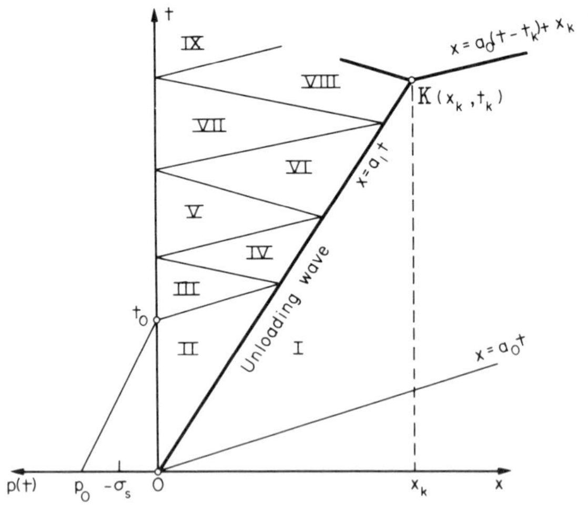

Fig. 33

A special case of (11.18) has been obtained in [120], namely, a semi-infinite bar described by Prandtl's model (Fig. 32) again with the assumption that the pressure is suddenly applied at the end of the bar and then decreases with time. On account of the fact that the unloading wave may intersect only these Riemann waves with speed $a_1 = \sqrt{E_1/\varrho}$ (the elastic waves of speed a_0 carry strains $\epsilon(x) \leqslant \epsilon_s$), in the case considered the unloading wave can only take the form of a straight line

$$t = f(x) = \frac{x}{a_1}. \tag{11.21}$$

In this case the unloading wave is a wave of strong discontinuity. It can be proved that in the case discussed the unloading wave is always followed by another unloading wave.

In the special case when the pressure at the end is suddenly applied at the instant $t = 0$, and then decreases linearly with time, i.e.

$$p(t) = p_0\left(1 - \frac{t}{t_0}\right), \tag{11.22}$$

we obtain from (11.18), making use of (11.2), the strain on the unloading wave

$$\varepsilon_0(x) = -\frac{p_0}{E_1}\left(1 - \frac{\mu^2 - 1}{2\mu} - \frac{x}{a_0 t_0}\right) + \varepsilon_s(1 - \mu^2), \tag{11.23}$$

where $\mu = a_0/a_1$ (for steel $10 \leqslant \mu \leqslant 15$, for dry sandy soils $2.5 \leqslant \mu \leqslant 4$).

It results from formula (11.23) that $\epsilon_0(x)$ is a decreasing function of x, therefore at a certain cross-section $x = x_k$ it will diminish to value ϵ_s. Point (x_k, t_k) (Fig. 33) denotes the end of the unloading wave. Thus the length of the segment of the bar that is plastically deformed x_k is obtained from (11.23) provided we substitute $\epsilon_0(x_k) = \epsilon_s$. For $x > x_k$ only elastic waves will propagate into the bar.

The stress on the unloading wave is uniquely determined by the relation $\sigma_0(x) = \sigma_0[\epsilon_0(x)]$ resulting from the model assumed for medium; thus we have

$$\sigma_0(x) = -p_0\left(1 - \frac{\mu^2 - 1}{2\mu} - \frac{x}{a_0 t_0}\right). \tag{11.24}$$

The unloading wave $x = a_1 t$ is a wave of strong discontinuity. In order to determine the velocity of the particles behind the front of the wave one should make use of the condition of the dynamic continuity.

The condition of the dynamic continuity (7.18) for the longitudinal waves in a bar takes the form:

for *forward-facing* waves ($dx = +a(\epsilon)dt$):

$$[\sigma] = -\varrho a[v], \tag{11.25}$$

for *backward-facing* waves ($dx = -a(\epsilon)dt$):

$$[\sigma] = +\varrho a[v]. \tag{11.26}$$

Making use of formula (11.25) on the unloading wave $x = a_1 t$, we obtain an expression for speed v_0,

$$v_0(x) = -\frac{1}{\varrho a_0}\sigma_s(1-\mu) + \frac{p_0}{\varrho a_1}\left[1 - \frac{\mu^2-1}{2\mu} - \frac{x}{a_0 t_0}\right]. \tag{11.27}$$

The solution in region II is constructed by the method of characteristics using (10.12), where $C_{1,2}$ are determined from the boundary condition (11.22) and from the prescribed values σ and v on the unloading wave (formulae (11.24) and (11.27)). The solution in region II, after determining constants $C_{1,2}$, will take the forms

$$\begin{aligned}\sigma(x,t) &= -p_0\left(1-\frac{t_1}{t_0}\right) + \frac{1}{2}(1-\mu)\left[\sigma_0(a_1 t_2) - \sigma_0\left(\frac{\mu}{1+\mu}a_1 t_1\right)\right],\\ v(x,t) &= \frac{1}{\varrho a_0}p_0\left(1-\frac{t_1}{t_0}\right) + \frac{1}{2\varrho a_0}(1-\mu)\left[\sigma_0(a_1 t_2) + \sigma_0\left(\frac{\mu}{1+\mu}a_1 t_1\right)\right] + \frac{1}{\varrho a_1}\left(1-\frac{1}{\mu}\right)\sigma_s,\end{aligned} \tag{11.28}$$

where we have used the following notation:

$$t_1 = t - \frac{x}{a_0}, \qquad t_2 = \frac{\mu}{1+\mu}\left(t + \frac{x}{a_0}\right);$$

stresses $\sigma_0(a_1 t_2)$ and $\sigma_0(\mu a_1 t_1/(1+\mu))$ are determined on the unloading wave $x = a_1 t$ by (11.24).

The solution in region III is obtained directly from (11.28) by taking $p(t) \equiv 0$, i.e.

$$\begin{aligned}\sigma(x,t) &= \frac{1}{2}(1-\mu)\left[\sigma_0(a_1 t_2) - \sigma_0\left(\frac{\mu}{1+\mu}a_1 t_1\right)\right],\\ v(x,t) &= \frac{1}{2\varrho a_0}(1-\mu)\left[\sigma_0(a_1 t_2) + \sigma_0\left(\frac{\mu}{1+\mu}a_1 t_1\right)\right] + \frac{1}{\varrho a_1}\left(1-\frac{1}{\mu}\right)\sigma_s.\end{aligned} \tag{11.29}$$

The solution in regions IV, VI, and VIII (Fig. 33) can be determined by means of (11.28) while that applying to regions III, V, VII, and IX can be found on the basis of (11.29).

We shall now present a method of determining the unloading wave by means of the successive approximation method [17]. We shall be looking for the unloading wave in the case considered above; we assume the boundary condition (11.1″) (the load suddenly applied at the end of the bar and then monotonically decreasing) and relation $\sigma = \sigma(\epsilon)$ as in Fig. 29. We shall return to the solution of (11.7) in region D_2 (Fig. 31), for the boundary condition (11.1″) and for the conditions on the unloading wave $t = f(x)$ in the form (11.6), i.e. we shall return to the system of (11.16).

Introducing the symbol $z = x/\alpha$ we represent (11.16) in the form

$$\begin{aligned}\Phi_1'[z(1+\alpha)] &= \varepsilon_0(0) - \frac{\sigma_0(0)}{E} - e\left[\frac{z}{a_0}(1+\alpha)\right] + \frac{1}{2}\left[\frac{\sigma_0(\varepsilon_0)}{E} - \int_0^{\varepsilon_0}\frac{a(\varepsilon)}{a_0}\,d\varepsilon\right],\\ \Phi_1'[z(1-\alpha)] &= -\frac{1}{2}\left[\frac{\sigma_0(\varepsilon_0)}{E} + \int_0^{\varepsilon_0}\frac{a(\varepsilon)}{a_0}\,d\varepsilon\right].\end{aligned} \tag{11.30}$$

Subtracting (11.30) one from the other we obtain

$$\Phi_1'[z(1+\alpha)]-\Phi_1'[z(1-\alpha)] = \varepsilon_0(0)-\frac{\sigma_0(0)}{E}-e\left[\frac{z}{a_0}(1+\alpha)\right]+\frac{\sigma_0(\varepsilon_0)}{E} \tag{11.31}$$

Confining the discussion to those cases in which $a(\epsilon) \ll a_0$, i.e. to materials for which $a(\epsilon)$ is small (materials that do not exhibit significant hardening effects), we reduce (11.31) to the form

$$\sigma_0[\varepsilon_0(x)] = \sigma_0(0)-E\varepsilon_0(0)+E\,e\left(\frac{x}{a[\varepsilon_0(x)]}\right), \tag{11.32}$$

where $x/a = t$. Solving (11.32) we find simultaneously the zeroth approximation for $\epsilon_0(x)$ and for the equation of the unloading wave $t = x/a(\epsilon_0)$.

Next we determine the unloading wave by the method of successive approximations. Namely, we substitute into the right-hand side of $(11.16)_1$ the zeroth approximation $\epsilon_0^0(x)$ determined from (11.32):

$$\begin{aligned} &\Phi_1'[z(1+\alpha^0)] = f_0(z;\alpha^0),\\ &\Phi_1'[z(1-\alpha)] = -\frac{1}{2}\left[\frac{\sigma_0(\varepsilon_0)}{E}+\int_0^{\varepsilon_0}\frac{a(\varepsilon)}{a_0}\,d\varepsilon\right], \end{aligned} \tag{11.33}$$

where we denote

$$f_0(z,\alpha^0) = \varepsilon_0(0)-\frac{\sigma_0(0)}{E}-e\left[\frac{z}{a_0}(1+\alpha^0)\right]+\frac{1}{2}\left[\frac{\sigma_0(\varepsilon_0^0)}{E}-\int_0^{\varepsilon_0^0}\frac{a(\varepsilon)}{a_0}\,d\varepsilon\right].$$

Changing the variables in (11.33),

$$z = \frac{z_1(1-\alpha^0)}{1+\alpha^0},$$

we obtain

$$\begin{aligned} &\Phi_1'[z_1(1-\alpha^0)] = f_0\left(\frac{z_1(1-\alpha^0)}{1+\alpha^0};\ \alpha^0\right),\\ &\Phi_1'[z(1-\alpha)] = -\frac{1}{2}\left[\frac{\sigma_0(\varepsilon_0)}{E}+\int_0^{\varepsilon_0}\frac{a(\varepsilon)}{a_0}\,d\varepsilon\right]. \end{aligned} \tag{11.34}$$

On account of the assumption $a(\epsilon) \ll a_0$, α is a small quantity and so we can set $\alpha \approx \alpha^0$, i.e. $z = z_1$. We can then equate the left-hand sides of (11.34) and obtain

$$\frac{\sigma_0(\varepsilon_0)}{E}+\int\limits_0^{\varepsilon_0}\frac{a(\varepsilon)}{a_0}\,d\varepsilon=-2f_0\left(\frac{z_1(1-\alpha^0)}{1+\alpha^0};\quad\alpha^0\right). \tag{11.35}$$

From this equation we obtain the first approximation $\epsilon_0^1(x)$. The recurrence formula for the nth approximation has the following form:

$$\frac{\sigma_0(\varepsilon_0^{(n)})}{E}+\int\limits_0^{\varepsilon_0^{(n)}}\frac{a(\varepsilon)}{a_0}\,d\varepsilon=-2f_{(n-1)}\left(\frac{z(1-\alpha^{(n-1)})}{1+\alpha^{(n-1)}};\quad\alpha^{(n-1)}\right). \tag{11.36}$$

The convergence of the method can be proved for particular cases. In the case of the Prandtl model and for the boundary condition (11.22), we obtain from the recurrence formula (11.36), for $\alpha=1/2$, good agreement between the exact solution (11.24) and the third approximation.

The unloading wave problem can also be solved by the inverse method. Namely, the shape of the unloading wave is assumed *a priori*. We then solve the Cauchy problem in region D_2. We take into account the medium described by Fig. 29, confining ourselves to the case when the unloading wave is a wave of weak discontinuity.

At an arbitrary point M of region D_2 (Fig. 30) we can determine the solution by drawing from this point the negative and the positive elastic characteristics up to their intersection with the unloading wave (points $M_1(x_1, t_1), M_2(x_2, t_2)$, respectively). The equations of these characteristics have the forms

$$x-x_1=-a_0[t-f(x_1)],\qquad x-x_2=a_0[t-f(x_2)], \tag{11.37}$$

where already the equation of the unloading wave has been utilized.

From the relations on the characteristics (10.12) we obtain the following equations:

$$\begin{aligned}v+\frac{1}{\varrho a_0}\sigma&=v[x_1,f(x_1)]+\frac{1}{\varrho a_0}\sigma[x_1,f(x_1)],\\ v-\frac{1}{\varrho a_0}\sigma&=v[x_2,f(x_2)]-\frac{1}{\varrho a_0}\sigma[x_2,f(x_2)],\end{aligned} \tag{11.38}$$

from which we can eliminate consecutively stress σ and velocity v. At point $M(x, t)$ the solution is as follows:

$$\begin{aligned}\sigma(x,t)&=\frac{1}{2}\varrho a_0\{v[x_1,f(x_1)]-v[x_2,f(x_2)]\}+\frac{1}{2}\{\sigma[x_1,f(x_1)]+\sigma[x_2,f(x_2)]\},\\ v(x,t)&=\frac{1}{2}\{v[x_1,f(x_1)]+v[x_2,f(x_2)]\}+\frac{1}{2\varrho a_0}\{\sigma[x_1,f(x_1)]-\sigma[x_2,f(x_2)]\},\end{aligned} \tag{11.39}$$

where x_1 and x_2 are determined by (11.37). Assuming $x = 0$ we obtain the solution for the end of the bar, i.e. we obtain the stress distribution at the end such that it corresponds to the shape of the unloading wave $t = f(x)$ assumed beforehand. By a trial-and-error method, assuming various different shapes for the unloading wave, we can obtain a stress distribution at the end which is close to the prescribed one.

We consider now the method of local linearization of the unloading wave. The unloading wave is determined (assuming relation $\sigma = \sigma(\epsilon)$ as in Fig. 29) by the method of characteristics. We consider the case of the pressure change at the end of the bar $x = 0$ as in Fig. 34, and the boundary condition in the form (11.1). The unloading wave starts propagating from point M_0, which corresponds to the maximum value p_{max} of the applied pressure $p(t)$.

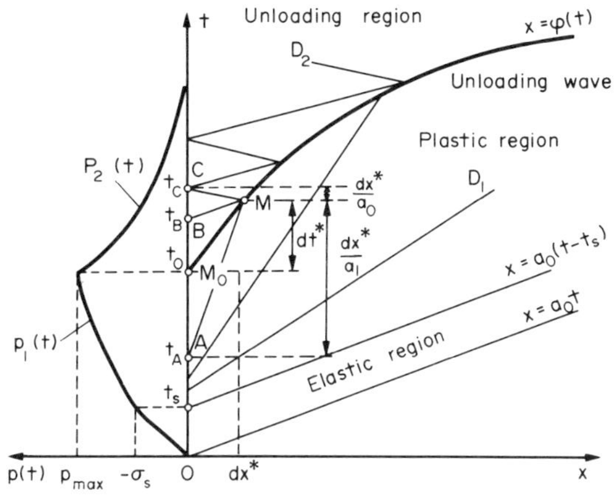

Fig. 34

Let us assume that the unloading wave in the time dt^* (Fig. 34) moves to point M of abscissa dx^*. From this point we draw a plastic, positive characteristic into region D_1, while in the unloading region D_2 we draw the elastic positive and negative characteristics up to the intersection with $x = 0$, at points A, B, and C respectively. We assume that the pressure at $x = 0$ has a discontinuous derivative at the instant $t = t_0$. Denoting by $k_1 = \left.\frac{dp_1(t)}{dt}\right|_{t=t^0}$, $k_2 = \left.\frac{dp_2(t)}{dt}\right|_{t=t_0}$ we can assume in the vicinity of point $(0, t_0)$ that

(11.40)
$$\begin{aligned} \sigma_0 &= -p_{max} - k_1(t - t_0) \quad \text{for} \quad t < t_0, \\ \sigma_0 &= -p_{max} - k_2(t - t_0) \quad \text{for} \quad t > t_0. \end{aligned}$$

The distance between points B and C and point M_0 is small; thus introducing $j = \partial v/\partial t$ we can assume for $t \geqslant t_0$ that the velocity of the points at the end of the bar $x = 0$ is equal to

(11.41)
$$\begin{aligned} v_B - v_{max} &= j\,(t_B - t_0) = j\left(dt^* - \frac{dx^*}{a_0}\right), \\ v_C - v_{max} &= j\,(t_C - t_0) = j\left(dt^* + \frac{dx^*}{a_0}\right), \end{aligned}$$

where v_{max} denotes the maximum velocity of the end $x = 0$.

Dividing one of the above equations by the other we obtain

$$\frac{v_B - v_{max}}{v_C - v_{max}} = \frac{dt^* - \dfrac{dx^*}{a_0}}{dt^* + \dfrac{dx^*}{a_0}} = \frac{1 - \dfrac{c}{a_0}}{1 + \dfrac{c}{a_0}}, \tag{11.42}$$

where $c = dx^*/dt^*$ denotes the initial velocity of the unloading wave.

Making use of (10.12) we can determine the characteristics the stress and the velocity at points B, C, and M from the corresponding values at point M_0. Consequently we can determine velocities v_B, v_C, and v_{max} in (11.42). Doing this we obtain

$$c = \sqrt{\frac{a_1^2 a_0^2 (k_1 - k_2)}{a_0^2 k_1 - a_1^2 k_2}}, \tag{11.43}$$

where a_1 denotes the speed of propagation of the plastic strains for stress $\sigma = -p_{max}$.

In the particular case, when $k_1 = 0$, $k_2 \neq 0$, or when $k_1 = 0$, $k_2 = \infty$, it follows from (11.43) that $c = a_0$. Thus the direction of the initial segment of the unloading wave M_0M coincides with the direction of the elastic wave. If, on the other hand, $k_1 \neq 0$, $k_2 = 0$, or if $k_1 \neq 0$, $k_2 = \infty$, then $c = a_1$ and the initial velocity of propagation of the unloading wave coincides with the velocity of the plastic wave. If $k_1 = k_2 = 0$, i.e. if the stress derivative at $x = 0$ is continuous for $t = t_0$, then formula (11.43) cannot be used. In this case we have to take into account the terms containing the second powers of dx^* in the expansions for the stress and velocity with respect to time in the vicinity of point for which $\sigma = -p_{max}$. Next, proceeding exactly as in the case described above, we obtain the formula for the initial velocity of the unloading wave in the form

$$c = a_0 \left[\sqrt{\left(\frac{a_0}{a_1}\right)^2 + 3} - \frac{a_0}{a_1} \right]. \tag{11.44}$$

We assume that the initial segment M_0M of the unloading wave takes the form of a straight-line segment. Its angle of inclination is found from the initial velocity which is determined by formula (11.43) or (11.44) depending upon whether the derivative $d^2\sigma/dt^2|_{x=0}$ is continuous or discontinuous. The values of stress and velocity are known on this segment from the solution in the plastic region D_1. On the elastic, negative characteristics in the unloading region D_2 in triangle M_0MC, we have

$$v = -\frac{\sigma}{\varrho a_0} + \psi(t_1), \tag{11.45}$$

where t_1 denotes the time for which the prescribed negative characteristic begins on the unloading wave $x = \varphi(t)$,

$$t_1 = t - \frac{\varphi(t_1) - x}{a_0} = t - \frac{c(t_1 - t_0) - x}{a_0} = \frac{a_0 t + x + c t_0}{a_0 + c}, \tag{11.46}$$

and the function

$$\Phi(t_1) = v[\varphi(t), t] + \frac{1}{\varrho a_0}\sigma[\varphi(t), t],$$

which is known on the wave $x = \varphi(t)$ from the solution in region D_1.

From (11.45), utilizing the boundary condition (11.1), we can determine the velocity v at the end $x = 0$ for $t_0 \leqslant t \leqslant t_C$:

$$v(0, t) = \frac{p_2(t)}{\varrho a_0} + \Phi\left(\frac{a_0 t + c t_0}{a_0 + c}\right). \tag{11.47}$$

On the positive characteristics in region D_2, leaving the end $x = 0$ during $t_B \leqslant t \leqslant t_C$, we obtain the relation

$$v = \frac{\sigma}{\varrho a_0} + \frac{2p_2\left(t - \dfrac{x}{a_0}\right)}{\varrho a_0} + \Phi\left(\frac{a_0 t - x + c t_0}{a_0 + c}\right). \tag{11.48}$$

The constant of integration C_1 is determined from the condition $\sigma(0, t) = -p_2(t)$ and condition (11.47).

In the case considered the unloading wave is a wave of weak discontinuity. The following compatibility conditions have to be satisfied on its front:

$$v_2[\varphi(t), t] = v_1[\varphi(t), t], \qquad \sigma_2[\varphi(t), t] = \sigma_1[\varphi(t), t], \tag{11.49}$$

where indices 1 and 2 denote the solution from the side in regions D_1 and D_2 respectively. Thus we obtain from (11.48) the equation

$$-v_1[\varphi(t), t] + \frac{\sigma[\varphi(t), t]}{\varrho a_0} + \frac{2p_2\left(t - \dfrac{\varphi(t)}{a_0}\right)}{\varrho a_0} + \Phi\left(\frac{a_0 t - \varphi(t) + c t_0}{a_0 + c}\right) = 0, \tag{11.50}$$

from which the next segment MM_1 of the unloading wave $x = \varphi(t)$ can be determined (v_1, σ_1, and Φ are determined from the solution in region D_1). For the consecutive segments of the unloading wave the procedure is repeated.

Besides the approximate analytical methods, graphical methods for the determination of the unloading wave and the distribution of stress and strain along the unloading wave are in frequent use. These methods are presented in details in [37], [34], and [125]. They consist in the simultaneous graphical construction of the solution on the phase plane (x, t) and on the hodograph plane (σ, v). Use is made of the equations of characteristics (10.10) as well as of the relations on the characteristics (10.11). Furthermore, if the unloading wave is a weak discontinuity (in the case of the boundary condition of the form (11)), the initial velocity of the unloading wave c is determined from (11.43) or (11.44). In the case, however, when the pressure is suddenly applied at the bar end and then decreases (the boundary condition (11.1′)), the equation of the unloading wave is known and the wave is a strong discontinuity. The dynamic continuity condition is utilized on the wave front (11.25). One can also find the end of the unloading wave and the strain distribution on the unloading wave.

There exists an extensive literature devoted to problems of the propagation of the plane stress waves, but the scope of this book does not allow us to discuss these papers. We mention only the papers by Rakhmatulin (compare [125]), where the effect of a change in the elastic limit on the propagation of loading and unloading waves was investigated. This kind of inhomogeneity occurs, for example, in the bars of materials that exhibit hardening since plastic waves are already propagated in it. Problems of loading and unloading wave propagation in a medium with general inhomogeneity in the elastic region as well as in the plastic region were investigated in [49] and [109]. Solutions were obtained in the form of integro-differential equations which were then solved by the method of successive approximations. A number of problems of wave propagation in bars of varying cross-section was also considered. Problems concerning the dynamics of a rigid plate resting on an elastic–plastic foundation constitute a separate group of papers. An interesting problem on stress wave propagation produced by a moving source of mechanical disturbance in an elastic–plastic medium was solved in [121]. Three cases were considered, depending on the velocity of the moving source: subsonic velocity, the velocity equal to the speed of sound in the medium, and supersonic velocity. The exact form of the solution was obtained by means of the method of characteristics.

12. Reflection of an unloading plane wave from an obstacle in an elastic–plastic medium

Now we consider the problem of the propagation of plane stress waves in a finite bar, made of an elastic–plastic material. The problem of the reflection of an unloading wave from the end of the bar is fairly complicated, particularly in the case of the reflection of an unloading wave which is a discontinuity since the loading waves reach faster than the unloading wave the bar end $x = l$ (e.g. fixed end), reflect from it, and on their way back encounter the unloading wave that can be refracted.

Usually, after a certain time, a new unloading wave starts propagating from the fixed end, moving towards the bar end $x = 0$. The problem of the reflection of a strong discontinuity unloading wave is a little bit simpler, particularly in the case when the front of the strong discontinuity wave is simultaneously an unloading wave. In this case the unloading wave reaches the end of the bar first and reflects from it.

Let us consider the problem of the reflection of a strong discontinuity wave (an arbitrary wave of strong discontinuity not necessarily an unloading wave) from the end of the bar that can be rigidly fixed, elastic, or free. In the general case we assume that at the bar end $x = l$ (Fig. 35) an undeformable mass M is attached with a damper of constant viscosity c and with a

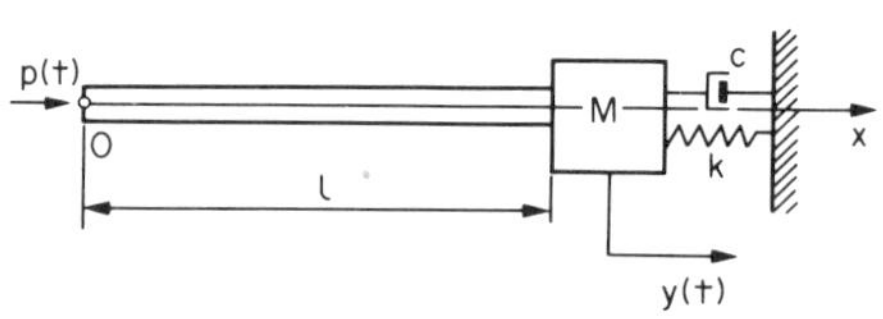

Fig. 35

spring of rigidity coefficient k. After the reflection of the strong discontinuity wave from the end $x = l$ a new strong discontinuity wave (called the reflected wave) appears in the bar. Denoting by $y(t)$ the displacement of mass M in the x-axis direction, by R the reaction of the combined system of mass M, spring k, and damper c_1, measured in the units of stress, we obtain (assuming permanent contact between mass M and the bar)

$$R = -\sigma_{\text{ref}}, \quad \frac{dy}{dt} = v_{\text{ref}}. \tag{12.1}$$

Since the reflected wave is a wave of strong discontinuity, the condition of the dynamic continuity (11.26) should hold on its front

$$\sigma_{\text{ref}} - \sigma_{\text{in}} = +\varrho a_{\text{ref}} (v_{\text{ref}} - v_p), \tag{12.2}$$

where the indices in and ref designate the values of the stress and velocity on the incident and reflected waves respectively; a_{ref} is the velocity of propagation of the reflected wave. We obtain from (12.1),

$$R = -\sigma_{\text{in}} + \varrho a_{\text{ref}} v_{\text{in}} - \varrho a_{\text{ref}} \frac{dy}{dt}. \tag{12.3}$$

On the other hand, mass M can be treated as a vibrating system with a single degree of freedom excited by force $R(\bar{t})$. The equation of its vibrations takes the form

$$M\ddot{y}(\bar{t}) + c_1 \dot{y}(\bar{t}) + ky(\bar{t}) = R(\bar{t}). \tag{12.4}$$

The vibrations are initiated from the instant when the incident wave reaches the end $x = l$. Thus for $\bar{t} = 0$, taking into account (12.3), we can obtain from (12.4)

$$M\ddot{y}(0) + (c_1 + \varrho a_{\text{ref}}) \dot{y}(0) + ky(0) = \varrho a_{\text{ref}} v_{\text{in}} - \sigma_{\text{in}}. \tag{12.5}$$

If the incident wave propagates into an undisturbed medium then we obtain from the condition of the dynamical continuity (11.25) that

$$\sigma_{\text{in}} = -\varrho a_{\text{in}} v_{\text{in}}, \tag{12.6}$$

where a_{in} is the velocity of the incident wave. Thus, from (12.5)

$$M\ddot{y}(0) + (c_1 + \varrho a_{\text{ref}}) \dot{y}(0) + ky(0) = -(1+\varkappa)\sigma_{\text{in}}, \tag{12.7}$$

where $\varkappa = a_{\text{ref}}/a_{\text{in}}$.

If we assume for (12.7) homogeneous, zero boundary conditions, i.e. $y(0) = \dot{y}(0) = 0$, we obtain

$$M\ddot{y}(0) = -(1+\varkappa)\sigma_{\text{in}} = R(0). \tag{12.8}$$

At the instant of reflection of the strong discontinuity wave the stress on its front is equal to $\sigma_{\text{ref}} = (1 + \varkappa)\sigma_{\text{in}}$. If the speed of propagation of the incident wave is the same as that of the reflected wave, i.e. when $\varkappa = 1$ at the instant that the wave is reflected from the system composed of the elements $M + k + c_1$, the stress on the front of the reflected wave is equal to $2\sigma_{\text{in}}$.

If we assume that $M = 0$, then we can assume one initial condition only, namely $y(0) = 0$, and the reaction R at the instant $t = 0$ is equal to

$$R(0) = -(1+\varkappa)\frac{c_1}{c_1+\varrho a}\sigma_{\text{in}}. \tag{12.9}$$

(subscript "ref" also printed beneath the fraction)

If we likewise assume that $c_1 = 0$, then

$$R(0) = 0. \tag{12.10}$$

The value of the reaction R in the case when the end of the bar $x = l$ is rigidly fixed can be obtained from the assumptions that $M = 0$, $k \to \infty$, and $\dot{y}(0) \to 0$, $y(0) \to 0$:

$$R(0) = -(1+\varkappa)\sigma_{\text{in}}, \quad \text{i.e. that} \quad \sigma_{\text{ref}} = (1+\varkappa)\sigma_{\text{in}}, \tag{12.11}$$

while, for the free-ended case we obtain

$$R(0) = 0. \tag{12.12}$$

We make some remarks connected with the propagation of plane stress waves in layered media. Let us assume that in a semi-infinite bar, at a distance $x = l$ from the end, there exists an undeformable mass M (Fig. 36); further we assume that for $x < l$ the density of the bar is ϱ_1

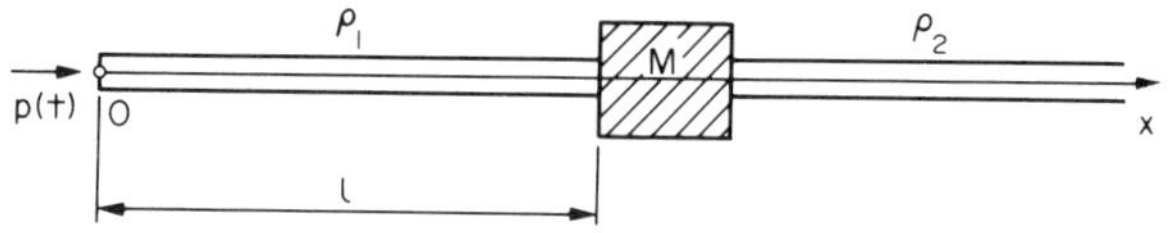

Fig. 36

while for $x > l$ it is equal to ϱ_2. It is also assumed that during the vibrations the bars cannot be separated from the mass M.

The following conditions have to be satisfied at the cross-section $x = l$:
compatibility condition of particle velocities

$$v_l(l, \bar{t}) = v_{\text{in}}(l, \bar{t}), \tag{12.13}$$

the condition of equilibrium of the forces applied to the concentrated mass M

$$M\frac{\partial v(l, \bar{t})}{\partial \bar{t}} + \sigma_{\text{in}}(l, \bar{t}) - \sigma_l(l, \bar{t}) = 0, \tag{12.14}$$

where the indices l and in denote the values of stress for $x < l$ and $x > l$ respectively. Time $\bar{t}$ is measured from the instant when the incident wave reaches the cross-section $x = l$. The incident wave will be reflected from mass M and will also penetrate into the second bar.

The solution of this problem is slightly simpler when we are dealing with waves of strong discontinuity. We should note that when the waves of strong discontinuity reach mass M, then the reflected wave is also a wave of strong discontinuity while only waves of weak discontinuity will propagate into the region $x > l$. This results from the inertia of mass M. The initial condition for (12.14) takes the form $v(l, 0) = 0$. In the case when $M = 0$, condition (12.14) is simplified to the form

$$\sigma_{in}(l, t) = \sigma_l(l, t). \tag{12.15}$$

When a wave of strong discontinuity strikes the boundary between media, then the wave reflected from the boundary $x = l$, as well as the wave moving into the second bar, are the waves of strong discontinuity. Because of their complicated numerical form, we shall not cite the solutions in the particular regions of the phase plane.

There exists a number of solutions of this type for the case of homogeneous media as well as in the case of the non-homogeneous media, e.g. in media with variable yield limit [123]. In section 14 we shall present the solution of the propagation problem of an unloading wave in a semi-infinite bar for which there exists, at the cross section $x = l$, an undeformable mass M, i.e. conditions (12.13) and (12.14) are assumed for $x = l$ while the model for the material is an elastic–plastic body with rigid unloading.

In conclusion we present a particular case of wave reflection in an elastic–plastic bar whose end $x = l$ is fixed (Fig. 37). We assume that the pressure at the end $x = 0$ varies periodically and the frequency of the load oscillations depends on the material properties of the bar and its length. We assume that the amplitude of the load is constant and equal to the elastic limit, the period of the pressure changes is equal to $T = 2l\left(\dfrac{1}{a_0} + \dfrac{1}{a_1}\right)$. This problem was solved in [67],

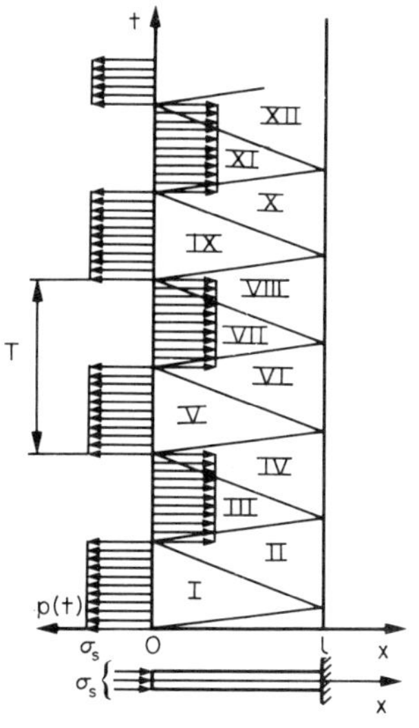

Fig. 37

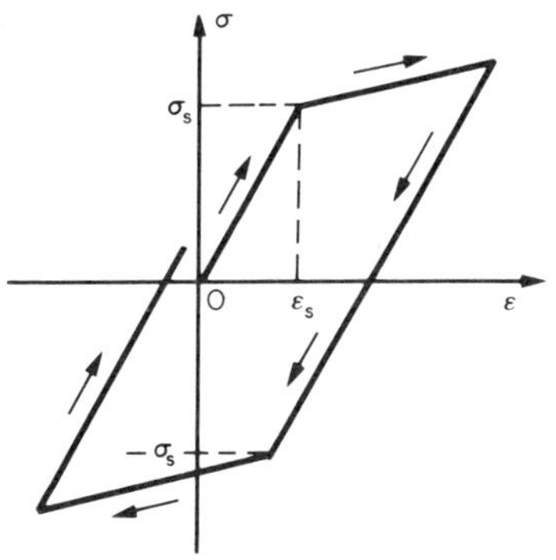

Fig. 38

analytically, and by a graphical method. The Prandtl model was assumed, neglecting the Bauschinger effect, and it was assumed that the elastic limit in tension is equal to that in compression (Fig. 38). Constructing the solutions in the separate region of the phase plane (Fig. 37), we obtain the recurrence formulae:

for the regions $2n+1$:

$$\sigma_{2n+1}(x,t) = (-1)^n \sigma_s,$$

(12.16)
$$v_{2n+1}(x,t) = (-1)^{n+1}\left(\mu^{-n} + 2\sum_{i=1}^{n} \frac{1}{\mu^{i-1}}\right) v_s,$$

$$\varepsilon_1(x,t) = \varepsilon_s, \qquad \varepsilon_{2n+3}(x,t) = \left[1 + (-1)^{n+1} \frac{\mu^{n+1} - \mu - 1}{\mu^n}\right] \varepsilon_s,$$

where $n = 0, 1, 2, 3, \ldots,$

for the regions $2n$:

$$\sigma_{2n}(x,t) = (-1)^{n+1}\left[(1+\mu^{-n}) + 2\sum_{i=2}^{n} \frac{1}{\mu^{i-1}}\right] \sigma_s, \qquad v_{2n}(x,t) = 0,$$

(12.17)
$$\varepsilon_{2n}(x,t) = (-1)^{n+1}\left[\frac{1+\mu^n}{\mu^{n-1}} + 2\sum_{i=2}^{n} \frac{\mu^{i-1}}{\mu^{n-1}}\right] \varepsilon_s,$$

where $n = 1, 2, 3, \ldots.$

For $n \to \infty$, we obtain the following values for the limiting cycle:

$$\sigma_{\max} = \lim_{n\to\infty} |\sigma_{2n}| = \left(1 + \frac{2}{\mu - 1}\right) \sigma_s,$$

(12.18)
$$v_{\max} = \lim_{n\to\infty} |v_{2n+1}| = 2\frac{\mu}{\mu - 1} v_s,$$

$$\varepsilon_{\max} = \lim_{n\to\infty} |\varepsilon_{2n}| = \mu\left(1 + \frac{2}{\mu - 1}\right) \varepsilon_s.$$

The coefficient of the increase of the stress amplitude $\alpha = \sigma_{\max}/\sigma_s$ is expressed as

(12.19)
$$\alpha = 1 + \frac{2}{\mu - 1}, \qquad \mu = \frac{a_0}{a_1}.$$

We can assert that for small values of μ the stress amplitude increases considerably; on the other hand, for high values (strongly plastic media) the increase of the amplitude decreases. For $\mu \to 1$ we obtain the limit of the theory of elasticity; it leads to the classical case of harmonic resonance (the period of the load change is then equal to $T = 4l/a_0$). For $\mu \to \infty$ two cases are possible: the case of ideal plasticity or rigid unloading. It has been proved that the phenomenon of "plastic resonance" also exists if we assume that the period of application of the load is equal to

$$T_m = (2n+1)\,T, \tag{12.20}$$

where T denotes the fundamental period, $n = 1, 2, 3, \ldots$ This is important from the practical standpoint since it is difficult to obtain experimentally load changes over the fundamental period T, which is very small, on account of the very great wave propagation velocities in the bar and due to the use of bars of small length in experimental practice.

13. Propagation of plane shock waves

So far we have considered the problem of the propagation of plane stress waves in elastic–plastic media in the case when $d^2\sigma/d\epsilon^2 < 0$. The waves of strong discontinuity which were considered were caused solely by the discontinuity of the boundary condition (pressure suddenly applied to the bar end, the impact of the bar at an obstacle, etc.). Now we shall discuss the problems associated with the propagation of plane shock waves for which the discontinuities of stress, velocity, and deformation (the first derivatives of the displacements) appear across the wave front independently of the boundary condition. In the case of plane waves, shock waves occur when $d^2\sigma/d\epsilon^2 > 0$ during the loading process (i.e. when $\sigma = \sigma(\epsilon)$ represents a convex curve with respect to the ϵ-axis (Fig. 39)). In the case of the material characteristic for which $\sigma''(\epsilon) > 0$ (where we denoted $\sigma''(\epsilon) = d^2\sigma/d\epsilon^2$) greater stresses

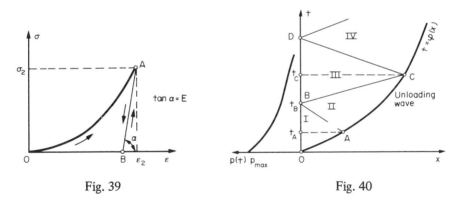

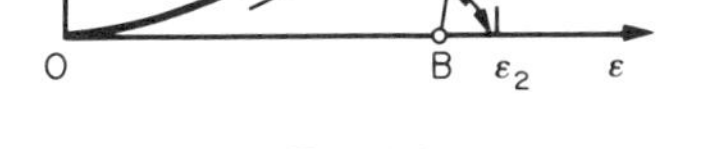

Fig. 39

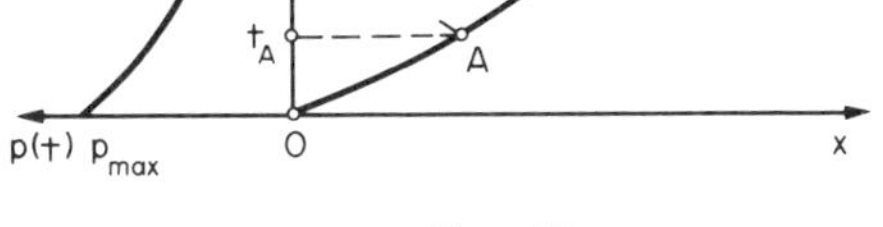

Fig. 40

propagate with greater velocities than do smaller stresses. For example, when at the bar end the pressure acts, monotonically increasing from zero, then waves will propagate in the medium. The higher the stress at the end of the bar the faster the waves travel; they will overtake the slower waves produced by smaller stresses. The result of such an accumulation will be that the front of a discontinuity wave will be generated. A *shock wave* is the name given to such a wave. In the general case the analysis of a shock wave is very complicated.

We consider the case of shock wave propagation in a bar of elastic–plastic material when a sudden pressure of value p_0, which monotonically decreases with time, has been applied at the end of the bar $x = 0$. The boundary condition takes the form

(13.1) $$\sigma(0,t) = -p(t), \quad p(t) > 0, \quad \frac{dp}{dt} \geqslant 0,$$

or in terms of strains

(13.1') $$\varepsilon(0,t) = e(t);$$

$e(t)$ is uniquely determined from the $\sigma = \sigma(\epsilon)$ diagram by $p(t)$. We assume the curvilinear material characteristic (Fig. 39), also that in the loading process the condition $\sigma''(\epsilon) > 0$ holds, whereas we assume that the unloading process is elastic.

Equation (10.8) holds in the loading region,

(13.2) $$\frac{\partial^2 u}{\partial t^2} - a^2(\varepsilon)\frac{\partial^2 u}{\partial x^2} = 0 \quad \text{for} \quad \frac{d\varepsilon}{dt} \geqslant 0,$$

while in the unloading region (11.7) is valid,

(13.3) $$\frac{\partial^2 u}{\partial t^2} - a_0^2\frac{\partial^2 u}{\partial x^2} = F(x) \quad \text{for} \quad \frac{d\varepsilon}{dt} < 0,$$

where

$$F(x) = \frac{1}{\varrho}\frac{d\sigma_0(x)}{dx} - a_0^2\frac{d\varepsilon_0(x)}{dx}, \qquad a(\varepsilon) = \sqrt{\frac{1}{\varrho}\frac{d\sigma}{d\varepsilon}}, \qquad a_0 = \sqrt{\frac{E}{\varrho}}.$$

In the case of the assumed loading of the bar end (condition (13.1)) the shock wave will start propagating from the origin of the coordinate system in the phase plane (x, t) (Fig. 40). This wave will be denoted by $t = \varphi(x)$. The shape of the wave is unknown until the problem is solved.

The general solution of (13.3) takes the form of (11.10), i.e.

(13.4) $$u(x,t) = \Phi_1(a_0 t - x) + \Phi_2(a_0 t + x) - \frac{1}{E}\int_0^x [\sigma_0(\xi) - E\,\varepsilon_0(\xi)]\,d\xi.$$

The following conditions have to be satisfied on the shock wave:

(13.5) $$\left.\frac{\partial u}{\partial x}\right|_{t=\varphi(x)} = \varepsilon_2(x), \qquad \left.\frac{\partial u}{\partial t}\right|_{t=\varphi(x)} = v_2(x),$$

and the continuity condition (7.8) as well as the condition of the dynamic continuity (7.17). In the case of plane waves they have the form:

(13.6) $$\varrho_1(v_1 - G) = \varrho_2(v_2 - G), \qquad \sigma_2 - \sigma_1 = \varrho_1(v_1 - G)(v_2 - v_1),$$

where the indices 1 and 2 denote the values before and behind the shock wave front respectively,

$G = \dfrac{1}{\varphi'(x)}$ denotes the velocity of the shock wave. If the shock wave propagates into an undisturbed medium then, obviously, $\sigma_1 = v_1 = 0$.

Making use of the initial conditions on the shock wave (13.5), the continuity conditions (13.6), and the boundary condition (13.1′) we are in position to determine the unknown functions $\Phi_1, \Phi_2, \epsilon_2(x), v_2(x)$, and the shape of the shock wave $t = \varphi(x)$. Making use of the boundary condition (13.1′) we obtain

$$-\Phi_1'(a_0 t) + \Phi_2'(a_0 t) - \frac{\sigma_2(0)}{E} + \varepsilon_2(0) = e(t), \tag{13.7}$$

where $\sigma_0(x) = \sigma_2(x)$, $\epsilon_0(x) = \epsilon_2(x)$.

Satisfying conditions (13.5) on the shock wave $t = \varphi(x)$ we have

$$\begin{aligned} &\Phi_1'(a_0\varphi(x) - x) + \Phi_2'(a_0\varphi(x) + x) = \frac{v_2(x)}{a_0}, \\ &-\Phi_1'(a_0\varphi(x) - x) + \Phi_2'(a_0\varphi(x) + x) - \frac{\sigma_2(x)}{E} = 0. \end{aligned} \tag{13.8}$$

Next, utilizing (13.7) and $\sigma_2(x) = \sigma_2[\epsilon_2(x)]$, we obtain the following system of two equations:

$$\begin{aligned} &\Phi_1'(a_0\varphi(x) - x) + \Phi_1'(a_0\varphi(x) + x) + \frac{\sigma_2(0)}{E} - \varepsilon_2(0) + e\left(\varphi(x) + \frac{x}{a^0}\right) = \frac{v_2(x)}{a_0}, \\ &-\Phi_1'(a_0\varphi(x) - x) + \Phi_1'(a_0\varphi(x) + x) + \frac{\sigma_2(0)}{E} - \varepsilon_2(0) + e\left(\varphi(x) + \frac{x}{a_0}\right) = \frac{\sigma_2[\varepsilon_2(x)]}{E}, \end{aligned} \tag{13.9}$$

whence, denoting $\alpha = \dfrac{x}{a_0\varphi(x)}$, $z = \dfrac{x}{\alpha}$, we obtain

$$\begin{aligned} &\Phi_1'[z(1+\alpha)] = \frac{1}{2}\left[\frac{v_2(x)}{a_0} + \frac{\sigma_2[\varepsilon_2(x)]}{E} - 2\frac{\sigma_0(0)}{E} + 2\varepsilon_2(0)\right] - e\left[\frac{z}{a_0}(1+\alpha)\right], \\ &\Phi_1'[z(1-\alpha)] = \frac{1}{2}\left[\frac{v_2(x)}{a_0} - \frac{\sigma_2[\varepsilon_2(x)]}{E}\right]. \end{aligned} \tag{13.10}$$

Subtracting equations (13.10) we obtain

$$\Phi_1'[z(1+\alpha)] = \Phi_1'[z(1-\alpha)] + \frac{\sigma_2[\varepsilon_2(x)] - \sigma_2(0)}{E} + \varepsilon_2(0) - e\left[\frac{z}{a_0}(1+\alpha)\right]. \tag{13.11}$$

If we confine ourselves to the case when the speed of the shock wave is a small quantity of lower order than the velocities of elastic waves in the medium, i.e. if $\alpha \ll 1$ [17], we obtain from (13.11) the following relation:

$$\sigma_2[\varepsilon_2(x)] = \sigma_2(0) - E\varepsilon_2(0) + Ee[\varphi(x)]. \tag{13.12}$$

Solving (13.12) we determine simultaneously $\epsilon_2(x)$ and the equation of shock wave $t = \varphi(x)$ for the zeroth approximation. Next the unloading wave is determined by the method of successive approximations. $\epsilon_2^0(x)$, determined from (13.12), and the parameters of the solution $\sigma_2^0, v_2^0, \alpha^0, G^0$ are assumed as the zeroth approximation. Substituting the obtained zeroth approximation into the right-hand side of $(13.10)_1$ and eliminating, in $(13.10)_2$, σ_2 and v_2 by means of the continuity conditions (13.6), we then obtain the system of equations:

$$\Phi_1'[z(1+\alpha)] = f_0(z;\alpha^0),$$

(13.13)

$$\Phi_1'[z(1-\alpha)] = \frac{1}{2}\left\{\frac{G\varepsilon_2(G)}{a_0} - \frac{\sigma_1}{E} + \frac{\varrho_1 v_1}{a_0\varrho_2} + \frac{\varrho_1\varepsilon_2(G)(G-v_1)^2}{E}\right\},$$

where

$$f_0(z;\alpha^0) = \frac{1}{2}\left[\frac{v_2^0}{a_0} + \frac{\sigma_2[\varepsilon_2^0(x)]}{E} - 2\frac{\sigma_2(0)}{E} + 2\varepsilon_2(0)\right] - e\left[\frac{z}{a_0}(1+\alpha^0)\right].$$

In this, the relation $\epsilon = 1 - \varrho_1/\varrho_2$ has also been used.

By changing the variables in $(13.13)_1$

$$z = \frac{z_1(1-\alpha^0)}{1+\alpha^0}$$

and assuming that, for small α^0 ($\alpha^0 \ll 1$) we have $z = z_1$, we deduce from the system (13.13) (substituting $G = 1/\varphi'(x)$), a differential equation of the first order, whence we determine the shock wave front $t = \varphi(x)$ to the first approximation. The recurrence formula for the determination of the nth approximation will have the form

(13.14)

$$\frac{1}{2}\left\{\frac{G^n\varepsilon_2(G^n)}{a_0} - \frac{\sigma_1}{E} + \frac{\varrho_1 v_2}{a_0\varrho_2} + \frac{\varrho_1\varepsilon_2(G^n)(G^n-v_1)^2}{E}\right\} = f_{(n-1)}\left[\frac{z(1-\alpha^{(n-1)})}{1+\alpha^{(n-1)}};\ \alpha^{(n-1)}\right].$$

The convergence of solutions obtained in this way can be verified for particular cases only [134].

We shall not discuss the cases when function $\sigma = \sigma(\epsilon)$ possesses points of inflexion. Then the solution is much more complicated.

Now we present a simple and very effective method of finding the shock wave [162] for the case considered above. Let us assume that the shape of the first segment of the shock wave is known for the time interval $0 \leqslant t \leqslant t_A$ – curve OA in Fig. 40. The equation of curve OA we denote by $x = \varphi_1(t)$. In region I, bounded by $x = 0$ and $x = \varphi_1(t)$, we can pose the Goursat problem.

We expand the function describing the shock wave over the segment OA into a Maclaurin's series in the vicinity of its initial point,

(13.15)
$$\varphi_1(t) = \frac{\varphi_1'(0)}{1!}t + \frac{\varphi_1''(0)}{2!}t^2 + \dots .$$

The coefficients of the expansion are computed from the equations describing the problem and from the conditions of the dynamic and kinematic continuity across the shock wave front (7.18) and (8.9). In the case of plane waves they assume the forms

$$[\sigma] = -\varrho\varphi'(t)[v], \quad [v] = -\varphi'(t)[\varepsilon]. \tag{13.16}$$

If the shock wave $x = \varphi(t)$ propagates into an undisturbed medium, then relations (13.16) in front of the wave front take the forms

$$\sigma_0 = -\varrho\varphi'(t)\, v_0, \quad v_0 = -\varphi'(t)\,\varepsilon_0, \tag{13.17}$$

where the indices zero denote the values on the shock wave from the side of the unloading region.

From the continuity condition (13.17) and the condition $\epsilon_0 = \epsilon_0(\sigma_0)$ we deduce, for $t \to 0$, that

$$\varphi_1'(0) = \sqrt{\frac{-p_{\max}}{\varrho\varepsilon_0(-p_{\max})}}\,. \tag{13.18}$$

Differentiating $\sigma_0(t)$ and $v_0(t)$ on the shock wave $x = \varphi_1(t)$

$$d\sigma_0(t) = \left.\frac{\partial\sigma}{\partial x}\right|_{x=\varphi_1(t)} \varphi_1'(t) + \left.\frac{\partial\sigma}{\partial t}\right|_{x=\varphi_1(t)}, \qquad dv_0(t) = \left.\frac{\partial v}{\partial x}\right|_{x=\varphi_1(t)} \varphi_1'(t) + \left.\frac{\partial v}{\partial t}\right|_{x=\varphi_1(t)},$$

making use of (13.17) $\sigma_0(t) = (\varphi_1'(t))^2\varrho\varepsilon_0\,[\sigma_0(t)]$, differentiating it with respect to time

$$\sigma_0'(t) = \frac{2\varrho\varphi_1'(t)\,\varphi_1''(t)\,\varepsilon[\sigma_0(t)]}{1-\varrho\,(\varphi_1'(t))^2\,\varepsilon[\sigma_0(t)]}\,,$$

taking into account the equations of motion in the unloading region (11.4) as well as condition $(13.17)_1$ also differentiated with respect to time, and passing to the limit $x \to 0$, $t \to 0$, we obtain the expression for the second coefficient of the expansion

$$\varphi_1''(0) = -\frac{\left[1-\dfrac{[\varphi_1'(0)]^2}{a_0^2}\right]\left[1-\varrho\,(\varphi_1'(0))^2\,\varepsilon_0'(-p_{\max})\right]\varphi_1'(0)\,p'(0)}{4\varrho\,(\varphi_1'(0))^2\,\varepsilon_0(-p_{\max})+p_{\max}\left[1-\varrho\,(\varphi_1'(0))^2\,\varepsilon_0'(-p_{\max})\right]}\,.$$

Taking into account (13.18) we obtain

$$\varphi_1''(0) = -\frac{\left[1+\dfrac{p_{\max}}{\varrho a_0^2\,\varepsilon_0(-p_{\max})}\right]\left[1+p_{\max}\dfrac{\varepsilon_0'(-p_{\max})}{\varepsilon_0(-p_{\max})}\right]p'(0)\sqrt{\dfrac{-p_{\max}}{\varrho\varepsilon_0(-p_{\max})}}}{p_{\max}\left[p_{\max}\dfrac{\varepsilon_0'(-p_{\max})}{\varepsilon_0(-p_{\max})}-3\right]} \tag{13.19}$$

where $a_0 = \sqrt{E/\varrho}$ is the speed of elastic waves in the unloading region.

In the same way we can determine the subsequent coefficients of the expansion in formula (13.15). The radius of convergence of series (13.15) depends on the characteristic of the medium $\sigma = \sigma(\epsilon)$ and on the nature of the pressure change at $x = 0$. The error in the calculations resulting from the finite number of terms taken into account in expansion (13.15) can be estimated in the following way. Assuming the shape of the shock wave in the form of, for example, two terms of the expansion (13.15), we calculate by the inverse method (similarly to the way it was done in section 11 for the case of an unloading wave when $\sigma''(\epsilon) < 0$) the change of pressure $p(t)$ at the end of the bar, corresponding to the assumed shock wave. The difference between the pressure calculated in this way and the real pressure constitutes the calculation error. On the basis of the relations on the characteristics in the unloading region (11.38) and the dynamic and kinematic continuity conditions on the shock wave front (13.17), we obtain the solution in region I (Fig. 40):

$$\begin{aligned}
\sigma(x,t) &= \frac{1}{2}\left\{-p(t_1) - \varrho a_0 v_0(t_1) + \sigma_0(t_2)\left[1 - \frac{a_0}{\varphi_1'(t_2)}\right]\right\},\\
v(x,t) &= \frac{1}{2\varrho a_0}\left\{p(t_1) + \varrho a_0 v_0(t_1) + \sigma_0(t_2)\left[1 - \frac{a_0}{\varphi_1'(t_2)}\right]\right\},
\end{aligned} \tag{13.20}$$

where

$$v_0(t) = \frac{1}{\varrho a_0}\left\{p(t) + \sigma_0(t_3)\left[1 - \frac{a_0}{\varphi_1'(t_3)}\right]\right\}, \qquad \sigma_0(t) = \varrho\,[\varphi_1'(t)]^2\,\varepsilon_0(\sigma_0),$$

$$t_1 = t - \frac{x}{a_0}, \qquad t_2 = t - \frac{\varphi_1(t_2) - x}{a_0}, \qquad t_3 = t - \frac{\varphi_1(t_3)}{a_0}.$$

Now let us construct the next segment of the shock wave $x = \varphi_2(t)$ (Fig. 40). From the positive characteristics drawn from the bar end for $0 \leqslant t \leqslant t_B$, we have on account of $(11.38)_1$,

$$v = \frac{\sigma}{\varrho a_0} + v_0(t_1) + \frac{p(t_1)}{\varrho a_0}.$$

Eliminating v in the above equation from conditions (13.17) we obtain the following relation on the shock wave $x = \varphi_2(t)$:

$$\frac{1}{\varrho}\,\varepsilon_0\,\sigma_0(\varepsilon_0) = \left[\frac{\sigma_0(\varepsilon_0)}{\varrho a_0} + v_0\left(t - \frac{\varphi_2(t)}{a_0}\right) + \frac{p\left(t - \dfrac{\varphi_2(t)}{a_0}\right)}{\varrho a_0}\right]^2, \tag{13.21}$$

where $t_1 = t - \varphi_2(t)/a_0$. The above equation is an implicit equation for function ϵ_0.

If the form of the function $\sigma_0(\epsilon_0)$ is such that (13.21) can be solved with respect to ϵ_0 then we can write

$$\varepsilon_0 = \Phi\,[\varphi_2(t), t]. \tag{13.22}$$

From the continuity conditions (13.17) we deduce that

(13.23)
$$\varphi_2'(t) = \sqrt{\frac{\sigma_0[\Phi(\varphi_2(t), t)]}{\varrho\Phi[\varphi_2(t), t]}},$$

and on integration, with the initial condition $\varphi_2(t_A) = \varphi_1(t_A)$, we obtain the integral equation

(13.24)
$$\varphi_2(t) = \varphi_1(t_A) + \int_{t_A}^{t} \sqrt{\frac{\sigma_0[\Phi(\varphi_2(\tau), \tau)]}{\varrho\Phi[\varphi_2(\tau), \tau]}}\, d\tau .$$

Equation (13.24) can be solved by the method of successive approximations to a given degree of accuracy. On the other hand, when the form of the function $\sigma_0(\epsilon_0)$ is such that (13.31) cannot be reduced to an explicit form with respect to ϵ_0 then in (13.21), instead of function $\varphi_2(t)$, the zeroth approximation has to be assumed, e.g. $\varphi_2^{(0)}(t) = \varphi_1(t_A)+\varphi_1'(t_A)(t-t_A)$. Then we calculate, from (13.21), the zeroth approximation of ϵ_0, while from (13.24) we obtain the first approximation for the shock wave front $\varphi_2^{(1)}(t)$. Repeating this operation we obtain the successive approximations for the wave front $\varphi_2^{(n)}(t)$. It was shown in specific examples in [162] that the process of successive approximations converges fast. The solution in region II is constructed in a similar fashion as in region I on the basis of the relations along the characteristics. Replacing $\varphi_1(t)$ by $\varphi_2(t)$ in formulae (13.20) we obtain the solution in region II. The construction of the consecutive segments of the shock wave (Fig. 40) as well as the solution in the successive regions of the phase plane is obtained in exactly the same way as in regions I and II.

14. Propagation of plane stress waves in an elastic–plastic medium with rigid unloading

Let us consider a homogeneous, semi-infinite bar of constant cross-section for which the $\sigma - \epsilon$ relation is as shown in Fig. 41. At the beginning we assume that on the segment of loading OA we have $d^2\sigma/d\epsilon^2 \leqslant 0$, i.e. we exclude the possibility of the existence of a shock wave. Unloading takes place at constant strain. At $x = 0$ we assume the condition in the form of (11.1) (Fig. 42). For the assumed boundary condition and the relation as shown in Fig. 41 only waves

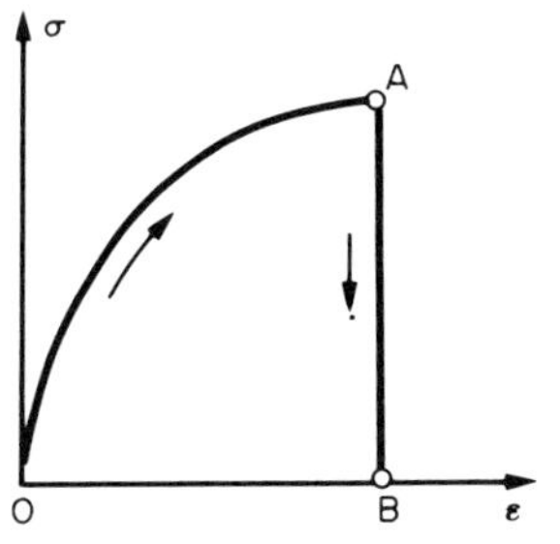

Fig. 41

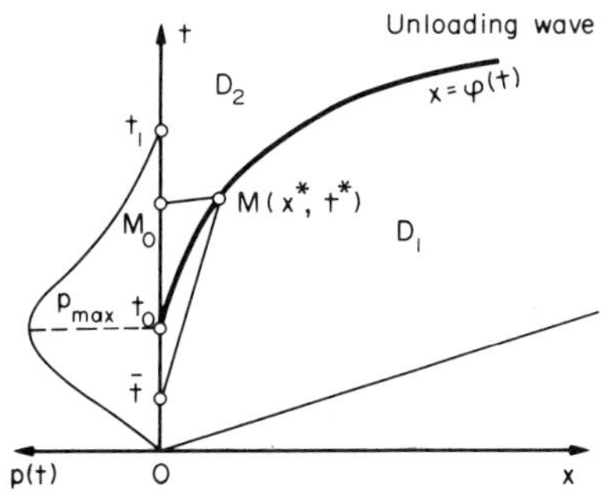

Fig. 42

of weak discontinuity propagate in the phase plane (x, t). The unloading wave $x = \varphi(t)$ (Fig. 42) divides the phase plane (x, t) into two regions: the loading region D_1 and the unloading region D_2. The problem reduces exactly as in the case of the unloading wave for elastic unloading (section 11) – to the determination of the shape of the wave $x = \varphi(t)$ and to finding the stress and velocity on this wave. In region D_1 we deal with the loading process, discussed already in detail in section 10. In this region Riemann waves propagate. The straight lines with equations

(14.1) $$x = a(\sigma)(t-\bar{t}), \qquad a^2 = \frac{1}{\varrho}\frac{d\varrho}{d\varepsilon},$$

where $\bar{t}$ denotes the time measured at the bar end $x = 0$ constitute the positive characteristics along which stress σ and velocity v are constant.

In this region we have (compare (10.11))

(14.2) $$dv = \frac{1}{\varrho a(\sigma)} d\sigma.$$

Now we shall derive the relations describing the unloading process. The equilibrium equation (10.1) and the continuity condition (10.4)

(14.3) $$\varrho\frac{\partial v}{\partial t} = \frac{\partial \sigma}{\partial x}, \qquad \frac{\partial v}{\partial x} = \frac{\partial \varepsilon}{\partial t},$$

are valid for an arbitrary $\sigma - \epsilon$ relation, consequently also for rigid unloading. Since in the unloading region D_2 the strain is constant at each cross-section x (i.e. $\epsilon(x)$ = const), then we deduce from the continuity condition $(14.3)_2$ that

(14.4) $$v = v(t) = v_0(t).$$

This means that in the unloading zone the bar segment M_0M (Fig. 42) is an ideally rigid body, the velocity of the particles is independent of x and is constant at any given instant in each cross-section of the bar [63], [170].

The equilibrium equation, on integration with respect to variable x, and taking into account the boundary condition (11.1), will take the form

(14.5) $$\sigma(x, t) = \varrho\frac{\partial v}{\partial t}x - p_2(t).$$

Assuming the continuity of velocity and stress across the unloading wave we are in a position to write

(14.6) $$dv_1 = dv_2 \qquad \text{along} \qquad x = \varphi(t),$$

where the indices 1 and 2 denote the values from the side of regions D_1 and D_2 respectively. Keeping in mind that $\partial v_2/\partial x = 0$, we deduce from (14.6)

(14.7) $$\frac{\partial v_2}{\partial t} = \frac{\partial v_1}{\partial t} + \frac{\partial v_1}{\partial x}\varphi'(t).$$

If we denote the stress and the velocity on the unloading wave by $\sigma_0 = \sigma_1[\varphi(t), t]$, $v_0 = v_1[\varphi(t), t]$, then

$$\frac{\partial v_2}{\partial t} = -\frac{1}{\varrho a(\sigma_0)}\frac{d\sigma_0}{dt}. \tag{14.8}$$

Substituting (14.8) into (14.5) and taking into account that $\sigma_2[\varphi(t), t] = \sigma_1[\varphi(t), t] = \sigma_0$ we obtain

$$\sigma_0 + \frac{d\sigma_0}{a(\sigma_0)\,dt} x^* + p_2(t^*) = 0, \tag{14.9}$$

where (x^*, t^*) are the coordinates of point M. Making use of the equation of the characteristic (14.1) passing through point $M(x^*, t^*)$ and replacing t^* by t, we derive from (14.9) the differential equation for $\sigma_0 = \sigma_0(t)$:

$$\frac{d\sigma_0}{dt} = -\frac{p_2(t)+\sigma_0}{t-\bar{t}(\sigma_0)} \quad \text{for} \quad t \geqslant t_0, \tag{14.10}$$

with the initial condition

$$\sigma_0(t_0) = -p_{\max}. \tag{14.11}$$

Equation (14.10) can be reduced to the form

$$d(t\sigma_0) + p_2(t)\,dt = \bar{t}(\sigma_0)\,d\sigma_0. \tag{14.12}$$

Integrating the above equation between the limits t_0 to t, we obtain

$$\sigma_0 t - \sigma_0(t_0)\,t_0 + \int_{t_0}^{t} p_2(\tau)\,d\tau = \int_{\sigma_0(t_0)}^{\sigma_0} \bar{t}(\sigma)\,d\sigma = \bar{t}(\sigma_0)\,\sigma_0 - t_0\,\sigma_0(t_0) + \int_{\bar{t}(\sigma_0)}^{t_0} p_1(t)\,dt, \tag{14.13}$$

or

$$\sigma_0[t-\bar{t}(\sigma_0)] = -\int_{t_0}^{t} p_2(\tau)\,d\tau - \int_{\bar{t}(\sigma_0)}^{t_0} p_1(t)\,dt. \tag{14.14}$$

Introducing the symbol

$$P(t) = -\int_{t_0}^{t} p_2(\tau)\,d\tau - \int_{0}^{t_0} p_1(t)\,dt \tag{14.15}$$

we obtain

$$[t-\bar{t}(\sigma_0)]\,\sigma_0 = P(t) + \int_{0}^{\bar{t}(\sigma_0)} p_1(t)\,dt. \tag{14.16}$$

Since $0<\bar{t}(\sigma_0)<t_0$, $0\leqslant p_1(t)\leqslant -\sigma_0(t_0)$ and $p_1(t)$ increase monotonically,

$$\int_{\bar{t}(\sigma_0)}^{t_0} p_1(t)\,dt > -[t_0-\bar{t}(\sigma_0)]\,\sigma_0 \tag{14.17}$$

whereas from (14.14) and (14.16) we obtain the estimation

$$-\frac{1}{t-t_0}\int_{t_0}^{t} p_2(\tau)\,d\tau < -\sigma_0(t) < -\frac{P(t)}{t-t_0}. \tag{14.18}$$

From the first of these inequalities we can write

$$p_2(t)+\sigma_0(t) < p_2(t)+\frac{1}{t-t_0}\int_{t_0}^{t} p_2(\tau)\,d\tau. \tag{14.19}$$

If $p_2(t)$ is a monotonically decreasing function of time, then $p_2(t)+\sigma_0(t)<0$. Consequently, on the basis of (14.10), we deduce that $\sigma_0(t)$ increases monotonically in time and tends to zero, provided $P(\infty)<+\infty$. By means of (14.16), taking into account (14.13), we obtain

$$\sigma_0(t)=\frac{P(t)}{t}+\frac{\sigma_0(t)\bar{t}(\sigma_0)}{t}+\frac{1}{t}\int_0^{\bar{t}(\sigma_0)} p_1(t)\,dt=\frac{P(t)}{t}+\frac{1}{t}\int_0^{\sigma_0}\bar{t}(\sigma)\,d\sigma, \tag{14.20}$$

whence

$$\sigma_0(t)=\frac{P(t)}{t}+o\left(\frac{1}{t}\right)=\frac{P(\infty)}{t}+o\left(\frac{1}{t}\right). \tag{14.21}$$

In [170] the asymptote of the unloading wave and the initial velocity of the unloading wave were also determined. The pressure distribution $p(t)$ is assumed in the vicinity of point $(0, t_0)$ to be of the form

$$\begin{aligned} p_2(t) &= p_{\max}-k_1(t-t_0)+\ldots, \\ \bar{t}(\sigma_0) &= t_0+k_2(\sigma_0+p_{\max})+\ldots, \\ \sigma_0 &= -p_{\max}-k_3(t-t_0)+\ldots, \end{aligned} \tag{14.22}$$

where $k_1>0$, $k_2>0$ are known while k_3 is an unknown coefficient. Making use of the above expressions (14.22) and passing to the limit $t\to t_0$, moreover assuming that $k_3\geqslant 0$, we obtain

$$k_3=\frac{k_1}{1+\sqrt{1+k_1 k_2}}. \tag{14.23}$$

The equation of the tangent to the wave at point $(0, t_0)$ takes the form

$$x=(\sigma_0)\sqrt{1+k_1 k_2}\,(t-t_0). \tag{14.24}$$

For certain forms of the loading $p_1(t)$ and for the simplified $\sigma - \epsilon$ relation in the unloading zone we can find the solution of the problem of the propagation of an unloading wave in closed form. Alternatively, after finding the explicit equation with respect to the derivative $\varphi'(t)$ it can be reduced to a problem involving integral equations. The unloading wave can be determined by means of successive iterations, and the stress, velocity and strain in each region of the phase plane are then determined.

Let us consider the case in which during the loading process the Prandtl model is valid whilst during unloading $\epsilon = \epsilon_0(x)$ (Fig. 43). We also assume that the loading $p_1(t)$ varies linearly, i.e. $p_1(t) = p_{max}\dfrac{t}{t_0}$. In region D_1 elastic waves first propagate followed by plastic Riemann waves the velocity of which is constant ($a_1 = \sqrt{E_1/\varrho}$). The stress on these waves is given by the formula

$$\sigma = -p_{max}\frac{\left(t-\dfrac{x}{a_1}\right)}{t_0}.$$

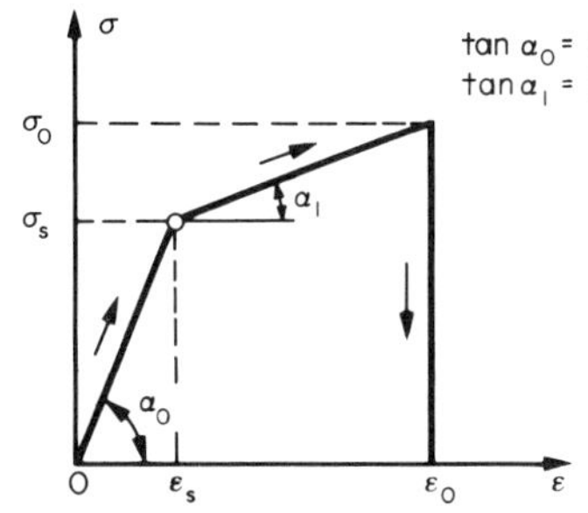

Fig. 43

On the unloading wave $x = \varphi(t)$ the conditions of the continuity of stress and velocity have to be satisfied, hence

$$\text{(14.25)}\quad \begin{aligned} \sigma_0(t) &= -\frac{p_{max}}{t_0}\left(t-\frac{\varphi(t)}{a_1}\right), \\ v_0(t) &= -\frac{\sigma_s}{\varrho a_0}+\frac{1}{\varrho a_1}\left[\sigma_s+\frac{p_{max}}{t_0}\left(t-\frac{\varphi(t)}{a_1}\right)\right]. \end{aligned}$$

Calculating the derivative $v_0(t)$, and taking into account $(14.25)_1$, in (14.5) we obtain (for $x = \varphi(t)$)

$$\text{(14.26)}\quad \begin{aligned} &\varphi(t)\varphi'(t) = a_1^2\left[t - t_0\frac{p_2(t)}{p_{max}}\right] = \Phi(t) \quad &&\text{for} \quad t_0 \leqslant t \leqslant t_1, \\ &\varphi(t)\varphi'(t) = a_1^2 t &&\text{for} \quad t \geqslant t_1. \end{aligned}$$

On integration and determination of the constants of integration from condition $\varphi(t_0) = 0$ and the condition of continuity on the unloading wave for $t = t_1$, we obtain the equation of the unloading wave in the forms:

(14.27)
$$\varphi(t) = \left\{2\int \Phi(t)\,dt - 2\int \Phi(t)\,dt\Big|_{t=t_0}\right\}^{1/2} \quad \text{for} \quad t_0 \leqslant t \leqslant t_1,$$
$$\varphi(t) = a_1\left\{t^2 - t_0^2 - \frac{2t_0}{p_{max}}\int_{t_0}^{t_1} p_2(t)\,dt\right\}^{1/2} \quad \text{for} \quad t \geqslant t_1.$$

Assuming a definite form for the loading change $p_2(t)$, we determine from (14.27) the unloading wave $x = \varphi(t)$. In the unloading region D_2 velocity v is equal to the velocity on the unloading wave, i.e. $v(t) = v_0(t)$, consequently it is known from formula $(14.25)_2$. The stress is determined from (14.5). For the time $t \geqslant t_1$ we obviously have to assume in formula (14.5) that $p_2(t) = 0$. The strain on the unloading wave results directly from the $\sigma - \epsilon$ relation applicable to the Prandtl model (Fig. 43) being considered.

Assuming, for example, that pressure $p_2(t)$ also varies linearly, $p_2(t) = p_{max}\dfrac{t_1 - t}{t_1 - t_0}$, we obtain from (14.27):

(14.28)
$$\varphi(t) = a_1\sqrt{\frac{t_1}{t_1 - t_0}}\,(t - t_0) \quad \text{for} \quad t_0 \leqslant t \leqslant t_1,$$
$$\varphi(t) = a_1\sqrt{t^2 - t_0 t_1} \quad \text{for} \quad t \geqslant t_1.$$

If we assume linear variations in pressures $p_1(t)$ and $p_2(t)$ for the case of the Prandtl model, the first segment of the unloading wave is a straight line (for $t_0 \leqslant t \leqslant t_1$).

If the shape of the unloading wave is known, the stress and the velocity on the unloading wave front can be determined from formulae (14.25):

(14.29)
$$\sigma_0(t) = -\frac{p_{max}}{t_0}\left[t - \sqrt{\frac{t_1}{t_1 - t_0}}\,(t - t_0)\right],$$
$$v_0(t) = -\frac{\sigma_s}{\varrho a_0}(1 - \mu) + \frac{p_{max}}{\varrho a_1 t_0}\left[t - \sqrt{\frac{t_1}{t_1 - t_0}}\,(t - t_0)\right] \quad \text{for} \quad t_0 \leqslant t \leqslant t_1,$$

(14.30)
$$\sigma_0(t) = -\frac{p_{max}}{t_0}\left(t - \sqrt{t^2 - t_0 t_1}\right),$$
$$v_0(t) = -\frac{\sigma_s}{\varrho a_0}(1 - \mu) + \frac{p_{max}}{\varrho a_1 t_0}\left(t - \sqrt{t^2 - t_0 t_1}\right) \quad \text{for} \quad t \geqslant t_1.$$

The stress in region D_2 is calculated from the equilibrium equation (14.5) after eliminating $v_0'(t)$ on differentiation with respect to time of velocity $v_0(t)$ determined in (14.29) and (14.30),

$$(14.31)\quad \begin{aligned} \sigma(x,t) &= \frac{p_{\max}}{a_1 t_0}\left(1-\sqrt{\frac{t_1}{t_1-t_0}}\right)x - p_{\max}\frac{t_1-t}{t_1-t_0} \quad &\text{for}\quad t_0 \leqslant t \leqslant t_1,\\ \sigma(x,t) &= \frac{p_{\max}}{a_1 t_0}\left(1-\frac{t}{\sqrt{t^2-t_1 t_0}}\right)x \quad &\text{for}\quad t \geqslant t_1. \end{aligned}$$

The strain $\epsilon(x,t) = \epsilon(x)$ in region D_2 is determined from the definition of the Prandtl model (Fig. 43):

$$(14.32)\quad \begin{aligned} \varepsilon(x) &= \frac{\sigma_s}{E_0}\left(1-\frac{E_0}{E_1}\right)+\left.\frac{\sigma_0(t)}{E_1}\right|_{t=\bar{\varphi}(x)}\\ &= \frac{\sigma_s}{\varrho a_0^2}(1-\mu^2)-\frac{p_{\max}}{t_0\varrho a_1^2}\left\{t_0+\frac{x}{a_1}\left(\frac{1}{\sqrt{\frac{t_1}{t_1-t_0}}}-1\right)\right\} \quad &\text{for}\quad t_0 \leqslant t \leqslant t_1\\ \varepsilon(x) &= \frac{\sigma_s}{\varrho a_0^2}(1-\mu^2)-\frac{p_{\max}}{t_0\varrho a_1^2}\left\{\sqrt{\frac{x^2}{a_1^2}+t_0 t_1}-\frac{x}{a_1}\right\} \quad &\text{for}\quad t \geqslant t_1. \end{aligned}$$

Now we consider the case when the pressure at the end of the bar described by Prandtl's model (Fig. 43) is applied suddenly and then decreases monotonically with time (Fig. 44). It can be proved, exactly as was done in the case of the unloading wave for the Prandtl model

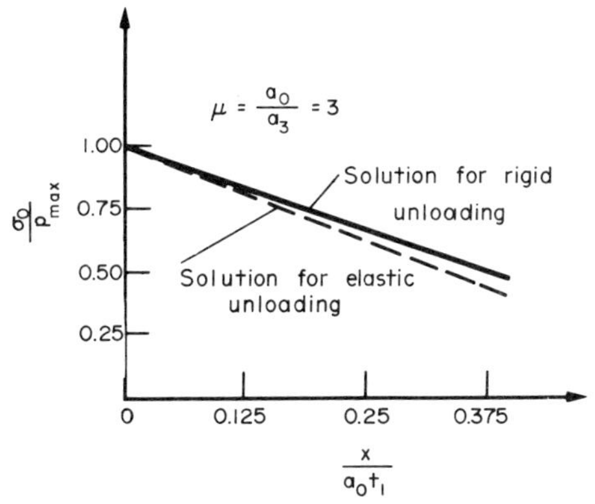

Fig. 44

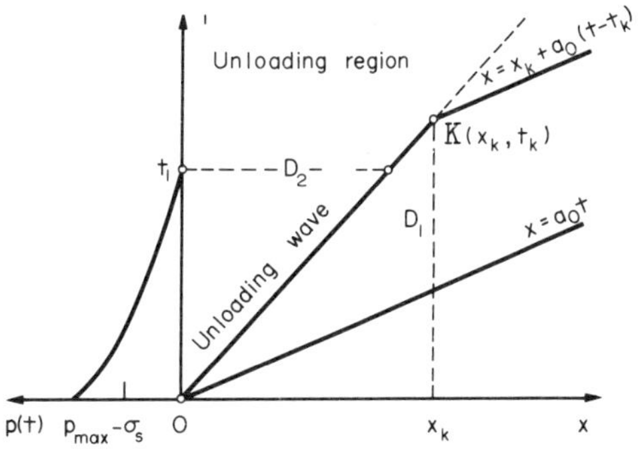

Fig. 45

with elastic unloading, that the unloading wave coincides with the characteristic, the equation of which is $x = a_1 t$ (11.21), and constitutes a strong discontinuity wave. The result of the solution in the phase plane is presented in Fig. 45. First the elastic wave, coinciding with the characteristic of equation $x = a_0 t$, the stress on which is equal to $\sigma = \sigma_s$, propagates into the undisturbed medium. It is followed by the front of a strong discontinuity wave, the unloading wave $x = a_1 t$. In region D_1 we have $\sigma = \sigma_S$, $v = -\frac{1}{\varrho a_0}\sigma_s$. On the unloading wave the condition of the dynamic continuity (11.25) has to be satisfied in the following form:

$$\sigma_0 - \sigma_s = -\varrho a_1 (v_0 - v_s), \tag{14.33}$$

where the index zero denotes the value on the unloading wave.

Making use of (14.33) we deduce from the equation of motion (14.5) on the unloading wave $x = a_1 t$ that

$$\sigma_0(t) = t\frac{d\sigma_0(t)}{dt} - p_2(t). \tag{14.34}$$

Integrating the above equation, and taking into account the initial condition

$$\sigma_0(0) = -p_2(0),$$

we obtain the stress on the unloading wave,

$$\begin{aligned} \sigma_0(t) &= -\frac{1}{t}\int_0^t p_2(\tau)\,d\tau, \quad \text{for} \quad t \leqslant t_1, \\ \sigma_0(t) &= -\frac{1}{t}\int_0^{t_1} p_2(t)\,dt, \quad \text{for} \quad t > t_1, \end{aligned} \tag{14.35}$$

where it is assumed, for $t > t_1$, that $p_2(t) \equiv 0$.

Velocity v_0, on the unloading wave, is obtained directly from (14.33), namely

$$\begin{aligned} v_0(t) &= -(1-\mu)\frac{\sigma_s}{\varrho a_0} + \frac{1}{\varrho a_1 t}\int_0^t p_2(\tau)\,d\tau \quad \text{for} \quad t \leqslant t_1, \\ v_0(t) &= -(1-\mu)\frac{\sigma_s}{\varrho a_0} + \frac{1}{\varrho a_1 t}\int_0^{t_1} p_2(t)\,dt \quad \text{for} \quad t > t_1. \end{aligned} \tag{14.36}$$

The stress in the unloading region is determined from (14.25) taking into account the condition $v(t) = v_0(t)$, thus

$$\begin{aligned} \sigma(x,t) &= -\left(1 - \frac{x}{a_1 t}\right)p_2(t) - \frac{x}{a_1 t^2}\int_0^t p_2(\tau)\,d\tau \quad \text{for} \quad t \leqslant t_1, \\ \sigma(x,t) &= -\frac{x}{a_1 t^2}\int_0^{t_1} p_2(t)\,dt \quad \text{for} \quad t > t_1. \end{aligned} \tag{14.37}$$

The strain in the unloading region, in the case of the Prandtl model, assumes the forms

$$
(14.38)\quad
\begin{aligned}
\varepsilon(x) &= (1-\mu^2)\frac{\sigma_s}{\varrho a_0^2} - \frac{1}{\varrho a_1 x}\int_0^{\frac{x}{a_1}} p_2(\tau)\,d\tau \qquad \text{for} \qquad t \leqslant t_1,\\
\varepsilon(x) &= (1-\mu^2)\frac{\sigma_s}{\varrho a_0^2} - \frac{1}{\varrho a_1 x}\int_0^{t_1} p_2(t)\,dt \qquad \text{for} \qquad t > t_1.
\end{aligned}
$$

It can be readily observed on the basis of (14.35) and (14.38) that the stress, as well as the strain, on the unloading wave decreases along the wave. The end of the unloading wave is determined from the condition

$$\sigma(x_k) = \sigma_s.$$

Consequently, the bar is plastically deformed over the segment $(0, x_k)$. For $x > x_k$ only elastic waves propagate in a semi-infinite bar.

The above solutions for a strong discontinuity wave can also be obtained from the preceding solutions for a weak discontinuity wave (Fig. 42) by passing to the limit $t_0 \to 0$ in them. Thus for $t_0 \to 0$ the unloading wave, the equation of which is given by (14.28), is a straight line of equation $x = a_1 t$. Furthermore, from the solution for the unloading wave, as well as from the solutions in the unloading region ((14.29)–(14.32)) we obtain equations (14.35)–(14.38), respectively.

Comparing the values of stresses (14.35), on the unloading wave, in the case of the Prandtl model with rigid unloading for the linear variation of the pressure $p(t) = p_{max}(1-\frac{t}{t_1})$ at the bar end, with the stress values (11.24) obtained for a similar problem in the case of the elastic unloading (Fig. 32), we see [37] that the difference between the two solutions is minimal. The values of the unloading wave computed for both cases, for coefficient $\mu = 3$, are presented in Fig. 44. The difference between the two results does not exceed 10%. The divergence of the results decreases considerably with the increase of the coefficient μ, e.g. for $\mu = 10$ it does not exceed 1%.

So far in section 14 we have been dealing with the problems of wave propagation in a semi-infinite bar. Now we pass to a discussion of problems of wave propagation in a finite bar. As before, we assume that the bar is of elastic–plastic material with rigid unloading. In this case also the problem of reflection of a wave of weak discontinuity from the end is a very difficult one, therefore we confine ourselves to cases of waves of strong discontinuity. Further simplification of the wave reflection problem can be obtained in the case when the front of the strong discontinuity wave is an unloading wave as well. In this case the unloading wave is the first one that reaches the bar end and is then reflected. Such a case is obtained if it is assumed that, for a given material, the elastic limit is equal to zero $\sigma_s \equiv 0$.

We now consider the case of the propagation of a plane unloading wave of strong discontinuity in a layered medium with an undeformable mass M (Fig. 46) at the boundary of the media. The unloading wave impinging on mass M will be reflected from it; simultaneously waves start propagating into the second medium. We assume that the first medium is plastic, the $\sigma - \epsilon$ relation is assumed in accordance with Fig. 47, the loading process acts along the straight line OC while the unloading takes place at constant strain. The second medium is ideally elastic.

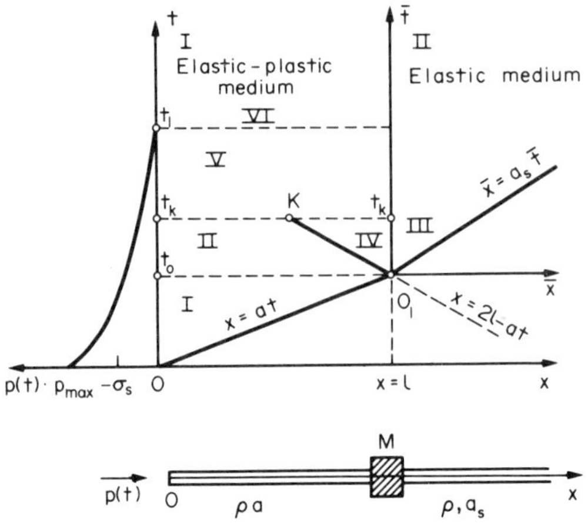

Fig. 46

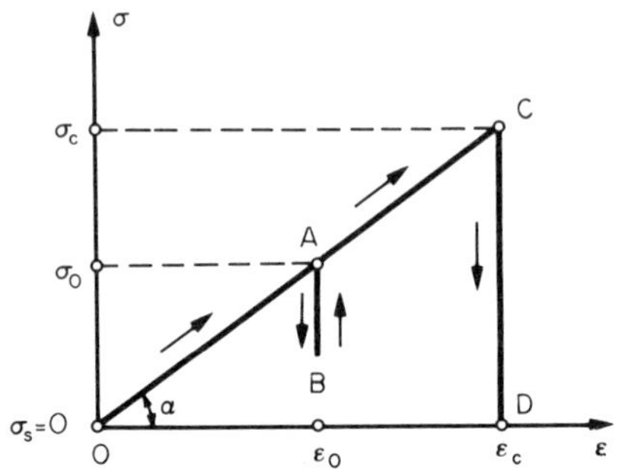

Fig. 47

Moreover we assume that pressure $p(t)$ suddenly appears at the end $x = 0$ and then decreases monotonically with time, i.e. the boundary condition is in the form (11.1′). This problem was solved in [64]. The solution in the phase plane (x, t) is presented in Fig. 46. The solution in region I is obtained from the solutions to the problem considered already ($(14.35)_1$, $(14.36)_1$, $(14.37)_1$, and $(14.38)_1$), assuming that $\sigma_s \equiv 0$, i.e.

$$\sigma_0(t) = -\frac{1}{t}\int_0^t p(\tau)\,d\tau \quad \text{or} \quad \sigma_0(x) = -\frac{a}{x}\int_0^{\frac{x}{a}} p(\tau)\,d\tau,$$

(14.39)

$$v_0(t) = \frac{1}{\varrho a t}\int_0^t p(\tau)d\tau,$$

$$\sigma_1(x, t) = -\left(1 - \frac{x}{at}\right)p(t) - \frac{x}{at^2}\int_0^t p(\tau)\,d\tau,$$

$$\varepsilon_1(x) = -\frac{1}{\varrho a x}\int_0^{\frac{x}{a}} p(\tau)\,d\tau,$$

where index 1 denotes the solution in region I (Fig. 46) and $a = \sqrt{E/\varrho}$.

At the instant when the incident wave is reflected from mass M an accumulation of stress occurs; it follows† from (12.8) that if the stress at the front of the incident wave was equal to σ_0 then at the instant of reflection of the wave from the mass $t_0 = l/a$ the stress on the front of the reflected wave would be equal to $\sigma_c = 2\sigma_0$. The reflected wave of equation $x = 2l - at$ also

†Assuming that the elastic medium II can be represented as a spring with a suitable coefficient of elasticity (model presented in Fig. 35).

constitutes a wave of strong discontinuity that propagates into the medium already deformed by the incident wave. At a particular cross-section $x (0 \leqslant x \leqslant l)$ the medium has been deformed according to OAB (Fig. 47). The increase of stress first occurs along line BA up to value σ_0 at constant strain, i.e. to the value that existed in the medium before the unloading; after this there then occurs an increase of stress along line AC with simultaneous increase of strain. Consequently, the stress has to increase up to value $\sigma_0(x)$ in front of the reflected wave at the considered cross-section x, then it increases up to value $\sigma_c(x)$ on the wave front. Behind the wave front it decreases – it can be proved that the reflected wave with equation $x = 2l - at$ is an unloading wave.

We consider the solution in region II. For a fixed time $t (t_0 \leqslant t \leqslant t_k)$ the medium is ideally rigid, the process of stress variation takes place along the straight line AB without any change of strain, velocity v_2 is function of time only. The equilibrium equation takes the form of (14.5), i.e.

$$\sigma_2(x, t) = \varrho v_2'(t)\, x - p(t). \tag{14.40}$$

Since for a considered cross-section x in front of the reflected wave front

$$x = 2l - at \tag{14.41}$$

the stress takes the value $\sigma_0(t)$, then on account of (14.39) we have

$$\sigma_0(x) = -\frac{a}{2l - at} \int_0^{\frac{2l}{a} - t} p(\tau)\, d\tau. \tag{14.42}$$

From (14.40), taking into account (14.42), we obtain for $x = 2l - at$

$$-\frac{a}{2l - at} \int_0^{\frac{2l}{a} - t} p(\tau)\, d\tau = \varrho v_2'(t)(2l - at) - p(t), \tag{14.43}$$

and on integration

$$v_2(t) = \frac{1}{\varrho} \int_{t_0}^{t} \frac{p(\tau)}{2l - at}\, d\tau - \frac{1}{\varrho} \int_{t_0}^{t} \left[\frac{a}{(2l - a\eta)^2} \int_0^{\frac{2l}{a} - \eta} p(\tau)\, d\tau \right] d\eta + C, \tag{14.44}$$

where the constant of integration C is determined from the initial condition for $t = t_0$,

$$v_2(t_0) = v_0(t_0) = \frac{1}{\varrho a\, t_0} \int_0^{t_0} p(t)\, dt,$$

or

$$(14.45)\quad v_2(t) = \frac{1}{\varrho}\int\limits_{t_0}^{t}\frac{p(\tau)}{2l-a\tau}\,d\tau - \frac{1}{\varrho}\int\limits_{t_0}^{t}\left[\frac{a}{(2l-a\eta)^2}\int\limits_{0}^{\frac{2l}{a}-\eta} p(\tau)\,dt\right]d\eta + \frac{1}{\varrho a t_0}\int\limits_{0}^{t_0} p(\tau)\,d\tau.$$

The stress distribution in region II is determined from (14.40) upon elimination of $v'(t)$ from (14.43):

$$(14.46)\qquad \sigma_2(x,t) = -p(t)\left(1-\frac{x}{2l-at}\right) - \frac{ax}{(2l-at)^2}\int\limits_{0}^{\frac{2l}{a}-t} p(\tau)\,d\tau,$$

the strains are identical with those in region I, namely

$$(14.47)\qquad \varepsilon_2(x) \equiv \varepsilon_1(x).$$

Now, we pass to the solution in region III. Assuming that medium II is elastic and that the velocity of wave propagation in that medium is equal to a_s, the equation of motion (10.8) in the region can be represented in the form

$$(14.48)\qquad \frac{\partial^2 u}{\partial \bar{t}^2} - a_s^2\frac{\partial^2 u}{\partial \bar{x}^2} = 0,$$

where $\bar{x} = x - l,\ t = \bar{t} - l/a$. The general solution of (14.48) is as follows:

$$(14.49)\qquad u(\bar{x},\bar{t}) = \varphi_1(\bar{x}-a_s\bar{t}) + \varphi_2(\bar{x}+a_s\bar{t}).$$

Functions φ_1 and φ_2 are determined from the boundary conditions for $x = 0$, and from the conditions on the characteristic $\bar{x} = a_s\bar{t}$

$$(14.50)\qquad \begin{aligned} \sigma_3(0,\bar{t}) &= \varrho_1 a_s^2\varepsilon_s(0,\bar{t}) = \varrho_1 a_s^2\left.\frac{\partial u}{\partial \bar{x}}\right|_{\bar{x}=0} = \varrho_1 a_s^2[\varphi_1'(-a_s\bar{t}) + \varphi_2'(a_s\bar{t})] = -\sigma_p(\bar{t}),\\ \sigma_3(x,\bar{t})|_{\bar{x}=a_s\bar{t}} &= \varrho_1 a_s^2\left.\frac{\partial u}{\partial \bar{x}}\right|_{\bar{x}=a_s\bar{t}} = \varrho_1 a_s^2[\varphi_1'(0) + \varphi_2'(2a_s\bar{t})] = \sigma_3(0,0) = 0, \end{aligned}$$

where $\sigma_p(\bar{t})$ denotes the stress acting on mass M from the side of the elastic medium. Condition $\sigma_3(0,0) = 0$ results from the fact that mass M cannot be set into motion suddenly by a force of finite value. The wave whose equation is $\bar{x} = a_s\bar{t}$ constitutes a wave of weak discontinuity. We obtain from condition $(14.50)_2$ that $\varphi_2'(2a_s\bar{t}) = -\varphi_1'(0)$, and next from $(14.50)_1$ we find that

$$(14.51)\qquad \varphi_1'(-a_s\bar{t}) = \varphi_1'(0) - \frac{\sigma_p(\bar{t})}{\varrho_1 a_s^2}.$$

Next, making use of the initial condition $v_3(0, 0) = 0$ we determine the constant $\varphi_1'(0)$,

$$\left.\frac{\partial u}{\partial \bar{t}}\right|_{\substack{\bar{x}=0\\ \bar{t}=0}} = -a_s\varphi_1'(0) + a_s\varphi_2'(0) = 0 \rightarrow \varphi_1'(0) = 0.$$

From (14.51), assuming that $z = -a_s\bar{t}$, we obtain

$$\varphi_1'(z) = -\frac{\sigma_p\left(-\dfrac{z}{a_s}\right)}{\varrho_1 a_s^2},$$

then, assuming $z = \bar{x} - a_s\bar{t}$, we obtain

(14.52) $$\varphi_1'(\bar{x} - a_s\bar{t}) = -\frac{\sigma_p\left(\bar{t} - \dfrac{\bar{x}}{a_s}\right)}{\varrho_1 a_s^2}.$$

Velocity $v(\bar{x}, \bar{t})$, in region III, can be represented by the following relation:

(14.53) $$v_s = -a_s\varphi'(\bar{x} - a_s\bar{t}) = \frac{1}{\varrho_1 a_s}\sigma_p\left(\bar{t} - \frac{\bar{x}}{as}\right),$$

where the stress σ_p over the contact of media I and II will be determined from the equation governing the dynamic coupling between the media (condition (12.19)).

It can be proved that, in region IV, unloading takes place. Taking into account that for unloading $\epsilon(t) = \text{const}$, we obtain, from $(14.3)_1$, the equation of dynamic equilibrium in region IV,

$$\sigma_4(x, t) = \varrho v_4'(t)x + C(t),$$

where the constant of integration is to be determined from the condition that $\sigma = \sigma_c$ (Fig. 47) on the reflected wave $x = 2l - at$, i.e. that

(14.54) $$\sigma_4(x, t) = \varrho v_4'(t)(x - 2l + at) + \sigma_c.$$

Hence, we obtain for $x = l$

(14.55) $$\sigma(l, t) = \varrho v_4'(t)(at - l) + \sigma_c = -\sigma_l(t),$$

where σ_1 denotes the reaction on mass M from the side of region I. Taking into account that $\bar{t} = t - l/a$ we have

(14.56) $$\sigma(l, t) = \varrho a v_4'(\bar{t})\bar{t} + \sigma_c(\bar{t}) = -\sigma_l(\bar{t}).$$

Making use of condition (12.14) and of the velocity compatibility condition (12.13), we obtain the following equation:

$$(M+\varrho a\bar{t})\,v_4'(\bar{t})+\varrho_1\,a_s\,v_4(\bar{t})=-\sigma_c(\bar{t}). \tag{14.57}$$

From the condition of dynamic continuity (12.2), at the front of the reflected wave $x = 2l - at$ we deduce that

$$\sigma_c=\varrho a v_4+\sigma_0-\varrho a v_2, \tag{14.58}$$

where σ_0 and v_2 are known functions of time, determined by formulae (14.42) and (14.45). Substituting (14.58) in (14.57), and using the notation

$$f_1(\bar{t})=\frac{\varrho a+\varrho_1\,a_s}{M+\varrho a\bar{t}},\qquad f_2(\bar{t})=\frac{-\sigma_0(\bar{t})+\varrho a v_2(\bar{t})}{M+\varrho a\bar{t}},$$

we obtain the following ordinary differential equation of the first order in terms of function $v_4(\bar{t})$:

$$v_4'(\bar{t})+f_1(\bar{t})\,v_4(\bar{t})=f_2(\bar{t}), \tag{14.59}$$

the solution of which is as follows:

$$v_4(\bar{t})=\int_0^{\bar{t}} f_2(\eta)\exp\left[-\int_\eta^{\bar{t}} f_1(\tau)\,d\tau\right]d\eta. \tag{14.60}$$

The constant of integration has been determined from the condition that $v_4(0) = 0$.

On the basis of formulae (14.53) and (14.55) stresses $\sigma_l(t)$ and $\sigma_p(t)$ take the forms

$$\sigma_l(\bar{t})=Mf_2(\bar{t})+[\varrho_1\,a_s-Mf_1(\bar{t})]\int_0^{\bar{t}} f_2(\eta)\exp\left[-\int_0^{\bar{t}} f_1(\tau)\,d\tau\right]d\eta,$$

(14.61)

$$\sigma_p(\bar{t})=\varrho_1\,a_s\int_0^{\bar{t}} f_2(\eta)\exp\left[-\int_0^{\bar{t}} f_1(\tau)d\tau\right]d\eta.$$

Formulae (14.61) are valid for the time $\bar{t} < t_k - t_0$ (assuming that $t_0 < t_k$), where t_k denotes the time at which the unloading wave fades away, i.e. when the condition

$$v_4(t_k)=v_2(t_k) \tag{14.62}$$

is satisfied.

The equation of dynamic equilibrium (14.5), in region V, takes the form

$$\sigma_5(x,\bar{t}) = \varrho x v_5'(\bar{t}) - p(\bar{t}). \tag{14.63}$$

Hence, for $x = l$, we have

$$\varrho l v_5'(\bar{t}) - p(\bar{t}) = -\sigma_l(\bar{t}). \tag{14.64}$$

Making use of the equilibrium equation for the forces applied to mass M (12.14)

$$M v_5'(\bar{t}) + \sigma_p(l,\bar{t}) - \sigma_l(l,\bar{t}) = 0,$$

and of (14.52) ($\nu_3 = \nu_5$),

$$v_5(\bar{t}) = \frac{1}{\varrho_1 a_s} \sigma_p(\bar{t}),$$

we obtain from (14.63) the following differential equation in terms of $\nu_5(\bar{t})$:

$$v_5'(\bar{t}) + f_3 v_5(\bar{t}) = f_4 p(\bar{t}), \tag{14.65}$$

where

$$f_3 = \frac{\varrho_1 a_s}{M + \varrho l}, \qquad f_4 = \frac{1}{M + \varrho l},$$

The solution of this equation takes the form

$$v_5(\bar{t}) = \mathrm{e}^{-f_3(\bar{t}-\bar{t}_k)} \left[f_4 \int_{\bar{t}_k}^{\bar{t}} p(\eta) \exp[f_3(\eta - \bar{t}_k)]\, d\eta + v_2(\bar{t}_k) \right], \tag{14.66}$$

where the constant of integration has been determined from the condition at $\bar{t} = \bar{t}_k$, i.e. that $\nu_5(\bar{t}_k) = \nu_2(\bar{t}_k)$. Finally, stresses $\sigma_l(\bar{t})$, $\sigma_p(\bar{t})$ can be put in the forms

(14.67)

$$\sigma_l(\bar{t}) = \{1 - \varrho l f_4 \exp[-f_3(\bar{t} - \bar{t}_k)]\}\, p(\bar{t}) + \varrho l f_3 \exp[-f_3(\bar{t} - \bar{t}_k)] \times \left[f_4 \int_{\bar{t}_k}^{\bar{t}} p(\eta) \exp[f_3(\eta - \bar{t}_k)]\, d\eta + v_2(\bar{t}_k) \right],$$

$$\sigma_p(\bar{t}) = \varrho_1 a_s \left\{ f_4 p(\bar{t}) - f_3 \left[f_4 \int_{\bar{t}_k}^{\bar{t}} p(\eta) \exp[f_3(\eta - \bar{t}_k)]\, d\eta + v_2(\bar{t}_k) \right] \right\} \exp[-f_3(\bar{t} - \bar{t}_k)].$$

The solution in region VI is obtained if we assume in equation (14.65) that $p(\bar{t}) \equiv 0$.

The strains in the respective regions take the forms

$$\varepsilon_2(x) \equiv \varepsilon_1(x), \qquad \varepsilon_5(x) \equiv \varepsilon_6(x)|_{x<x_k} \equiv \varepsilon_1(x),$$

(14.68)
$$\varepsilon_4(x) = \frac{1}{E}\,\sigma_c(\bar{t})\Big|_{\bar{t}=\frac{l-x}{a}},$$
$$\varepsilon_5(x) \equiv \varepsilon_6(x)|_{x>x_k} \equiv \varepsilon_4(x),$$
$$\varepsilon_3(x,t) = -\frac{1}{\varrho_1 a_s^2}\,\sigma_p\left(\frac{x}{a} - t\right).$$

In Fig. 48 the diagram is presented of the stress $\sigma_l(\bar{t})$ – the reaction acting on mass M from the side of region I [53] – in the case of a linear variation of pressure $p(t)$ on the boundary $\left(p(t) = p_{\max}\left[1 - \frac{t}{t_1}\right]\right)$, curve 1. In the case when there is no mass at the boundary of the

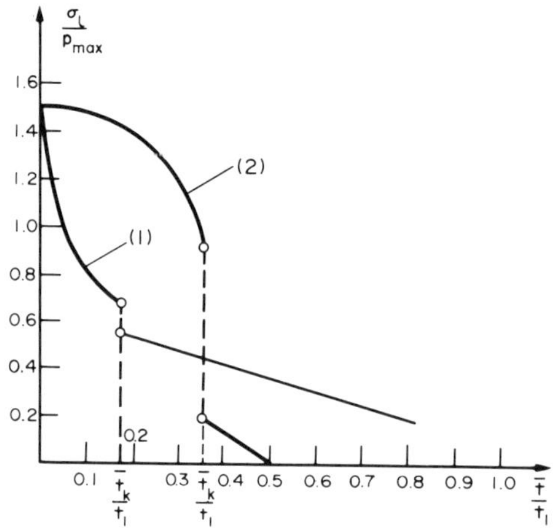

Fig. 48

media (while medium II is elastic) then, putting $M \equiv 0$ in (14.57), we obtain the solution for a layered medium, this being a particular case of the solutions presented by Kaliski in [56]. The solutions in regions I and II remain unchanged while the solutions in regions IV, V, and VI take the forms

$$f_1(\bar{t}) = \frac{\varrho a + \varrho_1 a_s}{\varrho a \bar{t}}, \quad f_2(\bar{t}) = \frac{-\sigma_0(\bar{t}) + \varrho a v_2(\bar{t})}{\varrho a \bar{t}}, \quad f_3 = \frac{\varrho_1 a_s}{\varrho l}, \quad f_4 = \frac{1}{\varrho l}.$$

Next, passing to the limit in the solutions of $M \to \infty$, we obtain the solutions for the case of the reflection of an unloading wave from an undeformable obstacle. Then we have $f_1 = f_2 = f_3 = f_4 = 0$, thus on the basis of (14.60) we obtain

(14.69)
$$v_4(x,t) = 0.$$

From the equation of dynamic equilibrium (14.54) one concludes that the stress is a function of time only $\sigma_4(x, t) = \sigma_c(t)$. Making use of the condition of dynamic continuity (14.58) we obtain the following solution:

$$\begin{aligned}\sigma_4(t) = \sigma_c(t) = &-\frac{a}{2l-at}\int_0^{\frac{2l}{a}-t} p(\tau)\,d\tau - \int_{t_0}^{t}\frac{a}{2l-a\tau}\,p(\tau)\,d\tau \\ &+\int_{t_0}^{t}\left[\frac{a^2}{(2l-a\eta)^2}\int_0^{\frac{2l}{a}-\eta} p(\tau)\,d\tau\right]d\eta - \frac{1}{t_0}\int_t^{t_0} p(t)\,dt.\end{aligned} \tag{14.70}$$

It is easy to observe that the stress $\sigma_4(t)$ is a decreasing function of time; consequently region IV is in fact an unloading region. The stress discontinuity across the front of the reflected wave vanishes with the increase of time. From (14.58) we have $[\sigma] = \sigma_c - \sigma_0 = -\varrho a v_2(t)$, where $v_2(t)$ is a decreasing function of time, determined from (14.45). In this case the end of the reflected unloading wave (14.62) is determined from the condition $v_2(t_k) = 0$. At that instant the motion of the bar as a rigid body has ceased ($v_5 \equiv 0$). It follows from (14.63) that for $\bar{t} = \bar{t}_k$ we have $\sigma_5(\bar{t}_k) = -p(\bar{t}_k)$, consequently there occurs a stepwise smoothing of the stresses to the value $-p(t_k)$. In region V we obtain from (14.67) that $\sigma_1(\bar{t}) = -\sigma_5(\bar{t}) = p(\bar{t})$. In region VI we obviously have $\sigma_6 = v_6 \equiv 0$. In Fig. 48 the variation of stress is also presented for $x = l$ from the side of region I, in the case when $M \to \infty$, i.e. in the case of the reflection of the unloading wave from a rigid obstacle [64] – curve 2. Time t_k corresponds to the time when the reflected unloading wave of strong discontinuity vanishes. At the initial instant the value of stress $\sigma_l(l, 0)$ is the same in the case of the reflection of the unloading wave from a rigid obstacle as it is in the case of an elastic obstacle. This is implied from a comparison of formulae (12.8) with (12.11). Next the stresses decrease in both cases; however, in the case of elastic fixing they decrease considerably faster. The gradient of stress $\sigma_l(l, \bar{t})$ decrease is the greater the smaller is mass M and the more flexible is medium II behind the mass M.

To conclude this section we present two cases of the propagation of plane shock waves in a semi-infinite bar with rigid unloading [58], [143].

The dynamic characteristics of soils for high values of pressure (compare p. 26, Fig. 16) can be approximated by the model shown in Fig. 49b, c. In the case of high pressures we can assume that the elastic limit of soil is equal to zero. The real curve $OBAD$ in Fig. 49a can be replaced either by curve OA (Fig. 49b) or by the straight line segment (Fig. 49c). When the stress increases above value $\sigma = \sigma^*$, loading takes place at a constant strain equal to $\epsilon = \epsilon^*$; for unloading, with $0 < \sigma \leqslant \sigma^*$ we have $\epsilon = \epsilon(x)$.

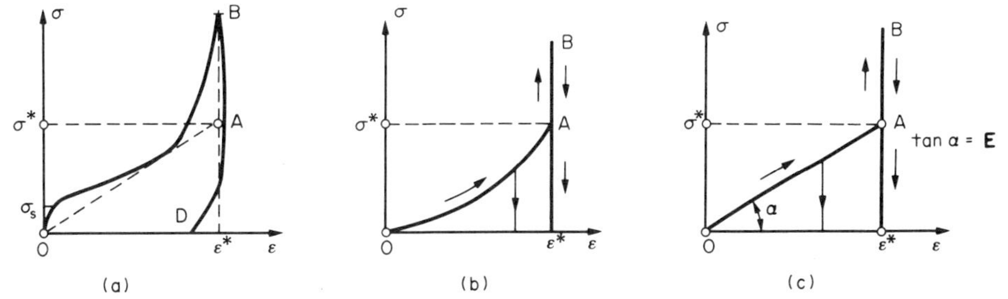

Fig. 49

We assume that the pressure at the end of the bar is suddenly applied and then it decreases monotonically with time, and that $p(0) = p_{max} > \sigma^*$. The case $p_{max} < \sigma^*$ for a linear variation of $\sigma = \sigma(\epsilon)$, Fig. 49c, has already been considered. The wave of strong discontinuity with equation $x = at$, being simultaneously an unloading wave, propagates in the semi-infinite bar. The solution for this case is given by (14.39). Also the wave process in a layered medium with a rigid mass at the boundary of the media has been considered for this case.

In the case referred to above of the end loading of a semi-infinite bar, a shock wave starts propagating from the bar end. The equation of this wave is indicated in Fig. 50. On account of the assumed loading this wave is simultaneously an unloading wave. Its shape will be determined from the equations of dynamic equilibrium (14.5) and from the conditions of dynamic and kinematic continuity (13.17), where we assume $\epsilon_0 = \epsilon^*$:

(14.71) $$\sigma(x,t) = \varrho v'(t)\,x - p(t), \qquad \sigma_0 = -\varrho\varphi'(t)\,v_0, \qquad v_0 = -\varphi'(t)\,\varepsilon^*.$$

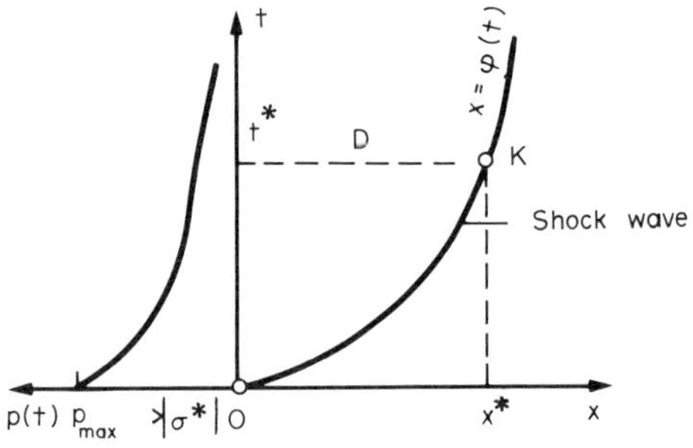

Fig. 50

For $x = \varphi(t)$, taking into account that $\sigma(x,t)|_{x=\varphi(t)} = \sigma_0$ and $v(t) = v_0(t)$, we obtain from the system of equations (14.71) the following ordinary differential equation of the second order:

(14.72) $$\varphi(t)\varphi''(t) + [\varphi'(t)]^2 = f(t).$$

where $f(t) = -p(t)/\varrho\varepsilon^*$. This equation can be transformed into

(14.73) $$\frac{\varphi(t)}{\varphi'(t)}\frac{d}{dt}\left[\frac{(\varphi'(t))^2}{2}\right] + [\varphi'(t)]^2 = f(t),$$

Integrating it twice and making use of the initial conditions:

(14.74) $$\varphi(0) = 0, \qquad \varphi'(0) = \sqrt{\frac{-p_{max}}{\varrho\varepsilon^*}},$$

we obtain the required equation of the shock wave

(14.75) $$\varphi(t) = \left[2\int_0^t\left(\int_0^{t_1} f(t_2)\,dt_2\right)dt_1\right]^{1/2}.$$

After some manipulation we obtain from formulae (14.71) the solution in the unloading region, namely

$$\sigma(x,t) = \varrho\varepsilon^* f(t) - \varrho\varepsilon^* x \frac{f(t)\left[2\int_0^t\left(\int_0^{t_1} f(t_2)\,dt_2\right)dt_1\right]^{1/2} - \left[\int_0^t f(t_1)\,dt_1\right]^2\left[2\int_0^t\left(\int_0^{t_1} f(t_2)\,dt_2\right)dt_1\right]^{-1/2}}{2\int_0^t\left(\int_0^{t_1} f(t_2)dt_2\right)dt_1}$$

(14.76)
$$v(t) = -\varepsilon^* \frac{\int_0^t f(t_1)\,dt_1}{\left[2\int_0^t\left(\int_0^{t_1} f(t_2)\,dt_2\right)dt_1\right]^{1/2}},$$

$$\sigma_0(t) = 2\varrho\varepsilon^* \int_0^t\left(\int_0^{t_1} (f(t_2)\,dt_2\right)dt_1.$$

The stress at the unloading wave front, determined by (14.76), decreases with time. For time $t = t^*$ (Fig. 50) it assumes the value $\sigma = \sigma^*$. For time $t > t^*$ in the case of the relation $\sigma = \sigma(\epsilon)$ shown in Fig. 49c, the solution can be obtained in closed form. For the case, however, of the relation shown in Fig. 49b the solution has to be constructed numerically.

The above solution can be readily generalized to the case of wave reflection in a finite bar for arbitrary boundary conditions at the end $x = l$. In the case of the reflection of a shock wave from a rigid obstacle, on account of the condition $v = 0$ at the fixed end, $x = l$, of the bar, the displacements tend to a uniform distribution while the stresses become stepwise smoothed.

Now we consider a similar problem of plane shock wave propagation in a slightly more general case [143]. We assume that the $\sigma = \sigma(\epsilon)$ relation for soils, in the range of medium pressures (the dashed line in Fig. 51a), can be approximated by straight line segments, as is shown in the same figure (continuous lines). Also we consider the case for which the pressure at the end $x = 0$ increases monotonically from zero during $0 \leqslant t \leqslant t_0$, and then monotonically decreases to zero.

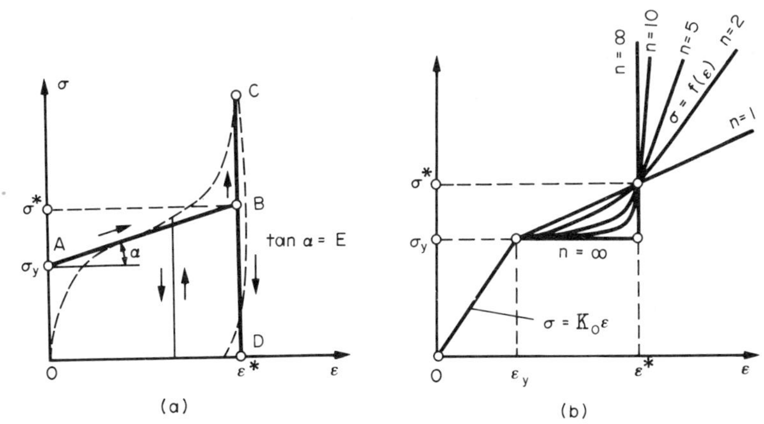

Fig. 51

The outline of the solution in the (x, t) plane is as follows (Fig. 52). For $t \leqslant t_y$ the bar is prestressed, then $\sigma = -p(t) \leqslant \sigma_y$, $v = \epsilon \equiv 0$. In region I the state is steady: $\sigma_1 = \sigma_y$, $v_1 = \epsilon_1 \equiv 0$.

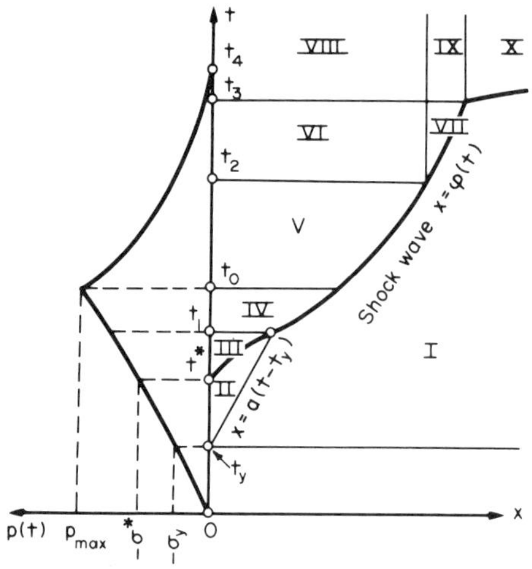

Fig. 52

In region II, bounded by the characteristic $x = a(t-t_y)$ and t-axis, Riemann waves propagate with a speed $a = \sqrt{E/\varrho}$. In this region we have

$$\sigma_2(x,t) = -p\left(t-\frac{x}{a}\right),$$

(14.77)
$$v_2(x,t) = -\frac{1}{\varrho a}\left[\sigma_y + p\left(t-\frac{x}{a}\right)\right],$$

$$\varepsilon_2(x,t) = \frac{1}{a}\, v_2(x,t).$$

If at time $t = t^*$ the pressure at $x = 0$ exceeds the value $\sigma = \sigma^*$ then a shock wave $x = \varphi(t)$ will start propagating from point $(0, t)$. The strain at the front of the wave is constant and equal to $\epsilon = \epsilon^* = \text{const}$. In regions III, IV, and V we have $\epsilon = \epsilon^*$. The shape of the shock wave is determined in the same way as it was done in the case considered in equations (14.71), only now, in the conditions of the dynamic and kinematic continuity $(14.71)_{2,3}$ we have to take into account the fact that the unloading wave propagates into a disturbed region. Thus the equations governing the problem in region III take the form

(14.78)
$$\sigma_3(x,t) = \varrho v_3'(t)\,x - p(t), \quad \sigma_3-\sigma_2 = -\varrho\varphi'(t)(v_3-v_2), \quad v_3-v_2 = -\varphi'(t)(\varepsilon^*-\varepsilon_2)$$
$$\text{for} \quad t^* \leqslant t \leqslant t_1.$$

By means of (14.77) we obtain from (14.78) an ordinary differential equation for determining the first segment of the shock wave front, which is

(14.79)
$$-p\left(t-\frac{\varphi(t)}{a}\right)+p(t)+\frac{\varphi(t)}{a}\left(1-\frac{\varphi'(t)}{a}\right)^2 p'\left(t-\frac{\varphi(t)}{a}\right)$$
$$+\frac{1}{2}\varrho\left\{\varepsilon^*+\frac{\left[\sigma_y+p\left(t-\dfrac{\varphi(t)}{a}\right)\right]}{\varrho a^2}\right\}[\varphi^2(t)]'' = 0,$$

with the initial conditions

(14.80) $$\varphi(t^*) = 0, \qquad \varphi'(t^*) = a.$$

Since, in general, (14.79) is a non-linear equation, it has to be solved numerically. Knowing the shape of the shock wave $x = \varphi(t)$ for $t \leqslant t \leqslant t_1$ we can determine $\sigma(x, t)$ and $v(t)$ in region III. In region IV we have $\epsilon = \epsilon^*$. The shock wave $x = \varphi(t)$ for $t_1 \leqslant t \leqslant t_0$ is determined from the equations

(14.81) $$\sigma_4(x,t) = \varrho v'_4(t)\,x - p(t), \qquad \sigma_4 - \sigma_y = -\varrho\varphi'(t)\,v_4, \qquad v_4 = -\varphi'(t)\,\varepsilon^* \qquad \text{for} \quad t_1 \leqslant t \leqslant t_0$$

by taking into account the initial conditions, the continuity of function $\varphi(t)$, and its first derivative for $t = t_1$.

Introducing the functions

(14.82)
$$F(t) = -\int_{t_1}^{t}\int_{t_1}^{\xi}[p(\eta)+\sigma_y]\,d\eta\,d\xi,$$
$$G(t) = F'(t) = -\int_{t_1}^{t}[p(\eta)+\sigma_y]\,d\eta$$

we obtain

(14.83)
$$\varphi(t) = \left[\frac{2F(t)}{\varrho\varepsilon^*} + 2c(t-t_1) + \varphi^2(t_1)\right]^{1/2},$$
$$v_4(t) = \frac{G(t)+\varrho c\varepsilon^*}{\varrho\varphi(t)},$$
$$\sigma_4(x,t) = -p(t) - \frac{x}{\varphi(t)}\left[\frac{\varrho v_4^2(t)}{\varepsilon^*} + p(t) + \sigma_y\right],$$

where

$$c = \varphi'(t_1)\,\varphi'(t_1) = \text{const.}$$

The solution in region V is identical to that in region IV. In this region, because of the boundary condition, the stress is a decreasing function of time on the shock wave front. Time t_2 is determined from the condition $\sigma_5(x, t_2)|_{x=\varphi(t)} = \sigma_y$. To simplify the problem we assume that $t_2 < t_4$.

In region VII, making use of the continuity conditions across the front of the shock wave, we obtain

(14.84) $$\sigma_7 - \sigma_y = \varrho(\varphi'(t))^2\varepsilon_7,$$

and since $\sigma_7 - \sigma_y = E\epsilon_7$, then

$$\varphi'(t) = a \quad \text{for} \quad t_2 \leqslant t \leqslant t_3. \tag{14.85}$$

Hence

$$\varphi(t) = a(t-t_2) + \varphi(t_2) \quad \text{for} \quad t_2 \leqslant t \leqslant t_3. \tag{14.86}$$

Since $\epsilon = \epsilon(x)$, then

$$v_6(t) = v_7(t). \tag{14.87}$$

From the equation of dynamic equilibrium

$$\sigma_c(x,t) = \sigma_7(x,t) = \varrho x v_7'(t) - p(t). \tag{14.88}$$

Making use of (14.84) and (14.87) we obtain the differential equation

$$\varrho[a(t-t_2) + \varphi(t_2)]v_7'(t) - \varrho a v_7(t) = -p(t) - \sigma_y, \tag{14.89}$$

whence, for the initial condition $v_7(t_2) = v_5(t_2)$, we have

$$v_7(t) = \frac{[G_1(t) + G(t_0) + \varrho\varepsilon^* c]}{\varrho[a(t-t_2) + \varphi(t_2)]}, \tag{14.90}$$

where

$$G_1(t) = -\int_t^{t_0} [p(\eta) + \sigma_y]\, d\eta.$$

The stress in regions VI and VII takes the form

$$\sigma_6(x,t) = \sigma_7(x,t) = -p(t) + \frac{x[-p(t) - \sigma_y - a\varrho v_7(t)]}{\varphi(t)}, \tag{14.91}$$

while the strain is given by

$$\varepsilon_6 = \varepsilon^*, \quad \varepsilon_7(x) = \left.\frac{G_1(t) + G(t) + \varrho c\varepsilon^*}{\varrho a x}\right|_{t = t_2 + \frac{x - \varphi(t_2)}{a}}. \tag{14.92}$$

Point $(x = \varphi(t_3),\ t = t_3)$ on the shock wave corresponds to point A in Fig. 51a. For $t > t_3$ in regions VIII, IX, and X we have

$$\begin{gathered} v_8 = v_9 = v_{10} = 0, \\ \sigma_8(t) = \sigma_9(t) = \sigma_{10}(t) = -p(t), \\ \varepsilon_8 = \varepsilon^*, \quad \varepsilon_9(x) = \varepsilon_7(x), \quad \varepsilon_{10} \equiv 0. \end{gathered} \tag{14.93}$$

On the straight line $t = t_3$ we have step-wise smoothing of the stress to the value prescribed at the boundary $x = 0$.

The problem of plane shock wave propagation was solved in [143] for the case of a curvilinear material characteristic during the process of plastic loading. The following relationship was assumed for the loading process:

$$\sigma = \bar{f}(\varepsilon) \quad \text{for} \quad \sigma \geqslant \sigma_y, \tag{14.94}$$

where $\bar{f}(\epsilon)$ belongs to C^2 with respect to σ and satisfies the following conditions: $\bar{f}''(\epsilon) \geqslant 0$, $\bar{f}'(\epsilon) \equiv K(\epsilon) > 0$. The function $f(\varepsilon) = \sigma + (\sigma^* - \sigma)\left(\dfrac{\varepsilon - \varepsilon_y}{\varepsilon^* - \varepsilon_y}\right)^n$, $(n \geqslant 1)$ is shown in Fig. 51b.

For an unloading process, as before, it is assumed that $\epsilon(x, t) = \epsilon_m(x)$ for $\sigma(x, t) \leqslant \sigma_m(x)$ for all $\epsilon > 0$, where $\epsilon_m(x)$ and $\sigma_m(x)$ denote the maximum strain and stress, respectively, at cross-section x during the loading process.

With these assumptions the problem was considered of the case of a suddenly applied pressure at the end of a semi-infinite bar, which then monotonically decreases in time. This solution was also obtained in closed form.

15. Propagation of plane, longitudinal waves in an elastic/viscoplastic medium

A number of papers have been devoted to the problem of the propagation of plane, longitudinal stress waves in an elastic/viscoplastic medium; see, among others, [65], [94], [147], [166]. The investigations in this field were commenced in 1948, and a full development took place in the sixties. Many problems were considered, namely the propagation of waves in homogeneous and non-homogeneous media, problems of reflection from undeformable and deformable obstacles, the propagation of waves in bars of varying cross-section, etc.

The constitutive equations for elastic/viscoplastic media (3.5) in a uniaxial stress state (Malvern's [83] equations) take the form (3.12):

$$\dot{\varepsilon} = \frac{\dot{\sigma}}{E} + \gamma^* \langle \Phi(F) \rangle, \tag{15.1}$$

where $F = \dfrac{\sigma}{f(\epsilon^p)} - 1$, $f(\epsilon p)$ denoting the static characteristic for simple tension and $\gamma^* = 2\gamma/\sqrt{3}$.

For elastic unloading the following equation has to be satisfied:

$$\sigma = \sigma_0(x) + E_0[\varepsilon - \varepsilon_0(x)] \quad \text{for} \quad \sigma \leqslant f(\varepsilon^p), \tag{15.2}$$

where $\sigma_0(x)$ and $\epsilon_0(x)$ denote the values of stress and strain respectively on the unloading wave front and E_0 is Young's modulus.

The equation of motion (10.1) and equation (10.4),

$$\varrho \frac{\partial v}{\partial t} = \frac{\partial \sigma}{\partial x}, \qquad \frac{\partial \varepsilon}{\partial t} = \frac{\partial v}{\partial x}, \tag{15.3}$$

together with the constitutive equation (15.1) or (15.2), constitute a closed system of equations for the problem. This system of equations has the following characteristics:

$$x = \pm at + \text{const}, \qquad x = \text{const}, \tag{15.4}$$

where $a = \sqrt{E_0/\varrho}$.

The following relationships are valid along the characteristics:

$$\begin{aligned} &d\sigma \mp \varrho a dv \pm \varrho a \gamma^* \langle \Phi(F) \rangle dx = 0, && \text{for} \quad x = \pm at + \text{const}, \\ &d\sigma - \varrho a^2 d\varepsilon + \varrho a^2 \gamma^* \langle \Phi(F) \rangle dt = 0 && \text{for} \quad x = \text{const}. \end{aligned} \tag{15.5}$$

In the regions of elastic straining and of unloading ($\sigma \leqslant f(\epsilon^p)$) we have to assume in (15.5) that $\langle \Phi(F) \rangle \equiv 0$. One should notice that in both – in the range of elastic strain as well as in the range of viscoplastic strain – the disturbances propagate with the same velocity, this being constant in the case of a homogeneous medium.

In the case of the propagation of plane stress waves in a semi-infinite bar, for the boundary conditions

$$\sigma(0, t) = -p(t), \qquad p(t) \geqslant 0, \qquad p(0) = p_0 > \sigma_s, \tag{15.6}$$

the solution in the phase plane (Fig. 53) is as follows. For the boundary condition (15.6) a wave of strong discontinuity will propagate in the bar, coinciding with the characteristic of the equation $x = at$. The conditions of dynamic continuity (7.18) and kinematic continuity (8.9), which for the plane waves assume the form

$$[\sigma] = -\varrho a [v], \qquad [v] = -a[\varepsilon], \tag{15.7}$$

have to be satisfied across the wave front.

If we have zero initial conditions then wave $x = at$ propagates into an undisturbed medium, therefore in front of the wave $\sigma = v = \epsilon = 0$, and conditions (15.7) are reduced to the form

$$\sigma = -\varrho a v, \qquad v = -a\varepsilon. \tag{15.8}$$

The solution on the wave front $x = at$ has been constructed based on $(15.5)_2$, along the positive characteristic, and on condition (15.8), namely

$$\frac{d\sigma}{dx} = -\frac{1}{2} \varrho a \gamma^* \langle \Phi(F) \rangle. \tag{15.9}$$

This equation can be reduced to the integral equation

$$\sigma(x) = -p_0 - \int_0^x \psi[\xi, \sigma(\xi)] \, d\xi, \tag{15.10}$$

where boundary condition (15.6) has been used,

$$\psi[x, \sigma(x)] = \frac{1}{2} \varrho a \gamma^* \langle \Phi(F) \rangle. \tag{15.11}$$

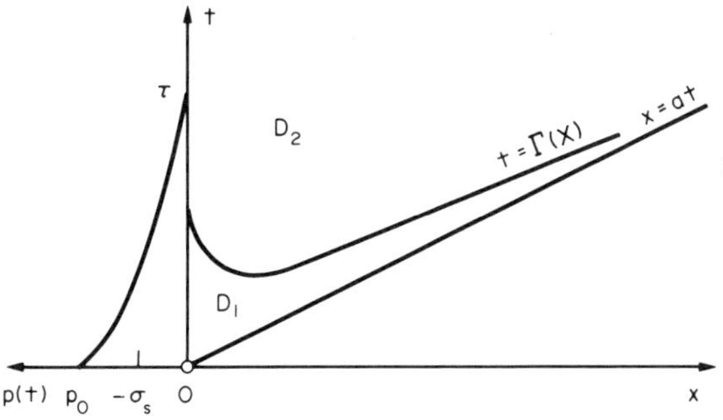

Fig. 53

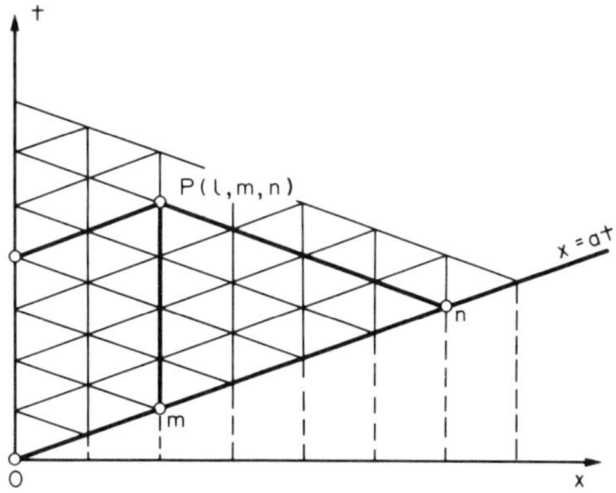

Fig. 54

If we take function $\gamma^*\langle\Phi(F)\rangle$ in the form of (3.30), then the integrand $\psi[x, \sigma(x)]$ becomes

(15.12)

$$\psi[x,\sigma(x)] = \left[\left(1-\frac{\varrho a^2(\sigma(x)-\sigma_s)}{\varepsilon_0-\varepsilon_s}\right)\frac{\sigma(x)-\sigma_{sn1}}{\sigma_{sn2}-\sigma_{sn2}}+\varrho a^2\frac{\sigma(x)-\sigma_s}{\varepsilon_w-\varepsilon_s}\,\frac{\sigma(x)-\sigma_{wn1}}{\sigma_{wn2}-\sigma_{wn1}}\right](\dot{\varepsilon}_{n2}-\dot{\varepsilon}_{n1}).$$

It is easy to show that the integrand in (15.10) represented in the form (15.12) is bounded and satisfies the Lipschitz condition; this implies the convergence of the method of successive approximations. Calculating a bound we obtain

$$|\sigma| \leqslant p_0 + \frac{M}{R}(e^{Rh}-1), \tag{15.13}$$

where M denotes the upper limit of the integrand (15.12), R is the value of derivative $\left|\frac{\partial\psi}{\partial\sigma}\right|$ for the parameters determining the upper limit of the integrand and h is the smaller of the two numbers

$$\frac{x^*}{2}, \quad \left|\frac{p_0-\sigma_s}{2M}\right|,$$

where x^* denotes the coordinate at which the stress $\sigma(x^*)$ at the front of the strong discontinuity $x = at$ reaches the static elastic limit.

Region D_1 (Fig. 53) is the region of viscoplastic deformations. In the general case, on account of the fact that the equations of the problem are semi-linear, we are not able to solve them in an exact form. The solution in region D_1 is constructed numerically by means of the method of characteristic nets (Fig. 54).

The recurrence formulae for discrete values of the parameters of the solution for the net of characteristics, based on formulae (15.5), take the following forms:

(15.14)

$$\varrho a\Phi(F)_{(l,m-1,n-1)}\Delta x-\varrho a[v(l,m,n)-v(l,m-1,n-1)] +[\sigma(l,m,n)-\sigma(l,m-1,n-1)]=0,$$

$$\varrho a\Phi(F)_{(l-1,m+1,n)}\Delta x-\varrho a[v(l,m,n)-v(l-1,m+1,n)] +[\sigma(l,m,n)-\sigma(l-1,m+1,n)]=0,$$

$$\varrho a^2 \Phi(F)_{(l-1,m,n-1)} \Delta t + [\sigma(l,m,n) - \sigma(l-1,m,n-1)] - \varrho a^2 [\varepsilon(l,m,n) - \varepsilon(l-1,m,n-1)] = 0.$$

The above equations constitute a system of equations in the three required functions σ, v, and ϵ. This system uniquely determines the solution in region D_1 for the prescribed boundary condition (15.6), and the conditions on the characteristic $x = at$; namely, stress $\sigma(x)$ is determined from (15.10), velocity $v(x) = -\frac{1}{(\varrho a)}\sigma(x)$, and the strain $\varepsilon(x) = \frac{1}{(\varrho a^2)}\sigma(x)$.

The curve $t = \Gamma(x)$ (Fig. 53), delineating the validity of the above equations, constitutes the "unloading wave" it is determined from the condition

$$\sigma[x, \Gamma(x)] = f[\varepsilon^p(x, \Gamma(x))]. \tag{15.15}$$

Curve $t = \Gamma(x)$ can also be determined numerically, making use simultaneously of the relations on the characteristics in the unloading region D_2 which are relations (15.5) with $\langle \Phi(F) \rangle \equiv 0$.

The problem of plane wave propagation in a bar of finite length has been solved in [94]; the problem of the reflection of waves from a rigidly fixed bar end has been investigated as well as the reflection from an elastically fixed end. Also wave propagation in a layered medium with an undeformable mass at the boundary of the medium has been examined. The function F in the constitutive equation has been taken in the forms (compare (3.13))

$$F = \frac{\sigma}{f(\varepsilon^p)} - 1 \quad \text{and} \quad F = \frac{|\sigma|}{\sigma_s} - 1. \tag{15.16}$$

A number of numerical examples have been solved for the special data applicable to a bar of mild steel. It is assumed that Young's modulus is $E_0 = 2.06 \times 10^{11}$ Pa, the hardening modulus $E_1 = 1.78 \times 10^9$ Pa, $\sigma_s = 2.75 \times 10^8$ Pa, $\sigma_w = 4.51 \times 10^8$ Pa, $\epsilon_s = 1.33 \times 10^{-3}$, $\epsilon_w = 0.1$, $\gamma = 7.8 \times 10^3$ kg m^{-3}. The pressure at the bar end is given in the form

$$p(t) = p_0\left(1 - \frac{t}{\tau}\right), \quad p_0 = 5.49 \times 10^8 \text{ Pa}, \ \tau_0 = 2.335 \times 10^{-4} \text{ s}.$$

In Fig. 55 we see the stress variation on the wave front $x = at$ for the above data. The strain variation at the bar end $x = 0$ for the case with hardening of the material $E_1/E_0 = 8.88 \times 10^{-3}$ is shown in Fig. 56a, and for the case without hardening in Fig. 56b. The differences in the permanent strains for both cases are of the order 9%. The difference evidently increases with the decrease of the pressure gradient $p(t)$ – curves (c) and (d) show the solution for a pressure gradient of half the original value.

The stress distribution is presented in Fig. 57 for the case of the reflection of the stress wave at the cross-section $x = l = 60$ cm for different kinds of supports:

(a) the reflection of the strong discontinuity from a rigid support;

(b) for the case of a layered medium with $\frac{\sigma_{sI}}{\sigma_{sII}} = 0.825$, $\frac{\varrho_{II} a_{II}}{\varrho_I a_I} = 2.13$;

(c) for the case of a layered medium with $\dfrac{\sigma_{sI}}{\sigma_{s2}} = 1.4, \quad \dfrac{\varrho_I a_I}{\varrho_{II} a_{II}} = 1.35$;

(d) for the case of a bar, for which at distance $x = 60$ cm there is an undeformable mass $M = 0.1\ \varrho l$ – the stress diagram referring to that in front of the mass;

(e) deformable support: $M = 0.1\ \varrho l$, spring constant $k = 0.05\ E_0 A l$, $c = 0$ (Fig. 35), where A denotes the cross-section of the bar, and l its length.

It is evident from the examples cited that in the case of dynamic problems at small strain rates ($\dot{\epsilon} \leqslant 300\ s^{-1}$) we cannot disregard the effect of material hardening. For higher values of the rate this effect decreases.

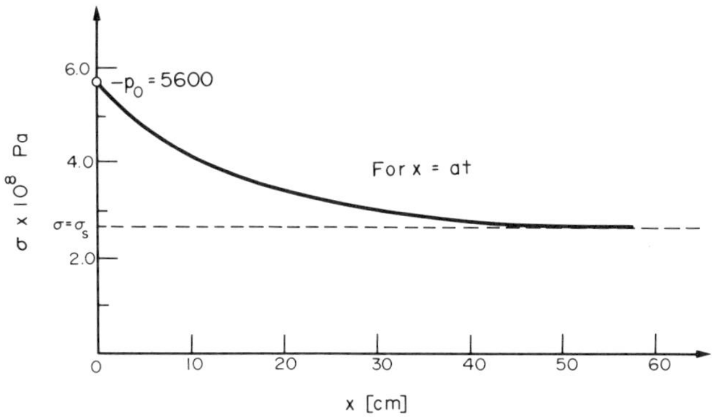

Fig. 55

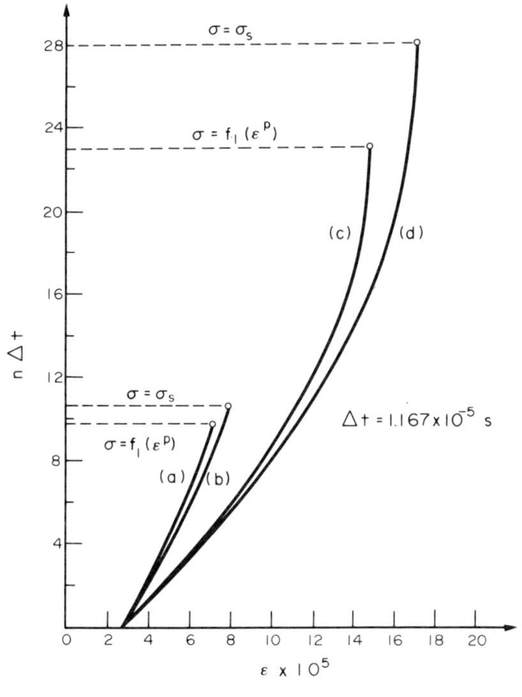

Fig. 56

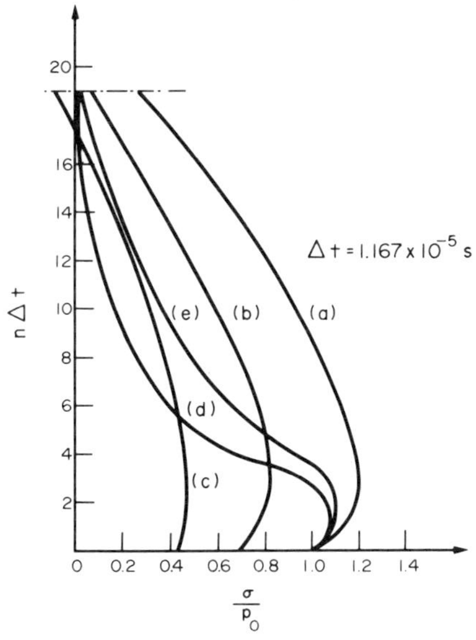

Fig. 57

The constitutive equations for elastic/viscoplastic media (3.5) in a uniaxial state of strain have the following forms:

$$\dot{\varepsilon} = \frac{1}{2\mu}(\dot{\sigma}_{11}-\dot{\sigma}_{22})+\sqrt{3}\gamma\langle\Phi(F)\rangle, \qquad \dot{\sigma}_{11}+2\dot{\sigma}_{22} = 3K\dot{\varepsilon}, \tag{15.17}$$

where

$$\varepsilon = \varepsilon_{11}, \qquad \sigma_{33} = \sigma_{22}, \qquad \varepsilon_{12} = \varepsilon_{13} = \varepsilon_{22} = \varepsilon_{33} = 0.$$

The system of (15.17) together with (15.3) constitutes a closed system of equations, the characteristics of which are the straight lines

$$x = \pm at+\text{const}, \qquad x = \text{const}, \tag{15.18}$$

where $a = \sqrt{(3K+4\mu)/3\varrho}$.

The relations along the characteristics (by means of (9.17)) take the forms

$$\begin{aligned} &d\sigma_{11}\mp\varrho a dv\pm\frac{4\sqrt{3}}{3a}\mu\gamma\langle\Phi(F)\rangle\, dx = 0 \quad \text{for} \quad x = \pm at+\text{const},\\ &d\sigma_{11}-\varrho a^2 d\varepsilon+\frac{4\sqrt{3}}{3}\mu\gamma\langle\Phi(F)\rangle\, dt = 0 \quad \text{for} \quad x = \text{const}. \end{aligned} \tag{15.19}$$

In the case of a uniaxial strain state the problems of the wave propagation can be solved in the same way as the corresponding problems for a uniaxial stress state.

Now we present a certain analytic solution of a problem concerned with propagation and reflection of stress waves in elastic/viscoplastic medium [65] using Sokolovskii's model (3.10). The constitutive equations can be rewritten in a different form, namely

$$E\dot{\varepsilon} = \dot{\sigma}+\varkappa k\langle\Phi(|\sigma|-\sigma_s)\rangle \tag{15.20}$$

where k is the viscosity coefficient $\langle\Phi(F)\rangle \equiv \begin{cases}\Phi(F) & \text{for } F>0\\ 0 & \text{for } F\leqslant 0\end{cases}$, $\varkappa = \operatorname{sgn}\sigma$. Taking into account $(15.3)_2$ and assuming that Φ is a linear function of its argument $\Phi(|\sigma|-\sigma_s) = |\sigma|-\sigma_s$ we obtain

$$Ev_{,x} = \sigma_{,t}+\varkappa k(|\sigma|-\sigma_s \tag{15.21}$$

The total values of the parameters of the problem can be represented in the form of the following sums:

$$\begin{aligned} &\sigma_c(x,t) = \varkappa\sigma_s+\sigma_p(x,t), \qquad \varepsilon_c(x,t) = \varkappa\varepsilon_s+\varepsilon_p(x,t),\\ &v_c(x,t) = v_s+v_p(x,t), \end{aligned} \tag{15.22}$$

where suffix c denotes the total values of the parameters of the problem, suffix s corresponds to the values at the elastic limit, and suffix p to the values exceeding that state.

Making use of (15.22) we obtain the equations of the problem for viscoplastic strains, with a linear function Φ :

$$Ev_{p,x} = \sigma_{p,t} + k\sigma_p, \quad \varrho v_{p,t} = \sigma_{p,x}; \tag{15.23}$$

hence

$$\sigma_{,tt} - a^2\sigma_{,xx} + k\sigma_{,t} = 0, \quad a = \sqrt{\frac{E}{\varrho}}. \tag{15.24}$$

For the velocity we obtain an equation of identical form,

$$v_{,tt} - a^2 v_{,xx} + kv_{,t} = 0, \tag{15.24'}$$

where the suffixes p have been omitted.

Performing (unilateral) Laplace transform on (15.24) we obtain for zero initial conditions

$$\bar{\sigma}\,\frac{s(s+k)}{a^2} - \bar{\sigma}_{,xx} = 0, \tag{15.25}$$

where s denotes the transform parameter.

The general solution of (15.25) takes the form

$$\bar{\sigma}(x, s) = A\exp\left[-\frac{x}{a}\sqrt{s(s+k)}\right] + B\exp\left[\frac{x}{a}\sqrt{s(s+k)}\right]. \tag{15.26}$$

In the case of wave propagation in a semi-infinite bar, on account of the condition that the stresses vanish as $x \to \infty$, we obtain $B = 0$. Constant A is determined from the boundary condition for $x = 0$.

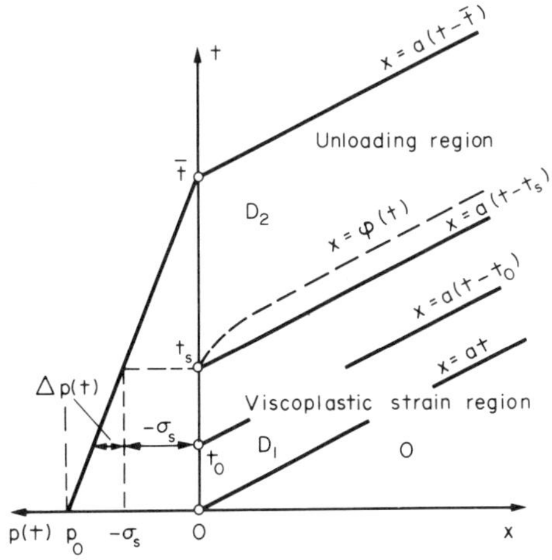

Fig. 58

Boundary condition (15.6) (Fig. 58), modified on the basis of $(15.22)_1$, assumes the form

$$\sigma(0, t) = -[p(t) - \sigma_s] = -\Delta p(t). \tag{15.27}$$

Transforming condition (15.27) we obtain from (15.26) for $B = 0$ the following equation:

$$\bar{\sigma}(x, s) = -\Delta\bar{p}(s) \exp\left[-\frac{x}{a}\sqrt{s(s+k)}\right]. \tag{15.28}$$

On performing the inverse transform and making use of the convolution theorem, (15.24) for the above boundary conditions takes the following form:

$$\sigma(x, t) = -\Delta p\left(t - \frac{x}{a}\right) e^{-\frac{kx}{2a}} - \frac{kx}{2a} \int_{\frac{x}{a}}^{t} \Delta p(t-\tau) e^{-\frac{k}{2}\tau} \frac{I_1\left[\frac{k}{2}\sqrt{\tau^2 - \left(\frac{x}{a}\right)^2}\right]}{\sqrt{\tau^2 - \left(\frac{x}{a}\right)^2}} d\tau, \tag{15.29}$$

where I_1 denotes the Bessel function of the first kind and the first order.

The strain in the viscoplastic range of strains can be obtained from the integrated constitutive equation (15.20), taking account of $(15.22)_1$ and $\Phi(F) = F$:

$$\varepsilon(x, t) = \frac{1}{E}\left[\sigma(x, t) + k \int_{\frac{x}{a}}^{t} \sigma(x, t)\, d\tau\right]. \tag{15.30}$$

The velocity, $v(x, t)$, is determined from the solution of (15.24′) in the same way as the stress. For the identical boundary conditions we have

$$\begin{aligned} v(x, t) = \frac{1}{\varrho a} \Delta p\left(t - \frac{x}{a}\right) e^{-\frac{kx}{2a}} + \frac{1}{\varrho a} \int_{\frac{x}{a}}^{t} \Delta p(t-\tau) \Bigg\{ e^{-\frac{k\tau}{0}} I_0\left[-\frac{k}{2}\sqrt{\tau^2 - \left(\frac{x}{a}\right)^2}\right] \\ + k \int_{\frac{x}{a}}^{\tau} e^{-\frac{k\xi}{2}} I_0\left[-\frac{k}{2}\sqrt{\xi^2 - \left(\frac{x}{a}\right)^2}\right] d\xi \Bigg\}_{,\tau} d\tau, \end{aligned} \tag{15.31}$$

where I_0 is the Bessel function of the first kind and zero order.

On the strong discontinuity wave $x = at$ (Fig. 58), from (15.22) and (15.29)–(15.31), the parameters of the solution are as follows:

$$\begin{aligned} \sigma_c(x) &= \varkappa\sigma_s - (p_0 - \sigma_s) \exp\left(-\frac{k}{2a}x\right), \\ \varepsilon_c(x) &= \varkappa\varepsilon_s - \frac{p_0 - \sigma_s}{E} \exp\left(-\frac{k}{2a}x\right), \\ v_c(x) &= v_c + \frac{p_0 - \sigma_s}{\varrho a} \exp\left(-\frac{k}{2a}x\right), \end{aligned} \tag{15.32}$$

whereas the stress on an arbitrary, positive characteristic $x = a(t-t_0)$ $(0 < t \leqslant t_s)$ in the viscoplastic region is given by the formula

$$\sigma_c(x) = -[p(t_0)-\sigma_s]\exp\left(-\frac{k}{2a}x\right) - \frac{k}{2a}x \int_{\frac{x}{a}}^{t_0+\frac{x}{a}} \left[p\left(t_0+\frac{x}{a}-\tau\right)-\sigma_s\right] \times e^{-\frac{k}{2}\tau} \frac{I_1\left[\frac{k}{2}\sqrt{\tau^2-\left(\frac{x}{a}\right)^2}\right]}{\sqrt{\tau^2-\left(\frac{x}{a}\right)^2}}\,d\tau. \tag{15.33}$$

For displacement boundary conditions on $x = 0$ of the forms

$$v(0,t) = \Delta w(t), \qquad w(0,0) = w_0, \tag{15.34}$$

where $v(t)$ is the given function, the solution in the viscoplastic region takes the forms [65]

$$v(x,t) = \Delta w\left(t-\frac{x}{a}\right)e^{-\frac{k}{2a}x} + \frac{k}{2a}x\int_{\frac{x}{a}}^{t} \Delta w(t-\tau)\,e^{-\frac{k}{2}\tau}\frac{I_1\left[\frac{k}{2}\sqrt{\tau^2-\left(\frac{x}{a}\right)^2}\right]}{\sqrt{\tau^2-\left(\frac{x}{a}\right)^2}}\,d\tau,$$

$$\sigma(x,t) = -\varrho a\left\{\Delta w\left(t-\frac{x}{a}\right)e^{-\frac{kx}{2a}} + \int_{\frac{x}{a}}^{t}\Delta w(t-\tau)\left[e^{-\frac{k}{2}\tau}I_0\left(\frac{k}{2}\sqrt{\tau^2-\left(\frac{x}{a}\right)^2}\right)\right]_{,\tau}d\tau\right\}, \tag{15.35}$$

$$\varepsilon(x,t) = \frac{1}{E}\left[\sigma(x,t)+k\int_{\frac{x}{a}}^{t}\sigma(x,\tau)\,d\tau\right].$$

On the wave of strong discontinuity $x = at$, we obtain for the above case

$$\sigma_c(x) = \varkappa\sigma_s - \varrho a(w_0-v_s)\,e^{-\frac{k}{2a}x},$$

$$\varepsilon_c(x) = \varkappa\varepsilon_s - \frac{w_0-v_s}{a}\,e^{-\frac{k}{2a}x}, \tag{15.36}$$

$$v_c(x) = v_s + (w_0-v_s)\,e^{-\frac{k}{2a}x}.$$

The stress on the positive characteristic $x = a(t-t_0)$ is determined by the formula

$$\sigma_c(x) = -\varrho a\left\{[w(t_0)-v_s]\exp\left(-\frac{k}{2a}x\right) + \int_{\frac{x}{a}}^{t_0+\frac{x}{a}}\left[v\left(t_0+\frac{x}{a}-\tau\right)-v_s\right] \times\left[e^{-\frac{k}{2}\tau} I_0\left(\frac{k}{2}\sqrt{\tau^2-\left(\frac{x}{a}\right)^2}\right)\right]_{,\tau} d\tau\right\}. \tag{15.37}$$

Discussing formulae (15.33) and (15.37) we can observe that the stress variation along the characteristics mainly depends on the viscosity coefficient k.

The stress, along the positive characteristics, in the case of boundary condition (15.6) is a decreasing function of x. Hence we can draw the conclusion that the velocity of the elastic unloading wave, for monotonically decreasing pressure at the end $x = 0$, is less than a.

The initial velocity of the unloading wave can be determined in the following way. We differentiate the stress along the characteristic $x = a(t-t_s)$, and along the unloading wave $x = \varphi(t)$:

$$\sigma_c'(t) = a\sigma_{,x} + \sigma_{,t}, \qquad \varphi'(t)\,\sigma_{,x} + \sigma_{,t} = 0, \tag{15.38}$$

where $\sigma_c(t)$ denotes the initial stress along the characteristic $x = a(t-t_s)$. The stress has been determined in (15.33),

$$\sigma_c(t) \approx -\frac{k}{2}(t-t_s)\int_0^{t_s}[p(t-\tau)-\sigma_s]\,e^{-\frac{k}{2}\tau}\,\frac{I_1\left[\frac{k}{2}\sqrt{\tau^2-(t-t_s)^2}\right]}{\sqrt{\tau^2-(t-t_s)^2}}\,d\tau, \tag{15.39}$$

where we have omitted the term containing factor $\exp(-k\tau/2)$ as a small quantity of low order (for metals the coefficient of viscosity is of the order 10^6 s^{-1}). From (15.38) and (15.39) we obtain for $x = 0$ and $t = t_s$

$$\varphi'(t) = a\,\frac{|p'(t)|_{t=t_s}}{|p'(t)|_{t=t_s} + \Delta}, \tag{15.40}$$

where

$$\Delta = \frac{k}{2}\int_0^{t_s}[p(t_s-\tau)-\sigma_s]\,e^{-\frac{k}{2}\tau}\,\frac{I_1\left(\frac{k}{2}\tau\right)}{\tau}\,d\tau.$$

Since k possesses a considerable value while the time in the considered dynamic cases is small, consequently we obtain $\Delta \approx 0$, i.e. $\varphi'(t) \approx a$. This, for practical purposes, the characteristic with equation $x = a(t-t_s)$ can be treated as the unloading wave.

The solution in the region of elastic unloading D_2 does not present many difficulties. In the case of impulse loading (boundary condition (15.6)) we have, in region D_2, the solution in the forms (Reimann waves):

$$\sigma(x,t) = -p\left(t-\frac{x}{a}\right), \qquad v_c(x,t) = v_c(t_s) + \frac{1}{\varrho a}\left[p\left(t-\frac{x}{a}\right)-\sigma_s\right], \tag{15.41}$$

where $v_c(t_s)$ denotes the velocity at the end $x = 0$ for time $t = t_s$. For kinematic loading (boundary condition (15.34)) we obtain

$$\sigma(x,t) = -\varrho a\left[w\left(t-\frac{x}{a}\right)-w(t_s)\right], \quad v_c(x,t) = w\left(t-\frac{x}{a}\right). \tag{15.42}$$

Now we pass to the problem shown in Fig. 59 of the reflection of plane waves from a fixed-end bar (end $x = l$) when the pressure at the bar end $x = 0$ is given in accordance with (15.6). The solution of this problem should be decomposed into two stages: the reflection of the stress waves produced by the constant pressure equal to $p_1(t) = -\sigma_s$, followed by the reflection of waves produced by the excess stress $\Delta p(t) = p(t) - \sigma_s$.

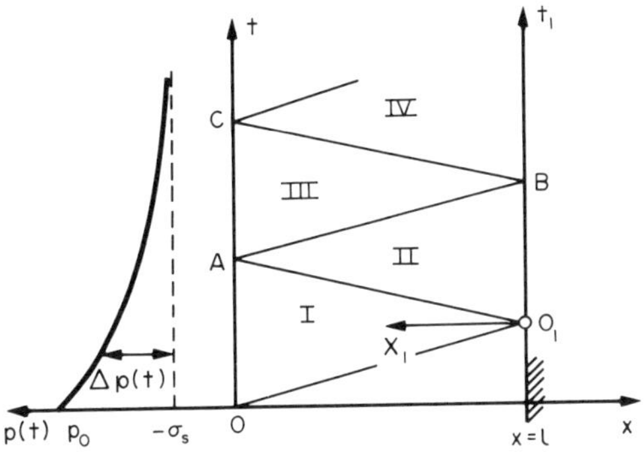

Fig. 59

The solution in region I of the phase plane (x, t) is the same as in the above case of wave propagation in a semi-infinite bar.

The wave with equation $x = at$ falling on the fixed end $x = l$ is reflected. Across the front of the reflected wave the following conditions hold: the condition on the negative characteristic $(15.5)_1$ and the condition of dynamic continuity,

$$\begin{gathered} d\sigma_2+\varrho a dv_2-\frac{k}{a}(\sigma_2-\sigma_s)\,dx = 0, \\ \sigma_2-\sigma_s = \varrho a(v_2-v_s). \end{gathered} \tag{15.43}$$

Solving the above system of equations we obtain

$$\begin{gathered} \sigma_2 = -\sigma_s\left[1+\exp\left(-\frac{k}{2a}\left(t-\frac{l}{a}\right)\right)\right], \\ v_2 = \frac{\sigma_s}{\varrho a}\left[1-\exp\left(-\frac{k}{2a}\left(t-\frac{l}{a}\right)\right)\right]. \end{gathered} \tag{15.44}$$

The reaction at the bar end $x = l$, denoted by

$$R(t_1) = \sigma_s+\Delta R(t_1), \quad t_1 = t-\frac{l}{a}, \tag{15.45}$$

is assumed, for the time being, to be known. Next making use of the solution of the problem of wave propagation in a semi-infinite bar (15.29), we find that

$$\sigma_{2s}(x,t) = -\sigma_s - \Delta R\left(t_1 - \frac{x_1}{a}\right) e^{-\frac{k}{2a}x_1} - \frac{k}{2a} x_1 \int_{\frac{x_1}{a}}^{t_1} \Delta R\,(t_1-\tau)\, e^{-\frac{k}{2}\tau} \times \frac{I_1\left[\frac{k}{2}\sqrt{\tau^2-\left(\frac{x_1}{a}\right)^2}\right]}{\sqrt{\tau^2-\left(\frac{x_1}{a}\right)^2}}\, d\tau. \tag{15.46}$$

Velocity $v_2(x, t)$ is determined from (15.21),

$$v_{2s,x} = \frac{1}{E}\left[\sigma_{2s,t} + k(\sigma_{2s}+\sigma_s)\right]. \tag{15.47}$$

Integration leads to

$$v_{2s}(x,t) = \frac{1}{E}\int_{2l-at}^{x} \left[\sigma_{2s,t} + k(\sigma_{2s}+\sigma_s)\right] d\xi + \frac{\sigma_s}{\varrho a}\left[1 - e^{-\frac{k}{2}\left(t-\frac{l}{a}\right)}\right], \tag{15.48}$$

where the constant of integration has been determined from the characteristic $x = 2l-at$ by making use of $(15.32)_3$.

From (15.46) and (15.48), making use of the condition at the fixed end of the bar,

$$v_{2s}(l,t) \equiv 0, \tag{15.49}$$

we deduce the Volterra type integral equation of the second kind for reaction $\Delta R\,(t)$:

$$\Delta R\left(t-\frac{l}{a}\right) - \frac{k}{2a}\int_{l}^{2l-at}\left(1+\frac{k}{4}\,\frac{l-\xi}{a}\right) e^{-\frac{k}{2}\left(\frac{l-\xi}{a}\right)} \Delta R\left(t-\frac{l}{a}-\frac{l-\xi}{a}\right) d\xi$$
$$-\frac{k}{2a}\int_{l}^{2l-at}\frac{l-\xi}{a}\int_{\frac{l-\xi}{a}}^{t-\frac{l}{a}}\left\{\frac{\partial}{\partial\tau}\left\{e^{-\frac{k\tau}{2}}\left[\tau^2-\left(\frac{l-\xi}{a}\right)^2\right]^{-1/2} I_1\left[\frac{k}{2}\sqrt{\tau^2-\left(\frac{l-\xi}{a}\right)^2}\right]\right.\right.$$
$$\left.+ k e^{-\frac{k}{2}\tau}\frac{I_1\left[\frac{k}{2}\sqrt{\tau^2-\left(\frac{l-\xi}{a}\right)^2}\right]}{\sqrt{\tau^2-\left(\frac{l-\xi}{a}\right)^2}}\right\} \Delta R\left(t-\frac{l}{a}-\tau\right) d\tau\, d\xi = \sigma_s. \tag{15.50}$$

In general (15.50) has to be solved by the successive approximation method. For large values of k we can neglect the term containing the factor $\exp\left[-\frac{\tau}{2}k\right]$, then (15.50) is reduced to the form

$$\Delta R(t_1)+\frac{k}{2a}\int\limits_0^{t_1}\left(1+\frac{k}{4}\tau\right)e^{-\frac{k}{2}\tau}\Delta R(t_1-\tau)\,d\tau=\sigma_s. \tag{15.51}$$

Performing the (unilateral) Laplace transform on (15.51) we obtain

$$\Delta\bar{R}=\frac{\sigma_s}{s\left[1+\frac{k}{2}\left(\frac{1}{s+\frac{k}{2}}+\frac{k}{4}\frac{1}{\left(s+\frac{k}{2}\right)^2}\right)\right]}. \tag{15.52}$$

Inverting the transform we find

$$\Delta R(t_1)=\sigma_s\left[1-\sqrt{\frac{2}{13}}\,e^{-3kt_1/2}\,\mathrm{sh}\left(\frac{1}{2}\sqrt{\frac{13}{2}}kt_1\right)-\frac{3}{4}\sqrt{\frac{2}{13}}k\right.$$
$$\left.\times\int\limits_0^{t_1}e^{-3/2k\tau}\,\mathrm{sh}\left(\sqrt{\frac{13}{8}}k\tau\right)d\tau\right]. \tag{15.53}$$

Making use, in (15.46) and (15.48), of (15.53), we obtain the solution for the stress σ_{s2} in region II.

We still have to find the solution for the excess stress $\Delta p\,(t)$ over the state σ_s. In the same way as in (15.26) we obtain

$$\bar{\sigma}_{2p}(x,s)=-\Delta\bar{p}(s)\exp\left[-\frac{x}{a}\sqrt{s(s+k)}\right]+B\exp\left[\frac{x}{a}\sqrt{s(s+k)}\right]. \tag{15.54}$$

On account of the boundary condition $v(l,t)=0$ we deduce from $(15.23)_2$ that

$$\bar{\sigma}_{2p,x}=0\qquad\text{for}\qquad x=l, \tag{15.55}$$

whence we have

$$B=-\Delta\bar{p}(s)\exp\left[-\frac{2l}{a}\sqrt{s(s+k)}\right],$$

or

$$\bar{\sigma}_{2p}(x,s)=-\Delta\bar{p}(s)\left\{\exp\left[-\frac{x}{a}\sqrt{s(s+k)}\right]+\exp\left[-\frac{2l-x}{a}\sqrt{s(s+k)}\right]\right\}. \tag{15.56}$$

The inverse transform yields

$$\sigma_{2p}(x,t) = -\Delta p\left(t-\frac{x}{a}\right)e^{-\frac{kx}{2a}} - \frac{kx}{2a}\int_{\frac{x}{a}}^{t}\Delta p(t-\tau)\,e^{-\frac{k\tau}{2}}$$

$$(15.57)\qquad \frac{I_1\left[\frac{k}{2}\sqrt{\tau^2-\left(\frac{x}{a}\right)^2}\right]d\tau}{\sqrt{\tau^2-\left(\frac{x}{a}\right)^2}} - \Delta p\left(t-\frac{2l-x}{a}\right)e^{-\frac{k}{2a}(2l-x)} - \frac{k}{2a}(2l-x)$$

$$\times\int_{\frac{2l-x}{a}}^{t}\Delta p(t-\tau)\,e^{-\frac{k\tau}{2}}\,\frac{I_1\left[\frac{k}{2}\sqrt{\tau^2-\left(\frac{2l-x}{a}\right)^2}\right]}{\sqrt{\tau^2+\left(\frac{2l-x}{a}\right)^2}}\,d\tau.$$

Finally, the total stresses in region II take the form

$$(15.58)\qquad \sigma_{2c}(x,t) = \sigma_{2s}(x,t) + \sigma_{2p}(x,t).$$

Velocity $v_{2p}(x,t)$ is determined in the same way as $v_{2s}(x,t)$ from (15.47), written down for the viscoplastic part.

In order to remove, in region III at the end $x = 0$, the plastic part of the stress (produced by the increment of reaction $\Delta R(t)$) carried to the free end of the bar by the reflected waves and expressed by formula (15.46) (for $x_1 = l$) we have to apply, at the end $x = 0$, the following stress:

(15.59)

$$p_3(t) = \Delta R\left(t-\frac{2l}{a}\right)e^{-\frac{kl}{2a}} + \frac{kl}{2a}\int_{l/a}^{t-l/a}\Delta R\left(t-\frac{l}{a}-\tau\right)e^{-\frac{k}{2}\tau}\,\frac{I_1\left[\frac{k}{2}\sqrt{\tau^2-\left(\frac{l}{a}\right)^2}\right]}{\sqrt{\tau^2-\left(\frac{l}{a}\right)^2}}\,d\tau$$

for $t \geqslant 2l/a$.

Thus we obtain in region III

$$\sigma_{3s}(x,t) = -\sigma_s - \Delta R\left(t-\frac{2l-x}{a}\right)e^{-\frac{k}{2a}(l-x)} - \frac{k}{2a}(l-x)\int_{\frac{l-x}{a}}^{t-\frac{l}{a}}\Delta R\left(t-\frac{l}{a}-\tau\right)$$

$$(15.60)\qquad \times\, e^{-\frac{k}{l}\tau}\,\frac{I_1\left[\frac{k}{2}\sqrt{\tau^2-\left(\frac{l-x}{a}\right)^2}\right]}{\sqrt{\tau^2-\left(\frac{l-x}{a}\right)^2}}\,d\tau + \Delta R\left(t-\frac{2l+x}{a}\right)e^{-\frac{k}{2a}(l+x)}$$

(15.60) [*cont.*]

$$+\frac{k}{2a}(l+x)\int\limits_{\frac{l+x}{a}}^{t-l/a}\Delta R\left(t-\frac{l}{a}-\tau\right)e^{-\frac{k}{2}\tau}\frac{I_1\left[\frac{k}{2}\sqrt{\tau^2-\left(\frac{l+x}{a}\right)^2}\right]}{\sqrt{\tau^2-\left(\frac{l+x}{a}\right)^2}}d\tau.$$

In exactly the same way we obtain the solution for $\Delta p\,(t)$:

$$\sigma_{3p}(x,t)=-\Delta p\left(t-\frac{x}{a}\right)e^{-\frac{k}{2a}x}-\frac{kx}{2a}\int\limits_{\frac{x}{a}}^{t}\Delta p(t-\tau)e^{-\frac{k\tau}{2}}$$

$$\frac{I_1\left[\frac{k}{2}\sqrt{\tau^2-\left(\frac{x}{a}\right)^2}\right]}{\sqrt{\tau^2-\left(\frac{x}{a}\right)^2}}d\tau-\Delta p\left(t-\frac{2l-x}{a}\right)e^{-\frac{k}{2a}(2l-x)}-\frac{k}{2a}(2l-x)\int\limits_{\frac{2l-x}{a}}^{t}\Delta p(t-\tau)e^{-\frac{k}{2}\tau}$$

(15.61)

$$\times\frac{I_1\left[\frac{k}{2}\sqrt{\tau^2-\left(\frac{2l-x}{a}\right)^2}\right]}{\sqrt{\tau^2-\left(\frac{2l-x}{a}\right)^2}}d\tau+\Delta p\left(t-\frac{2l+x}{a}\right)e^{-\frac{k}{2a}(2l+x)}$$

$$+\frac{k}{2a}(2l+x)\int\limits_{\frac{2l+x}{a}}^{t}\Delta p\,(t-\tau)\,e^{-\frac{k\tau}{2}}\frac{I_1\left[\frac{k}{2}\sqrt{\tau^2-\left(\frac{2l+x}{a}\right)^2}\right]}{\sqrt{\tau^2-\left(\frac{2l+x}{a}\right)^2}}d\tau.$$

The total stress in region III is expressed by the sum

$$\sigma_{3c}(x,\,t)=\sigma_{3s}(x,\,t)+\sigma_{3p}(x,\,t).$$

The solutions in the consecutive regions of the phase plane can be constructed in the same way as in region III, making use of the conditions at the bar end. The above solutions have been obtained under the assumption that the absolute values of the stresses in those regions exceed the elastic limit.

If the pressure at the bar end $x = 0$ decreases monotonically, smoothly passing through state $\sigma = \sigma_S$, then, in the general case, the unloading wave will not coincide with the characteristic. This problem, because of its complexity (there is interaction between the reflected waves), has not yet been solved completely. The above problem is considerably simplified when the pressure on the boundary $p(t) \geqslant \sigma_S$ suddenly disappears. In such a case a wave of elastic unloading coincides with the characteristic.

In [66] the problem of so-called "plastic resonance" has been solved for an elastic/visco-plastic material without hardening (in the case of the Sokolovskii model) by use of the above method of solving problems of wave reflection in a finite bar.

We now present the case of wave propagation in a generally non-homogeneous bar of slightly varying cross-section for the Sokolovskii material model which has been considered in [166].

Constitutive equation (15.20), in the case when the elastic modulus, the medium density, the viscosity coefficient, and the elastic limit of material depend on variable x, takes the following form:

$$E(x)\dot{\varepsilon} = \dot{\sigma} + \varkappa\gamma(x)[|\sigma| - \sigma_s(x)]. \tag{15.62}$$

Equation (10.1) of the dynamic equilibrium of an element, in the case of a bar of slightly varying cross-section $F(x)$, assumes the form

$$v_{,t} = \frac{1}{\varrho(x)}\sigma_{,x} + \frac{F'(x)}{\varrho(x)F(x)}\sigma. \tag{15.63}$$

If we consider the problem on the basis of small strains and make use of the mass continuity equation (10.4), then (15.62) takes the form

$$v_{,x} = \frac{1}{E(x)}\dot{\sigma} + \frac{\varkappa\gamma(x)}{E(x)}[|\sigma| - \sigma_s(x)]. \tag{15.64}$$

Eliminating velocity v between the above two equations, we obtain, as a result, the equation for the stress σ in the form

$$\begin{aligned}\sigma_{,tt} - \frac{E(x)}{\varrho(x)}\sigma_{,xx} + \frac{E(x)}{\varrho(x)}\left[\frac{\varrho'(x)}{\varrho(x)} - \frac{F'(x)}{F(x)}\right]\sigma_{,x} + \varkappa\gamma(x)\sigma_{,t} \\ - \frac{E(x)}{\varrho^2(x)F^2(x)}\{F''(x)\varrho(x)F(x) - [\varrho(x)F(x)]'F'(x)\}\sigma = 0.\end{aligned} \tag{15.65}$$

This equation can be reduced to an equation of the Euler–Darboux type. Proceeding as was done in reference [166] we introduce canonical coordinates (ξ, η):

$$\xi = t - F_1(x), \qquad \eta = t + F_1(x), \tag{15.66}$$

where

$$F_1(x) = \int \frac{dx}{a(x)}, \qquad a(x) = \sqrt{\frac{E(x)}{\varrho(x)}}.$$

Equation (15.66) in canonical coordinates (ξ, η) will assume the form

$$\begin{aligned}\sigma_{,\xi\eta} + \frac{1}{4}\left\{\frac{1}{F_1'(x)}\left[\frac{F_1''(x)}{F_1'(x)} + \frac{F'(x)}{F(x)} - \frac{\varrho'(x)}{\varrho(x)}\right] + \varkappa\gamma(x)\right\}\Bigg|_{\xi,\eta}\sigma_{,\xi} \\ - \frac{1}{4}\left\{\frac{1}{F_1'(x)}\left[\frac{F_1''(x)}{F_1'(x)} + \frac{F'(x)}{F(x)} - \frac{\varrho'(x)}{\varrho(x)}\right] - \varkappa\gamma(x)\right\}\Bigg|_{\xi,\eta}\sigma_{,\eta} \\ - \frac{E(x)}{4\varrho^2(x)F^2(x)}\{F''(x)F(x)\varrho(x) - [F(x)\varrho(x)]'F'(x)\}|_{\xi,\eta}\sigma = 0.\end{aligned} \tag{15.67}$$

This equation is of the Euler–Darboux type provided

$$\{F''(x)F(x)\varrho(x)-[F(x)\varrho(x)]'F'(x)\}|_{\xi,\eta}=0,$$

$$\frac{1}{4}\left\{\frac{1}{F_1'(x)}\left[\frac{F_1''(x)}{F_1'(x)}+\frac{F'(x)}{F(x)}-\frac{\varrho'(x)}{\varrho(x)}\right]+\varkappa\gamma(x)\right\}\Bigg|_{\xi,\eta}=-\frac{\beta}{\xi-\eta},\tag{15.68}$$

$$\frac{1}{4}\left\{\frac{1}{F_1'(x)}\left[\frac{F_1''(x)}{F_1'(x)}+\frac{F'(x)}{F(x)}-\frac{\varrho'(x)}{\varrho(x)}\right]-\varkappa\gamma(x)\right\}\Bigg|_{\xi,\eta}=-\frac{\alpha}{\xi-\eta},$$

where α and β are arbitrary constant parameters. If we assume that equations (15.68) are satisfied then (15.67) will take the form of the Euler–Darboux equation:

$$\sigma_{,\xi\eta}-\frac{\beta}{\xi-\eta}\sigma_{,\xi}+\frac{\alpha}{\xi-\eta}\sigma_{,\eta}=0,\tag{15.69}$$

the solution of which is known. In the elastic strain range the equation reduces to the form ($\gamma(x)\equiv 0$):

$$\sigma_{,\xi\eta}-\frac{\gamma}{\xi-\eta}(\sigma_{,\xi}-\sigma_{,\eta})=0,\tag{15.69'}$$

where $\gamma=(\alpha+\beta)/2$.

From the first relation of (15.68) we obtain on integration

$$F(x)=F_0\exp[C_1\textstyle\int\varrho(x)\,dx+C_2].\tag{15.70}$$

In the case when the cross-section of the bar is constant along the length, relation $(15.68)_1$ is, of course, satisfied identically. From the remaining parts of (15.68), on integration, the following two equations can be deduced:

$$E(x)=\varrho(x)\frac{F^4(x)}{C_3^2F'^2(x)}\left\{[1-(\alpha+\beta)]\left[\frac{C_3}{F(x)}+C_4\right]\right\}^{\frac{2(\alpha+\beta)}{\alpha+\beta-1}},\tag{15.71}$$

$$\gamma(x)=(\beta-\alpha)\left\{[1-(\alpha-\beta)]\left[\frac{C_3}{F(x)}+C_4\right]\right\}^{\frac{1}{\alpha+\beta-1}},$$

provided the condition $\alpha\neq\beta$ holds. The constants of integration are determined on the basis of the approximation of the mechanical properties of the medium under consideration. The variation of the elastic limit $\sigma_s(x)$ and the medium density can be arbitrary. The remaining quantities $E(x)$, $\gamma(x)$, and $F(x)$ are calculated from formulae (15.70) and (15.71). On account of the sufficiently large number of constants $C_i (i=1,\ldots,4)$ occurring in these formulae, practically we can approximate each inhomogeneity of the medium.

The problem of the propagation of viscoplastic waves of weak and strong discontinuity was solved in [166] for the case of the above-discussed inhomogeneous medium. In the case of weak discontinuity waves we encounter fundamental difficulties when determining the boundaries of the viscoplastic regions, i.e. when determining the front of the plastic loading wave and that of the unloading wave. According to the nature of the change in the mechanical properties of the medium $E(x)$, $\gamma(x)$, and $\sigma_S(x)$ and the variation of the cross-section of the bar $F(x)$, we obtain various positions of the wave of plastic loading and the unloading wave in the phase plane with respect to the characteristics of equations of the problem. Eventually it leads to solving a number of separate special cases. The solution of the problem of wave propagation in the non-homogeneous medium presented above is considerably simplified if the loads are suddenly applied at the bar end. Then the plastic wave front is a wave of strong discontinuity† and coincides with the positive characteristic which emanates from the origin of the coordinate system and has equation $t = F_1(x) - F_0$. Also here the position of the unloading wave depends on the inhomogeneity of the medium. The method discussed for the construction of the solutions – in the form of exact formulae – of the problem of stress wave propagation in inhomogeneous elastic/viscoplastic media can be generalized to treat the problem of wave reflection from an obstacle, the problems of the wave propagation in layered media, etc. These problems are of great practical value.

†Suck waves cannot propagate in an elastic/viscoplastic medium, since all disturbances propagate with the same speed.

CHAPTER IV

SPHERICAL AND CYLINDRICAL WAVES

16. Formulation of problems

This chapter is devoted to problems concerned with the dynamic loading of media whose geometry and boundary conditions exhibit either spherical or cylindrical symmetry. We shall consider, in turn, problems of spherical waves, of cylindrical radial waves, and cylindrical shear waves using the various constitutive equations discussed in Chapter I. In these problems, on account of the assumed symmetry, all the parameters determining the state of the medium under consideration are functions of a single spatial variable and time. In contradistinction to the problems presented in the preceding chapter we shall be dealing with complex states of stress and strain. The treatment will be confined to the case of small strains in the medium.

We shall formulate the geometrical conditions and the equations of motion, in turn, for the propagation of spherical waves and of cylindrical radial and cylindrical shear waves.

16.1. SPHERICAL WAVES

Let us consider an infinite medium with a spherical cavity of radius r_0, subjected to pressure $p(t)$, varying in time, and uniformly distributed over the cavity surface. In the spherical coordinate system (r, φ, θ) we have, subject to the above assumptions,

$$u_r = u(r, t), \qquad u_\varphi = u_\theta = 0, \tag{16.1}$$

where u_r, u_φ, and u_θ denote the spherical components of the displacement vector.

Under the assumption of spherical symmetry, the non-vanishing components of the strain tensor are the following ones:

$$\varepsilon_{rr} = \frac{\partial u}{\partial r}, \qquad \varepsilon_{\varphi\varphi} = \varepsilon_{\theta\theta} = \frac{u}{r}, \tag{16.2}$$

and the non-vanishing stress components

$$\sigma_{rr} = \sigma_{rr}(r, t), \qquad \sigma_{\varphi\varphi} = \sigma_{\varphi\varphi}(r, t), \qquad \sigma_{\theta\theta} = \sigma_{\theta\theta}(r, t). \tag{16.3}$$

If we denote by v the particle velocity $\partial u/\partial t$ and neglect body forces, then the equation of motion (5.4′), in spherical coordinates, reduces to the form

(16.4)
$$\frac{\partial \sigma_{rr}}{\partial r}+2\frac{\sigma_{rr}-\sigma_{\varphi\varphi}}{r}=\varrho\frac{\partial v}{\partial t}.$$

16.2. CYLINDRICAL RADIAL WAVES

Let us consider in an infinite medium a cavity of radius r_0, the boundary of which is subjected to a uniformly distributed pressure $p(t)$, varying in time. Under these assumptions we obtain in the cylindrical coordinate system (r, φ, z)

(16.5)
$$u_r = u(r,t), \qquad u_\varphi = u_z = 0,$$

where u_r, u_φ, and u_z denote the cylindrical components of the displacement vector.

The states of stress and strain are described by the components

(16.6)
$$\varepsilon_{rr}=\frac{\partial u}{\partial r}, \qquad \varepsilon_{\varphi\varphi}=\frac{u}{r}, \qquad \varepsilon_{zz}=0,$$

(16.7)
$$\sigma_{rr}=\sigma_{rr}(r,t), \qquad \sigma_{\varphi\varphi}=\sigma_{\varphi\varphi}(r,t), \qquad \sigma_{zz}=\sigma_{zz}(r,t),$$

respectively.

If we denote by v the particle velocity $\partial u/\partial t$, then the equation of motion (5.4′), in cylindrical coordinates, assumes the following form (for $X_i \equiv 0$):

(16.8)
$$\frac{\partial \sigma_{rr}}{\partial r}+\frac{\sigma_{rr}-\sigma_{\varphi\varphi}}{r}=\varrho\frac{\partial v}{\partial t}.$$

The equations of motion, (16.4) and (16.8), can be represented using a unified notation, namely

(16.9)
$$\frac{\partial \sigma_{rr}}{\partial r}+n_0\frac{\sigma_{rr}-\sigma_{\varphi\varphi}}{r}=\varrho\frac{\partial v}{\partial t}$$

where in the case of the spherical waves $n_0 = 2$ while in the case of the cylindrical radial waves we put $n_0 = 1$.

16.3. CYLINDRICAL SHEAR WAVES

Shear tractions $p(t)$ varying in time and uniformly distributed over the surface are applied to the surface of an infinitely long cylindrical cavity of radius r_0 in an unbounded medium. In this case in cylindrical coordinates we have

(16.10)
$$u_\varphi = u(r,t), \qquad u_r = u_z = 0,$$

where u_r, u_φ, and u_z denote the cylindrical components of the displacement vector. The stress component

(16.11)
$$\tau_{r\varphi}=\tau_{r\varphi}(r,t)$$

is the only non-vanishing one, similarly the strain component:

$$\varepsilon_{r\varphi} = \frac{\partial u}{\partial r} - \frac{u}{r}. \tag{16.12}$$

If we assume $v = \partial u/\partial t$, then the equation of motion (5.4′) in the cylindrical coordinate system (body forces being neglected) will take the form

$$\frac{\partial \tau_{r\varphi}}{\partial r} + \frac{2\tau_{r\varphi}}{r} = \varrho \frac{\partial v}{\partial t}. \tag{16.13}$$

We will consider, in turn, problems of spherical wave propagation in media described by the constitutive equations of the strain theory of plasticity under the assumption of elastic unloading or "rigid" unloading and also using the constitutive equations of the theory of viscoplasticity. Since the procedure of constructing the solution for the propagation of radial cylindrical waves is similar to that of the case of spherical waves we confine ourselves to a short description of problems involving the former waves only for the case of the viscoplastic constitutive equations. In the case of problems concerning cylindrical wave propagation in media insensitive to strain rate, we cite items of bibliography only (compare, e.g., [37] and [125]). The last section of this chapter will be devoted to problems of cylindrical shear wave propagation in an elastic/viscoplastic medium.

17. Spherical waves in an elastic–plastic medium with elastic unloading

We assume that inside a spherical cavity of radius r_0, in an elastic–plastic medium, a pressure has suddenly appeared, at instant $t = 0$, uniformly distributed over the surface of the cavity. The pressure of initial value $p(0) = p_0$ then varies monotonically with time. The boundary condition at $r = r_0$ takes the form

$$\sigma_{rr}(r_0, t) = -p(t), \tag{17.1}$$

where $p(t) > 0$, $p(0) = p_0$, $p_0 > \sigma_{is}$, σ_{is} denotes the pressure corresponding to the passage of the body into the plastic state.

We assume the homogeneous, zero initial conditions

$$u(r,t) = \frac{\partial u(r,t)}{\partial t} = 0 \quad \text{for} \quad t = 0. \tag{17.2}$$

This problem for the case of the equations of the Nádai–Hencky–Iliushin strain theory of plasticity was solved by Lunts [82].

If we take into account (16.2) and (16.3), then the constitutive equations of the strain theory (2.3) reduce, in the case of the spherical symmetry, to the following forms:

$$\begin{aligned} \sigma_{\varphi\varphi} - \sigma_{rr} &= \sqrt{3}\, 2m \left[\frac{2}{\sqrt{3}}\left(\frac{u}{r} - \frac{\partial u}{\partial r}\right)\right], \\ \sigma_{rr} + 2\sigma_{\varphi\varphi} &= 3K\left(\frac{\partial u}{\partial r} + 2\frac{u}{r}\right), \end{aligned} \tag{17.3}$$

where $\sigma_{\varphi\varphi} = \sigma_{\theta\theta}$.

The system of (17.3), together with the equation of motion (16.4), constitutes the complete system of equations for the problem under consideration, and will be solved taking into account boundary conditions (17.1) and (17.2).

To simplify the calculations, in [82], function $2m$ is replaced by function f defined as follows:

$$f(\theta)=\frac{2}{\sqrt{3}}\frac{1}{\varrho}2m\left(\frac{2}{\sqrt{3}}\theta\right)+\frac{K}{\varrho}\theta, \tag{17.4}$$

where $\theta=\frac{\sqrt{3}}{2}\varepsilon_i=\frac{u}{r}-\frac{\partial u}{\partial r}$.

The function $f(\theta)$ (Fig. 60) has the same shape as the function $2m(\epsilon_i)$ (Fig. 4). However, in the unloading region we have to impose the restriction that $f(\theta)\geqslant\frac{K}{\varrho}\theta$, since otherwise the function $2m(\epsilon_i)$ would become negative.

Introducing the function $f(\theta)$ defined as above, we can determine, from the constitutive equations (17.3), the components of the stress tensor

$$\begin{aligned}\sigma_{rr}&=-\varrho f(\theta)+3K\frac{u}{r},\\ \sigma_{\varphi\varphi}&=\frac{1}{2}\left[\varrho f(\theta)+3K\left(\frac{\partial u}{\partial r}+\frac{u}{r}\right)\right].\end{aligned} \tag{17.5}$$

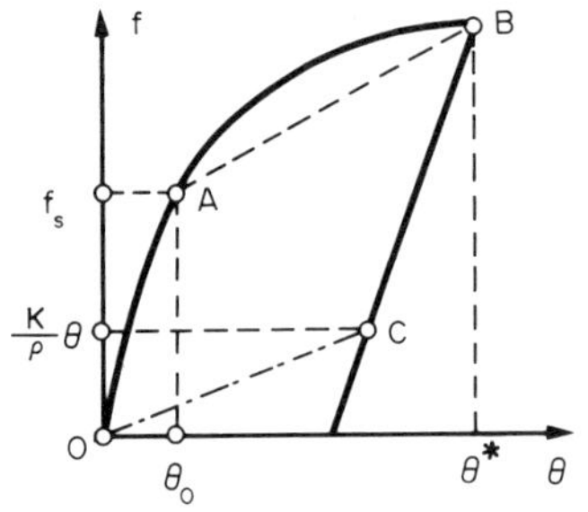

Fig. 60

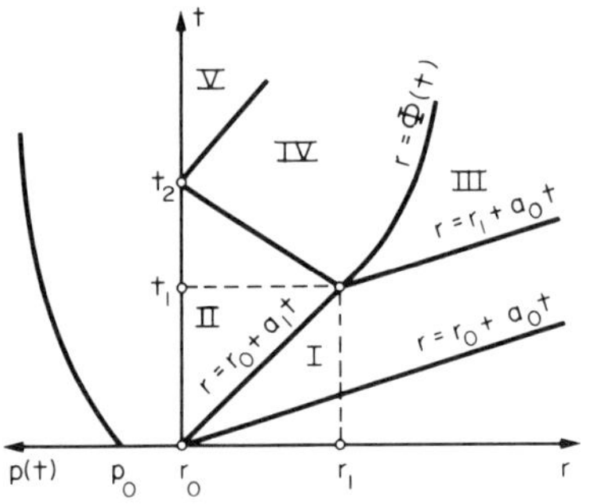

Fig. 61

Eliminating the components of stress tensor from the equations of motion (16.4) by means of formulae (17.5), and using $v=\partial u/\partial t$, we obtain the following differential equation of the second order:

$$\frac{\partial^2 u}{\partial t^2}=-\frac{\partial f(\theta)}{\partial r}-3\frac{f(\theta)}{r}. \tag{17.6}$$

The differential equations of the characteristics (compare 9.2) of the system of (17.5) and (16.4), or of the equivalent system (17.6), have the following form:

$$dr=\pm\, a(\theta)\,dt, \tag{17.7}$$

where $a(\theta)=\sqrt{f'(\theta)}$.

We restrict the discussion to the case when $f''(\theta) \leqslant 0$, i.e. we exclude, for the time being, the possibility of the occurrence of shock waves. First we consider the case when pressure $p(t)$ (boundary condition (17.1)) is a monotonically increasing function of time, $dp/dt > 0$ (Fig. 61).

A wave of strong discontinuity with equation $r = r_0 + a_0 t$ (where $a_0^2 = f(\theta)/\theta = \text{const}$ for $\theta \leqslant \theta_0$) propagates into the undisturbed region $r > r_0$. Behind this wave a region of elastic strain spreads out – region I. The motion of the medium in this region is governed by the equation

$$\frac{1}{a_0^2}\frac{\partial^2 u_1}{\partial t^2} = \frac{\partial^2 u_1}{\partial r^2} + \frac{2}{r}\frac{\partial u_1}{\partial r} - \frac{2u_1}{r^2}. \tag{17.8}$$

In general this problem of spherical wave propagation in an elastic–plastic medium for the curvilinear function $f(\theta)$ (on segment AB – Fig. 61) has to be solved numerically. In [82] it was assumed that segment AB is rectilinear (dashed line in Fig. 60), then we have

$$f(\theta) = a_1^2\theta + (a_0^2 - a_1^2)\theta_0 \quad \text{for} \quad \theta \geqslant \theta_0, \tag{17.9}$$

where a_1 denotes the velocity of plastic wave propagation. In the case of the above sectionally linear function $f(\theta)$, a plastic wave propagates into region $r > r_0$ with speed a_1. The wave coincides with the characteristic of equation $r = r_0 + a_1 t$. Therefore the region of elastic strain I is bounded by characteristics $r = r_0 + a_0 t$ and $r = r_0 + a_1 t$. In this region the Picard problem has to be solved for (17.8) with the data prescribed on the characteristics

$$\begin{aligned} &u_1(r,t) = 0 &&\text{for} \quad r = r_0 + a_0 t,\\ &\frac{u_1}{r} - \frac{\partial u_1}{\partial r} = \theta_0 &&\text{for} \quad r = r_0 + a_1 t. \end{aligned} \tag{17.10}$$

The solution of (17.8) in the case of the above initial conditions is sought in the form

$$u_1(r,t) = \frac{\varphi_1'(r - r_0 - a_0 t)}{r} - \frac{\varphi_1(r - r_0 - a_0 t)}{r^2}. \tag{17.11}$$

Substituting this form for the solution into the initial conditions (17.10) we obtain

$$\begin{aligned} &\varphi_1'(0) = \varphi_1(0) = 0,\\ &\frac{\varphi_1''(x)}{1 - \lambda x} - \frac{3\varphi_1'(x)}{(1+\lambda x)^2} + \frac{3\varphi_1(x)}{(1+\lambda x)^3} = -\theta_0, \end{aligned} \tag{17.12}$$

where the radius of the spherical cavity is of unit length, i.e. $r_0 = 1$, and the following notation has been used:

$$x = r - 1 - a_0 t, \qquad \lambda = \frac{a_1}{a_1 - a_0}.$$

Solving the differential equation of the second order $(17.12)_2$, for initial conditions $(17.12)_1$ we find the function $\varphi_1(x)$. Next, substituting this function into (17.11) we obtain an expression for the displacement field in the region of elastic strain I:

$$u_1(r,t) = \frac{\theta_0}{3(1-\lambda)(1-2\lambda)}\left\{\frac{\lambda}{r}\left[\xi^{\mu-1}\left[3\cos(\nu\ln\xi)+\frac{3\mu-\nu^2-\mu^2}{\nu}\right.\right.\right.$$
(17.13)
$$\left.\left.\left.\times\sin(\nu\ln\xi)\right]-3\xi^2\right]-\frac{1}{r^2}\left[\xi^{\mu}\left[\cos(\nu\ln\xi)+\frac{3-\mu}{\nu}\sin(\nu\ln\xi)\right]-\xi^3\right]\right\},$$

where

$$\mu = \frac{3+\lambda}{2\lambda}, \qquad \nu = -\frac{\sqrt{12-(3+\lambda)^2}}{2\lambda}, \qquad \xi = \lambda[(r-1-a_0t)+1].$$

Region II, as it has been said already, is a plastic strain region. The equation of motion in this region is obtained from (17.6) by means of expression (17.9):

$$\frac{1}{a_1^2}\frac{\partial^2 u_2}{\partial t^2} = \frac{\partial^2 u_2}{\partial r^2}+\frac{2}{r}\frac{\partial u_2}{\partial r}-2\frac{u_2}{r^2}+\frac{3(a_1^2-a_0^2)}{a_1^2 r}\theta_0. \tag{17.14}$$

This equation has to be solved for boundary condition (17.1), for $r = r_0 = 1$, which, by taking into account $(17.5)_1$ and (17.9), can be transformed into the following form:

$$a_1^2\varrho\frac{\partial u_2}{\partial r}+(3K-\varrho a_1^2)u_2-\varrho\theta_0(a_0^2-a_1^2) = -p(t) \quad \text{for} \quad r = 1, \tag{17.15}$$

and for the initial condition on characteristic $r = 1 + a_1 t$ is

$$u_2(r,t) = u_1(r,t). \tag{17.16}$$

The function $u_1(r, t)$ is obtained from the solution in region I (17.13), assuming there that $r = 1 + a_1 t$, thus the initial condition takes the form

$$u_2(r,t) = \frac{\theta_0}{3(1-\lambda)(1-2\lambda)}\left\{(1-3\lambda)r+r^{\mu-2}\left[(1-3\lambda)\cos(\nu\ln r)\right.\right.$$
(17.16′)
$$\left.\left.+\frac{3(\lambda^2-1)+4\lambda}{2\lambda\nu}\sin(\nu\ln r)\right]\right\} = g(r) \quad \text{for} \quad r = 1+a_1t.$$

The required solution of (17.14) for conditions (17.15) and (17.16′) is given in the form of the sum

$$u_2(r,t) = U_2(r,t)+hr\ln r, \tag{17.17}$$

where $h = \dfrac{a_0^2-a_1^2}{a_1^2}\theta_0$, while function $U_2(r, t)$ satisfies the following equation:

$$\frac{1}{a_1^2}\frac{\partial^2 U_2}{\partial t^2} = \frac{\partial^2 U_2}{\partial r^2}+\frac{2}{r}\frac{\partial U_2}{\partial r}-2\frac{U_2}{r^2}. \tag{17.18}$$

Subject to the initial conditions

$$\varrho a_1^2 \frac{\partial U_2}{\partial r} + (3K - \varrho a_1^2) U_2 = -p(t) \quad \text{for} \quad r = 1, \tag{17.19}$$
$$U_2(r,t) = g(r) - hr \ln r \quad \text{for} \quad r = 1 + a_1 t.$$

Equation (17.18) will be identically satisfied provided we assume that

$$U_2(r,t) = \frac{\partial}{\partial r}\left[\frac{\varphi_2(r-1+a_1 t) + \psi_2(r-1-a_1 t)}{r}\right], \tag{17.20}$$

where functions φ_2 and ψ_2 are determined from the initial conditions (17.19). We also make use of the fact that $u_2(r_0, t)\big|_{\substack{r_0=1\\ t=0}} = 0$, from which we obtain

$$\varphi_2(0) = \psi_2(0) = \psi_2'(0) = 0. \tag{17.21}$$

Substituting solution (17.20) into initial conditions (17.19) and taking into account (17.21), we obtain, after numerous operations, the expressions for functions φ_2 and ψ_2:

$$\varphi_2(x) = \frac{(x+2)^2}{2} \int_0^x \frac{1}{\tau+2}\left[g\left(\frac{\tau}{2}+1\right) - h\left(\frac{\tau}{2}-1\right)\ln\left(\frac{\tau}{2}+1\right)\right] d\tau,$$

$$\psi_2(x) = -\varphi_2(-x) + \frac{2}{\sqrt{4\varkappa-\varkappa^2}} \int_0^x \left\{\left[-\frac{1}{\varrho a_1^2}\, p\left(-\frac{x}{a_1}\right) - \varkappa^2 \varphi_2(-\tau)\right]\right. \tag{17.22}$$

$$\left.\times \sin\left(\frac{(x-\tau)\sqrt{4\varkappa-\varkappa^2}}{2}\right) - \varkappa\sqrt{4\varkappa-\varkappa^2}\,\varphi_2(-x)\cos\left(\frac{(x-\tau)\sqrt{4\varkappa-\varkappa^2}}{2}\right)\right\} \exp\left(\frac{\varkappa(x-\tau)}{2}\right) d\tau,$$

where $\varkappa = 3(\varrho a_1^2 - K)/\varrho a_1^2$.

The solution in region II for the displacement field $u_2(r, t)$ can be readily obtained by substituting expression (17.22) into (17.20) and next into (17.17). Analysing the solution we find that the function $\theta = (u_2/r)$–$(\partial u_2/\partial r)$ is a decreasing function of radius r. Along the plastic wave $r = r_0 + a_1 t$, function θ decreases reaching, at a certain point of coordinates (r_1, t_1) (Fig. 61), the value θ_0. For $r > r_1$ the rectilinear plastic wave becomes a curve $r = \Phi(t)$. The point (r_1, t_1) is determined from the condition

$$\theta = \frac{u_2}{r} - \frac{\partial u_2}{\partial r} = \theta_0 \quad \text{for} \quad r = r_0 + a_1 t. \tag{17.23}$$

Hence we obtain the following transcendental equation for value r_1 on the wave $r = 1 + a_1 t$:

$$\frac{10\lambda^3 - 22\lambda^2 + 15\lambda - 3}{2(1-\lambda)^2(2\lambda^2 - 6\lambda + 3)} + \frac{2\lambda^2 - 6\lambda + 3}{2(1-\lambda)^2}\, r_1 - r_1^{\mu-2}\left[\frac{6\lambda^2 - 8\lambda + 3}{2(1-\lambda)^2(2\lambda^2 - 6\lambda + 3)}\right. \tag{17.24}$$
$$\left.\times \cos(\nu \ln r_1) + \frac{6\lambda^3 + 2\lambda^2 - 9\lambda + 3}{4\nu(2\lambda^2 - 6\lambda + 3)(1-\lambda)^2}\sin(\nu \ln r_1)\right] = \frac{p_0}{\sigma_{is}}.$$

It is easy to see that the higher the value of the pressure p_0 is, the longer is the rectilinear segment of the plastic wave.

In spite of approximating the function $f(\theta)$ by the straight line (17.4), the solutions in the next regions of the phase plane III, IV cannot be obtained in all exact (or closed) form. The solutions in these regions are constructed numerically using the finite difference method and making use of the relationships along the characteristics (compare section 9, eqn. (9.17)):

$$(17.25) \quad dv = \pm a(\theta)\frac{\partial u}{\partial r} + \left[\frac{a^2(\theta)}{r}\left(\frac{u}{r} - \frac{\partial u}{\partial r}\right) - \frac{3f(\theta)}{r}\right] dt \quad \text{for} \quad dr = \pm a(\theta)\, dt,$$

where in region III (elastic strain region) we have $a(\theta) = a_0$, while in region IV (plastic strain region) $a(\theta) = a_1$. The displacement field $u_3(r, t), u_4(r, t)$, in regions III and IV, is determined from the solution of the equations of motion (17.8) with the initial conditions $u_3(r, t) = g_3(t)$ on the characteristic $r = r_1 + a_0 t$, and from the solution of (17.4) with the condition $u_4(r, t) = g_4(t)$ on the negative plastic characteristic drawn from point (r_1, t_1), and applying the continuity of displacement condition $u_3 = u_4$ on the curvilinear plastic wave $r = \Phi(t)$. Functions $g_3(t)$ and $g_4(t)$ are determined in regions I and II, respectively. The method of constructing solutions based on the differential relations along the characteristics for quasi-linear equations of the second order was discussed in section 9. In region V the field of displacements is determined from the solution of (17.14) by making use of boundary condition (17.15) for $r = 1$ and of condition $u_5(r, t) = g_5(t)$ on the positive characteristic starting at point (r_0, t_2), where $g_5(t)$ is a function known from the approximate solution in region IV. This solution is identical with that in region II.

Now let us pass to a discussion of the second case, namely of the problem of an unloading process. Taking the boundary condition in form (17.1) we suppose that $p(t)$ is a monotonically decreasing function of time, $dp/dt < 0$ (condition $f''(\theta) \leqslant 0$ is still valid).

The physical equations for the process of elastic unloading (2.4), for the problem under consideration, taking into account (16.2) and (16.3), written in terms of the function $f(\theta)$ will take the form (or directly from Fig. 60):

$$(17.26) \qquad f(\theta) = a_0^2\theta + (a_1^2 - a_0^2)[\theta^*(r) - \theta_0],$$

where $\theta^*(r)$ denotes the value of $\theta(r)$ on the unloading wave.

The equation of motion (16.4) for the medium, during unloading, will thus take the form

$$(17.27) \qquad \frac{\partial^2 \bar{u}}{\partial r^2} + \frac{2}{r}\frac{\partial \bar{u}}{\partial r} - \frac{2\bar{u}}{r^2} = \frac{\partial^2 \bar{u}}{\partial \tau^2} + h_0\left(\frac{d\bar{\theta}^*}{dr} + \frac{3(\bar{\theta}^* - 1)}{r}\right),$$

where the following notation has been introduced:

$$\bar{u} = \frac{u}{\theta_0}, \quad \bar{\theta} = \frac{\theta}{\theta_0}, \quad \bar{\theta}^* = \frac{\theta^*}{\theta_0}, \quad \tau = a_0 t, \quad h_0 = \frac{a_1^2 - a_0^2}{a_0^2}, \quad \bar{p}(\tau) = \frac{p(\tau)}{\theta_0}.$$

It has been proved in [82] and [169] that in the case considered the plastic wave coinciding with the characteristic $r = r_0 + a_1 t$ (Fig. 62) is an unloading wave. The solution in region I, i.e. in the region of elastic strain, is the same as before and is given in (17.13). In region II the unloading region, a Goursat problem, has to be solved for (17.27) (see section 9). If in an arbitrary small triangle 012 of the unloading region II (Fig. 62, region A) the solution is known, then we can find the solution of (17.27) in the entire region II. So far the exact form of the solution in region 012 is not known. An approximate method of finding the displacement field in this region was given in [82]. We shall not quote that solution here.

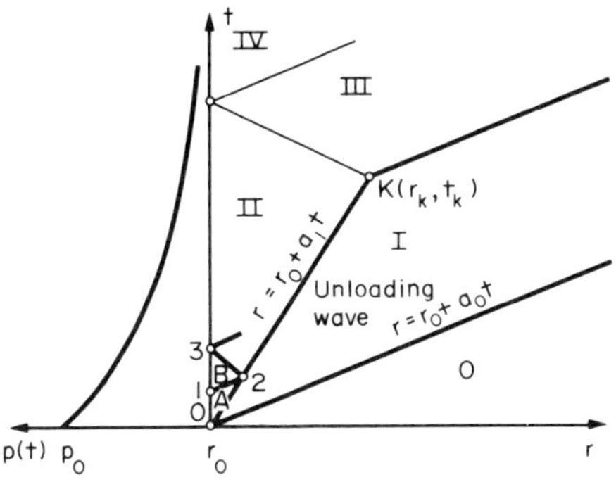

Fig. 62

The solution of (17.27) in the unloading region IIB can be represented in the form of the sum

(17.28) $$\bar{u} = U_B(r, \tau) + V_B(r),$$

where the above functions satisfy the equations

(17.29) $$\frac{d^2V_B}{dr^2} + \frac{2}{r}\frac{dV_B}{dr} - \frac{2V_B}{r^2} = h_0\left[\frac{d\bar{\theta}^*}{dr} + \frac{3(\bar{\theta}^*-1)}{r}\right],$$
$$\frac{\partial^2 U_B}{\partial r^2} + \frac{2}{r}\frac{\partial U_B}{\partial r} - \frac{2U_B}{r^2} = \frac{\partial^2 U_B}{\partial \tau^2}.$$

We define the initial conditions of (17.29). Assuming that $V_B(r)$ is defined as follows:

(17.30) $$\frac{dV_B}{dr} - \frac{V_B}{r} = h_0(\bar{\theta}^*-1), \qquad V_B(r_0)|_{r_0=1} = 0$$

then it constitutes a particular solution of $(17.29)_1$ and the condition of displacement continuity across the unloading wave $r = 1 + \dfrac{a_1}{a_0}\tau$ for function $U_B(r, t)$ can be written down in the form

(17.31) $$U_B(r, t)\Big|_{r=1+\frac{a_1}{a_0}\tau} = \frac{g(r)}{\theta_0} - V_B(r)$$

and

(17.32) $$\bar{\theta} = \bar{\theta}^*(r),$$

where the function $g(r)$ is determined from (17.16′). From the condition of displacement continuity across the unloading wave we have

$$\frac{U_B}{r} - \frac{\partial U_B}{\partial r} + \frac{V_B}{r} - \frac{dV_B}{dr} = \bar{\theta}^*,$$

whence, by means of (17.30), we obtain

$$\frac{U_B}{r} - \frac{\partial U_B}{\partial r} + \left(\frac{V_B}{r} - \frac{dV_B}{dr}\right)\frac{a_1^2}{a_1^2 - a_0^2} - 1 = 0. \tag{17.33}$$

Using the expression for $V_B(r)$ from (17.31) and eliminating $g(r)$ using (17.16′), we obtain, from (17.33), the initial condition for $(17.29)_2$ on the characteristic $r = 1 + \frac{a_1}{a_0}\tau$:

$$\frac{\partial U_B}{\partial r} + \frac{a_1}{a_0}\frac{\partial U_B}{\partial \tau} - \frac{U_B}{r} = h_0\left\{1 + \frac{\lambda}{1-2\lambda} r^{\mu-3}\left[\cos(\nu \ln r) + \frac{2\lambda^2 + \lambda - 1}{2\lambda\nu(2\lambda - 1)}\sin(\nu \ln r)\right]\right\}. \tag{17.34}$$

The boundary condition (17.1) for the unloading process, in contradistinction to condition $(17.19)_1$, will assume the form

$$\varrho a_0^2 \frac{\partial U_B}{\partial r} + (3K - \varrho a_0^2)\, U_B = -\bar{p}\left(\frac{\tau}{a_0}\right) \quad \text{for} \quad r_0 = 1, \tag{17.35}$$

where $V(1) = 0$ (compare (17.30)).

The solution in region IIB can be reduced to that of $(17.29)_2$ with the initial conditions (17.34) and (17.35). The general solution of this equation can be represented in the following form:

$$U_B(r,t) = \frac{\varphi_B'(r-1+\tau) + \psi_B'(r-1+\tau)}{r} - \frac{\varphi_B(r-1+\tau) + \psi_B(r-1-\tau)}{r^2}, \tag{17.36}$$

where φ_B and ψ_B are the functions of class C^3, determined on the basis of conditions (17.34) and (17.35):

$$\varphi_B(x) = \left(\frac{2 + x + \tau_1}{2}\right)^2 \int_{\tau_1}^{x} \frac{U_0\left[\frac{x-\tau_1}{2} + 1\right]}{\frac{x-\tau_1}{2} + 1}\, dx,$$

$$\begin{aligned} \psi_B(x) = {} & -\varphi_B(-x) - \frac{2U_0(1)}{\sqrt{4\varkappa_0 - \varkappa_0^2}} \exp\left[\frac{\varkappa_0(x+\tau_1)}{2}\right] \sin\left[\frac{2(x+\tau_1)}{\sqrt{4\varkappa_0 + \varkappa_0^2}}\right] \\ & + \frac{2}{\sqrt{4\varkappa_0 - \varkappa_0^2}} \int_{-\tau_1}^{x} \left\{\left[\frac{p(\tau)}{\varrho a_0^2 \theta_0} - \varkappa_0^2 \varphi_B(-\tau)\right] \sin\left(\frac{2(x-\tau)}{\sqrt{4\varkappa_0 - \varkappa_0^2}}\right)\right. \\ & \left. - \varkappa_0 \sqrt{4\varkappa_0 - \varkappa_0^2}\, \varphi_2(-\tau) \cos\frac{2(x-\tau)}{\sqrt{4\varkappa_0 - \varkappa_0^2}}\right\} \exp\frac{\varkappa_0(x-\tau)}{2}\, d\tau. \end{aligned} \tag{17.37}$$

Here U_0 denotes the solution in an arbitrary small region IIA in the vicinity of the point $(r_0, 0)$ on the positive characteristic 1–2 (Fig. 62) which has equation $\tau = r - 1 + \tau_1$, where τ_1 denotes the time at point 1 on the boundary $r_0 = 1$, $\varkappa_0 = 3\left(1 - \frac{K}{\varrho a_0^2}\right)$.

Taking into account (17.37) in expression (17.36) we obtain the displacement field $U_B(r, t)$ in region IIB. The construction of the solution in the remaining subregions of region II is analogous to that presented for subregion IIB.

The plastic wave $r = r_0 + a_1 t$ vanishes at point $K(r_k, t_k)$ (Fig. 62). The coordinates of the point are determined from the condition

$$\theta = \left(\frac{u_2}{r} - \frac{\partial u_2}{\partial r}\right)_{|r=r_0+a_1 t} = \theta_0. \tag{17.38}$$

In the medium, for $r > r_k$, only elastic waves propagate. The solution in region III is found by a solution, in the same region, of the Darboux problem for (17.27) (where for $r > r_k$ we have

$h_0\left(\frac{d\bar{\theta}^*}{dr} - \frac{3(\bar{\theta}^* - 1)}{r}\right) \equiv 0$) together with the displacements prescribed on the elastic

characteristics (positive and negative, leaving from point $K(r_k, t_k)$) determined from the solutions in regions I and II. In region II we have to solve a Goursat problem for (17.27).

The effect of material hardening on the form of stress field has been studied numerically [169] for the above case of any unloading wave. It was assumed that the pressure on the boundary of the spherical cavity appeared suddenly and then decreased linearly to zero. The hardening parameter was defined as the ratio $\lambda \to \mu_p/\mu_0$. The assumed value of Poisson's ratio was $\nu = 0.25$.

We are in a position to draw the following conclusions, on the basis of the numerical calculations performed over a wide range of the hardening parameter:

1. The maximum shear strain θ^* and the jump $\delta\theta^*$ on the plastic wave front decrease with increase of material hardening.
2. The width of the plastic strain region increases slowly with increase of the hardening parameter.
3. The permanent shear strain considerably decreases with increase of the hardening parameter.
4. The distance between the elastic and plastic wave fronts decreases with increase of the hardening parameter; the influence of this effect on the stress field is insignificant.

18. Spherical shock waves in a homogeneous elastic–plastic medium

We shall next consider the problem of spherical wave propagation in an elastic–plastic medium, similarly to that considered at the preceding point. The difference will consist in the assumption that the characteristic $f(\theta)$ is of positive curvature, i.e. $f''(\theta) > 0$. Here we refer to the solution presented in [162]. Function $f(\theta)$ will be assumed as shown in Fig. 63. For the sake of simplicity we also assume that $f_s \equiv 0$, and we consider the case of a pressure suddenly applied to the cavity boundary and then monotonically decreased to zero. This assumption simplifies the solutions since in such a case the shock wave front is simultaneously an unloading wave. If $f''(\theta) > 0$, then the front of the shock wave will propagate into the medium $r > r_0$ with a velocity which varies independently of the nature of the boundary condition. Further simplification of the solutions can be achieved by approximating the curve $f(\theta)$ by straight line segments.

The function $f(\theta)$ can be represented as follows,

$$(18.1)\qquad \begin{aligned} f(\theta) &= f_n + a_n^2(\theta-\theta_n) = a_n^2\theta + A_n && \text{for loading process,}\\ f(\theta) &= f^* - a_0^2(\theta^*-\theta) && \text{for unloading process,} \end{aligned}$$

where θ_n is the value of variable θ corresponding to the break point (Fig. 63) between the segments $n-1$ and n and θ^* is the value of the variable θ on the wave front from the side of the unloading region;

$$f^* = f(\theta^*), \qquad f_n = f(\theta_n), \qquad a_n^2 = \frac{f_{n+1}-f_n}{\theta_{n+1}-\theta_n}, \qquad a_0^2 = \frac{f^*}{\theta^*-\theta_k}, \qquad \theta_k = \varepsilon_{\varphi\varphi}^k - \varepsilon_{rr}^k,$$

$\varepsilon_{\varphi\varphi}^k$ and ε_{rr}^k denote the permanent transversal and radial strain components respectively.

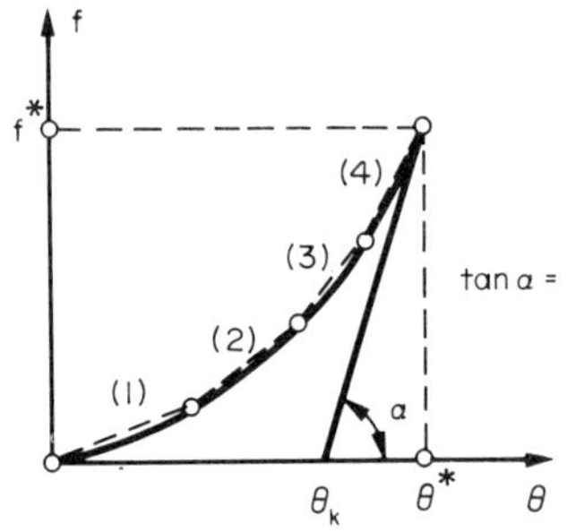

Fig. 63

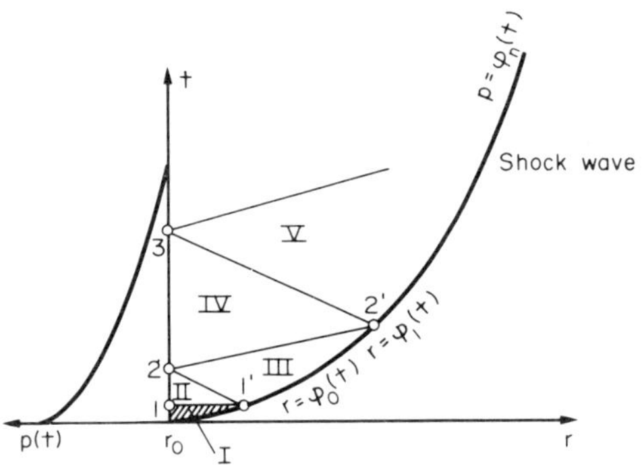

Fig. 64

The principle of constructing the solution of the problem posed is the same as in the plane shock wave case, presented in section 13. The solution in the phase plane (r, t) is presented in Fig. 64. The shock wave $r = \varphi(t)$ propagates into an undisturbed region (zero initial conditions have been assumed). On account of the assumed boundary condition the wave is simultaneously an unloading wave. The construction of the solution is as follows. We take a sufficiently small region I, bounded by the shock wave $r = \varphi_0(t)$ and by the positive elastic characteristic $1-1'$. We determine, in an approximate way, an initial segment of shock wave $r = \varphi_0(t)$ assuming that it coincides with the tangent to the real wave $\varphi(t)$ at point $r = r_0$, $t = 0$, i.e. that

$$(18.2)\qquad \varphi_0(t) = \varphi_0'(0)\, t.$$

The initial velocity of the shock wave $\varphi_0'(0)$ is determined from the conditions of dynamic and kinematic continuity across the shock wave front and from the boundary condition for $t = 0$.

The conditions of dynamic (7.18) and kinematic (8.9) continuity across the front of the shock wave, in the case of the spherical symmetry, take the forms

$$(18.3)\qquad \sigma_{rr} = -\varrho\varphi'(t)\, v, \qquad v = -\varphi'(t)\, \varepsilon_{rr}.$$

From the above conditions at $t = 0$, and taking $\sigma_{rr}(r_0, 0) = -p_0$, we deduce that

$$\varphi'(0) = \sqrt{\frac{p_0}{\varrho\theta^*}}, \tag{18.4}$$

where θ^* is determined from

$$f^* = f(\theta^*). \tag{18.5}$$

The initial segment of the shock wave can be determined more precisely by expanding function $\varphi(t)$ into a Maclaurin's series in the neighbourhood of the origin, exactly as was shown in section 13 in the case of a plane shock wave.

In region I the solution can be constructed numerically by the finite difference method by making use of the relations on the characteristics (17.25) (instead of function $f(\theta)$ relation $(18.1)_2$ ought to be used).

The equation of motion, governing the unloading process in regions II, III, IV, . . ., is obtained from (17.6) by eliminating function $f(\theta)$ in it, taking account of formula $(18.1)_2$:

$$\frac{\partial^2 u}{\partial r^2} + \frac{2}{r}\frac{\partial u}{\partial r} - \frac{2u}{r^2} = \frac{1}{a_0^2}\frac{\partial^2 u}{\partial t^2} + \left(\frac{a_n^2}{a_0^2} - 1\right)\left[\frac{d\theta^*}{dr} + \frac{3}{r}\left(\theta^* + \frac{A_n}{a_n^2 - a_0^2}\right)\right]. \tag{18.6}$$

In the same way as in the previously considered case of the propagation of spherical elastic–plastic waves, the solution of (18.6) will be represented in the form of the sum of two functions,

$$u(r, t) = U(r, t) + V(r). \tag{18.7}$$

These functions satisfy the following equations:

$$\begin{gathered}\frac{1}{a_0^2}\frac{\partial^2 U}{\partial t^2} = \frac{\partial^2 U}{\partial r^2} + \frac{2}{r}\frac{\partial U}{\partial r} - \frac{2U}{r^2},\\ \frac{d^2 V}{dr^2} + \frac{2}{r}\frac{dV}{dr} - \frac{2V}{r^2} = \left(\frac{a_n^2}{a_0^2} - 1\right)\left[\frac{d\theta^*}{dr} + \frac{3}{r}\left(\theta^* + \frac{A_n}{a_n^2 - a_0^2}\right)\right].\end{gathered} \tag{18.8}$$

The solution in region II is obtained in the same way as in region IIB of the preceding case in section 17.

The solution of the unloading problem in region III is slightly different since here we must simultaneously determine the next segment of the unloading wave $r = \varphi_1(t)$ contained between points 1′ and 2′ (Fig. 64). In order to determine the unloading wave $r = \varphi_1(t)$ we make use of the conditions of dynamic and kinematic continuity (18.3), still assuming that the shock wave propagates into an undisturbed medium. Hence we obtain an additional condition, namely

$$u = U + V = 0, \quad \text{whence} \quad U = -V \quad \text{for} \quad r = \varphi_1(t). \tag{18.9}$$

Conditions (18.3) should then be expressed in terms of the displacement $U(r, t)$. Next, by eliminating $\varphi_1'(t)$ in these equations, we obtain the relation that should be satisfied by function $U(r, t)$ on the front of the shock wave. The method of finding the unloading wave $r = \varphi_1(t)$ and the solution of the problem in region III are discussed with full particulars in [162]. The solution in the consecutive regions of the phase plane is constructed in the same way as it was done in regions II and III.

19. Propagation of a spherical unloading wave in an elastic–plastic medium with rigid unloading behaviour

We now consider the problem of the propagation of a spherical unloading wave in a homogeneous elastic–plastic medium assuming that the unloading takes place at constant strain intensity, i.e. ϵ_i = const for a given radius r (compare (4.22)). This condition, in the case considered, does not mean that the medium is "rigid" as it was in the case of the uniaxial state of stress or strain (see section 14); changes with time may exist in the strain tensor components ε_{rr}, $\varepsilon_{\varphi\varphi}$, and $\varepsilon_{\theta\theta}$.

We assume that the boundary condition for $r = r_0$ takes the form

$$\sigma_{rr}(r_0, t) = -p(t), \quad p(t) > 0, \quad \frac{dp}{dt} < 0, \quad p(0) = p_0, \quad p_0 > \sigma_{is}, \tag{19.1}$$

while as the initial conditions we take conditions (17.2).

The constitutive equations and the equations of motion are identical with those for the previously considered problem of the propagation of spherical elastic–plastic waves, (17.3) and (16.4)). To simplify the calculations we introduce, as previously, the function $f(\theta)$. For the loading process the linear approximation of function $f(\theta)$ is given by formula (17.9) and shown in Fig. 65a. For the unloading process we assume that θ is a function of radius r only, i.e. that

$$\theta = \frac{u}{r} - \frac{\partial u}{\partial r} = g(r). \tag{19.2}$$

Using the function $f(\theta)$, the equation of motion (16.4) is reduced to the form of (16.6).

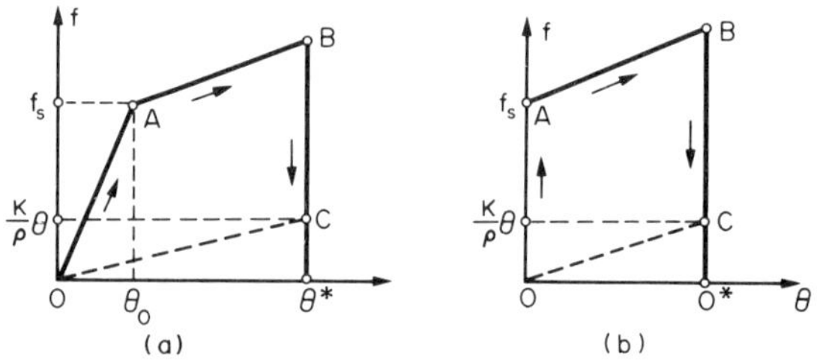

Fig. 65

For the sake of further simplification of the calculations the approximation of function $f(\theta)$, shown in Fig. 65b, is assumed in many cases referring to the propagation of spherical waves in a material with rigid unloading (see ref. [37]). Such an approximation yields substantial simplification in the solution of problems of the reflection of strong discontinuity waves.

The solution of the wave propagation problem is illustrated in Fig. 66. In region I the displacements and strains are equal to zero. This follows from the assumption of rigid loading

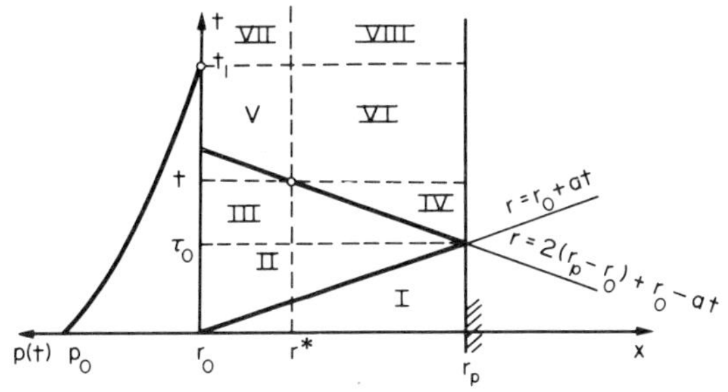

Fig. 66

(segment OA, Fig. 65b). Stress σ_{rr} in this region is determined from $(17.5)_1$ by taking into account that $u_1 \equiv 0$ and $f(\theta) = f_s$ (Fig. 65b). Hence

$$\sigma_{rr} = -\varrho f_s. \tag{19.3}$$

Region II is the unloading one. It results from the form of the assumed boundary condition and from the model for the medium. The region is bounded by a strong discontinuity wave which at the same time is an unloading wave. This wave coincides with the characteristic of equation $r = r_0 + at$ where $a = \sqrt{(3K+4\mu)/3\varrho}$, μ denoting the modulus of the linearly hardening material. Condition (19.2) is satisfied on account of the assumption that the unloading is "rigid" in region II. Integrating this condition we obtain an expression for displacement $u(r, t)$ in the form

$$u(r,t) = -r\left[\int \frac{g(r)}{r}\,dr - C(t)\right]. \tag{19.4}$$

Velocity $v(r, t) = \partial u/\partial t$ takes the form

$$v(r,t) = r\,C'(t). \tag{19.5}$$

We assume, for the time being, that the particle velocity on the boundary of the spherical cavity of radius r_0 is known, and we designate it by suffix zero, i.e. $v(r_0, t) = v_0(t)$. Then, by means of (19.5), we obtain

$$v(r,t) = \frac{r}{r_0}\,v_0(t). \tag{19.6}$$

Displacement $u(r, t)$ is determined by means of integration of (19.6), with respect to time, between the limits $(r-r_0)/a$ to t:

$$u(r,t) = \frac{r}{r_0}\int\limits_{\frac{r-r_0}{a}}^{t} v_0(\tau)\,d\tau. \tag{19.7}$$

Using the above expression for displacement $u(r, t)$ in equation of motion (17.6) and integrating it with respect to variable r, we obtain

$$f = -\frac{r^2}{5r_0} v_0'(t) + \frac{1}{r^2} \psi(t). \tag{19.8}$$

The arbitrary function $\psi(t)$ is determined from the boundary condition (19.1). We obtain after some algebra

$$\psi(t) = \frac{1}{\varrho}\left[p(t) + \varrho \frac{r_0}{5} v_0'(t) + \frac{3K}{r_0} \int_0^t v_0(\tau)\, d\tau\right]. \tag{19.9}$$

Thus

$$f = -\frac{r^2}{5r_0} v_0'(t) + \frac{1}{\varrho}\left[p(t) + \varrho \frac{r_0}{5} v_0'(t) + \frac{3K}{r_0} \int_0^t v_0(\tau)\, d\tau\right]. \tag{19.10}$$

The condition of dynamic continuity (7.18) has to be satisfied across the front of the strong discontinuity wave. In the case of spherical symmetry it takes the form

$$\sigma_{rr} = -\varrho a v. \tag{19.11}$$

Making use of the above condition and taking into account expressions $(17.5)_1$, (19.7), and (19.10) we obtain, after numerous manipulations, the following singular Volterra integral equation of the second kind for the unknown function $v_0(t)$:

$$v_0(t) = \Phi(t) + \frac{15Kr_0^3}{\varrho[(r_0+at)^5 - r_0^5]} \int_0^t \int_0^\xi v_0(\tau)\, d\tau\, d\xi, \tag{19.12}$$

where

$$\Phi(t) = \frac{5r_0^4\left[\int_0^t p(\tau)\, d\tau + \dfrac{(r_0+at)^4 - r_0^4}{4ar_0^3}\right]}{\varrho[(r_0+at)^5 - r_0^5]}.$$

Applying the method of successive approximations we obtain the general solution of (19.12) in the form

$$v_0(t) = \Phi(t) + \sum_{m=1}^{\infty} \lambda^m \frac{1}{(r_0+at)^5 - r_0^5} \int_0^t \int_0^\xi K_m(t,\tau)\Phi(\tau)\, d\tau\, d\xi, \tag{19.13}$$

where

$$K_m(t,\tau) = \frac{1}{(r_0+at)^5 - r_0^5} \int_0^t \int_\tau^\xi K_{m-1}(\zeta,\tau)\, d\zeta\, d\xi, \qquad K_1 = 1, \qquad \lambda = \frac{15r_0^3 K}{\varrho}.$$

It is shown in [37] that each term of the solution given by formula (19.13) is bounded. An estimation of the error is also given there for the approximate solution. In certain particular cases the solution of (19.12) can be found in an exact form.

If function $v_0(t)$ is known it is easy to find the solution in region II of the phase plane as follows. First, f is determined from the corresponding formula. Then, from (17.5), taking into account (19.7), we obtain the expressions for stresses $\sigma_{rr}(r, t)$ and $\sigma_{\varphi\varphi}(r, t)$, and velocity is found from (19.6). The permanent strain is determined from the formula

$$\theta = \frac{f_0 - f_s}{a^2}, \tag{19.14}$$

where f_0 denotes the value of function f on the unloading wave $r = r_0 + at$.

Now we consider the problem of the reflection of the unloading wave from an undeformable spherical obstacle, concentric with the spherical cavity of radius r_0, placed at a distance $r = r_p$. We have to assume that pressure $p(t)$ on the surface of the spherical cavity $r = r_0$ is sufficiently high that the stress intensity at $r = r_p$ on the wave reflected from the obstacle (of equation $r = 2(r_p - r_0)$) exceeds the value σ_{is} (in diagram f–θ, Fig. 65, value $f = f_s$). The reflected wave propagates into a non-homogeneous medium of variable yield limit.

In the case considered the validity of the solution in region II is limited by time $t = \tau_0$. In region III $\left(\text{for } \tau_0 \leqslant t \leqslant \dfrac{r_0 - r + 2(r_p - r_0)}{a}\right)$ the solution is constructed similarly to that in region II. On account of the fact that the assumed model for the medium contains the rigid unloading condition, (19.2) has to be satisfied in region III:

$$\frac{u_3}{r} - \frac{\partial u_3}{\partial r} = g(r), \tag{19.15}$$

where suffix 3 denotes the displacements in region III.

By the same procedure as before, integrating the above condition, we obtain

$$u_3(r, t) = -r\left[\int \frac{g(r)}{r}\, dr - C(t)\right], \tag{19.16}$$

and then the velocity

$$v_3(r, t) = rC'(t).$$

Assuming once again that the function $v_0(t)$ is known, we obtain, in region III, the same relations as (19.6), (19.7), and (19.8). Functions $v_0(t)$ and $\psi(t)$ are determined from the following conditions: the boundary condition on $r = r_0$ and the continuity condition of function $f(\theta)$ ahead of the front of the reflected wave $r = 2(r_p - r_0) + r_0 - at$:

$$\begin{aligned} \sigma_{rr}(r_0, t) &= -\varrho f_3(r_0, t) + 3K \frac{u_3(r_0, t)}{r_0} = -p(t) \quad \text{for} \quad r = r_0, \\ f_3(r, t) &= f_2\left(r, 2\frac{r_p - r_0}{a} - t\right) \quad \text{for} \quad r = 2(r_p - r_0) + r_0 - at. \end{aligned} \tag{19.17}$$

Substituting expressions (19.16), (19.18), and (19.5) into the above conditions, and assuming that $r = \varphi(t) = 2(r_p - r_0) + r_0 - at$, we obtain the system of equations

$$\varrho \frac{r_0^2}{5} C''(t) - \varrho \frac{\psi(t)}{r_0^3} + 3KC(t) - 3K \int \frac{g(r)}{r} dr \Big|_{r=r_0} = -p(t),$$

(19.18)
$$-\frac{\varphi^2}{5} C''(t) + \frac{\psi(t)}{\varphi^3(t)} = f_2 \left[\varphi(t), 2\frac{r_p - r_0}{a} - t \right],$$

whence on elimination of function $\psi(t)$ we find that

(19.19)
$$C''(t) + \alpha(t) C(t) = \Phi(t),$$

where

$$\Phi(t) = \frac{5\varphi^3(t)}{[\varphi^5(t) - r_0^5]\varrho} \left\{ \frac{r_0^3}{\varphi^3(t)} p(t) - \frac{r_0^3}{\varphi^3(t)} 3K \int \frac{g(r)}{r} dr \Big|_{r=r_0} - f_2 \left[\varphi(t), 2\frac{r_p - r_0}{a} - t \right] \right\},$$

$$\alpha(t) = \frac{15K[\varphi^3(t) - r_0^3]}{\varrho[\varphi^5(t) - r_0^5]}.$$

Equation (19.19) can be reduced to a Volterra integral equation of the second kind [37] by the substitutions

$$C''(t) = X(t), \qquad C'(t) = \int_{\tau_0}^{r} X(\tau) d\tau + A,$$

$$C(t) = \int_{\tau_0}^{t} (t - \tau) X(\tau) d\tau + At + B.$$

The integral equation takes the form

(19.20)
$$X(t) + \alpha(t) \int_{\tau_0}^{t} (t - \tau) X(\tau) d\tau = \Phi(t) - \alpha(t)(At + B),$$

where the constants A and B are determined from the initial conditions for $t = \tau_0$:

$$A = C'(\tau_0) = \frac{v_2(r_p, \tau_0)}{r_p},$$

$$B = C(\tau_0) - C'(\tau_0)\tau_0 = \int \frac{g(r)}{r} dr \Big|_{r=r_0} - \tau_0 \frac{v_2(r_0, \tau_0)}{r_p}.$$

If function $C(t)$ is known, the parameters of the solution in region III can be readily obtained from the solution of (19.19) by means of formulae (19.16), (19.5), and (17.5). The permanent strain for a given r is identical with that in region II, and is determined by formula (19.14).

Region IV of the phase plane (r, t) (Fig. 66) is an unloading region. For $r = r_p$, i.e. on the undeformed boundary we have to satisfy either the condition for displacement $u_4(r_p, t) = 0$ or the condition for velocity $v_4(r_p, t) = 0$. Taking into account the above condition we deduce from (19.4) that $C(t) = 0$, and next by means of (19.5) that

(19.21)
$$v_4(r, t) \equiv 0.$$

Consequently, the equation of motion (17.6) is considerably simplified, and we obtain

(19.22)
$$\frac{\partial f(\theta)}{\partial r} + 3\frac{f(\theta)}{r} = 0.$$

Integrating this equation we find

(19.23)
$$f(\theta) = \frac{1}{r^3}\,\psi(t),$$

where function $\psi(t)$ is determined from the condition of dynamic continuity across the front of the reflected wave:

(19.24)
$$\sigma_{rr4} - \sigma_{rr3} = \varrho a(v_4 - v_3) = -\varrho a v_3.$$

Making use of $(17.5)_1$ and of the fact that $u_4 = u_3$ on reflected wave $r = r_0 + 2(r_p - r_0) - at$, we deduce from (19.24) the following relation:

$$-\varrho f + 3K\frac{u_4}{r} = \sigma_{rr3} - \varrho a v_3,$$

from which function f can be determined:

$$f[\varphi(t), t] = \frac{1}{\varrho}\left\{3K\frac{u_3[\varphi(t), t]}{r} - \sigma_{rr3}[\varphi(t), t] + \varrho a v_3[\varphi(t), t]\right\},$$

where $\varphi(t) = 2(r_p - r_0) + r_0 - at$. In turn we obtain from (19.23) function $\psi(t)$:

(19.25)
$$\psi(t) = \frac{1}{\varrho}\varphi^3(t)\left\{3K\frac{u_3[\varphi(t), t]}{r} - \sigma_{rr3}[\varphi(t), t] + \varrho a v_3[\varphi(t), t]\right\}.$$

The stresses σ_{rr} and $\sigma_{\varphi\varphi}$ are determined from formulae (17.5), and the permanent strain from (19.14), where f_0 now denotes the value of the function f on the reflected unloading wave.

The reflected unloading wave of equation $r = 2(r_p - r_0) + r_0 - at$ vanishes at time $t = t^*$, where t^* is determined from the condition

(19.26)
$$v_0(t^*) = 0.$$

Time t^* simultaneously restricts the validity of the solutions in region IV.

The solution of the problem in the remaining regions of the phase plane does not present difficulties. On account of the boundary condition on $r = r_p$: $v(r_p, t) = 0$ we have $v_7(r, t) = v_8(r, t) = v_5(r, t) = v_6(r, t) \equiv 0$. Thus we have to deal with a static problem. In the consecutive regions we have the following displacements:

$$u_7(r, t) = u_5(r, t) = u_3(r, t^*), \qquad u_8(r, t) = u_6(r, t) = u_4(r, t) = u_3(r, \bar{\varphi}(r)),$$

strain:

$$\varepsilon_{rr7}(r, t) = \varepsilon_{rr5}(r, t) = \varepsilon_{rr3}(r, t^*), \qquad \varepsilon_{rr8}(r, t) = \epsilon_{rr6}(r, t) = \varepsilon_{rr4}(r, t),$$

and identical ones for components $\varepsilon_{\varphi\varphi}$, stress:

$$\sigma_{rr7}(r,t) = \sigma_{rr5}(r,t) = -\left(\frac{r_0}{r}\right)^3 p(t) - 3K\frac{r_0^2}{r^3}u_3(r_0,t^*) + 3K\frac{u_3(r,t^*)}{r},$$

$$\sigma_{rr8}(r,t) = \sigma_{rr6}(r,t) = -\left(\frac{r_0}{r}\right)^3 p(t) - 3K\frac{r_0^2}{r^3}u_3(r_0,t^*) + 3K\frac{u_3(r,\bar{\varphi}(r))}{r},$$

$$\sigma_{\varphi\varphi7}(r,t) = \sigma_{\varphi\varphi5}(r,t) = -\frac{1}{2}\sigma_{rr5} + \frac{3K}{2}\left(\frac{\partial u_3}{\partial r} + 2\frac{u_3}{r}\right),$$

$$\sigma_{\varphi\varphi8}(r,t) = \sigma_{\varphi\varphi6}(r,t) = -\frac{1}{2}\sigma_{\varphi\varphi6} + \frac{3K}{2}\left[\frac{\partial u_3(r,\varphi(r))}{\partial r} + 2\frac{u_3(r,\varphi(r))}{r}\right].$$

In the formulae for the stress components in regions VII and VIII we obviously have to assume that $p(t) \equiv 0$.

20. Spherical waves and radial cylindrical waves in a homogeneous elastic/viscoplastic medium

On account of the analogous procedures employed when solving problems of the propagation of spherical waves and of radial cylindrical waves in a medium which is sensitive to strain rate, we shall treat at this point the solutions for both types of wave simultaneously.

Let us consider an unbounded elastic/viscoplastic medium with either a spherical or a cylindrical cavity of initial radius equal to r_0. A uniformly distributed time-varying pressure $p(t)$ is applied to the surface of the cavity. We take the boundary condition in the form (17.1). If in the case of spherical or cylindrical symmetry we take into account the conditions given in (16.1) and (16.2), then the constitutive equations (3.5) for a strain rate sensitive medium take the forms

$$\frac{\partial v}{\partial r} - \frac{v}{r} = \frac{1}{2\mu}(\dot{\sigma}_{rr} - \dot{\sigma}_{\varphi\varphi}) + \gamma\left\langle\Phi\left[\frac{\sqrt{J_2}}{\varkappa} - 1\right]\right\rangle\frac{\sigma_{rr} - \sigma_{\varphi\varphi}}{\sqrt{J_2}},$$

(20.1)
$$-n_1\left(\frac{\partial v}{\partial r} + \frac{v}{r}\right) = n_1\frac{1}{2\mu}(2\dot{\sigma}_{zz} - \dot{\sigma}_{rr} - \dot{\sigma}_{\varphi\varphi}) + n_1\gamma\left\langle\Phi\left[\frac{\sqrt{J_2}}{\varkappa} - 1\right]\right\rangle\frac{2\sigma_{zz} - \sigma_{rr} - \sigma_{\varphi\varphi}}{\sqrt{J_2}},$$

$$\frac{\partial v}{\partial r} + n_0\frac{v}{r} = \frac{1}{3K}(\dot{\sigma}_{rr} + n_0\dot{\sigma}_{\varphi\varphi} + n_1\dot{\sigma}_{zz}),$$

where the small strains relations have been used:

$$\frac{\partial \varepsilon_{rr}}{\partial t} = \frac{\partial v}{\partial r}, \qquad \frac{\partial \varepsilon_{\varphi\varphi}}{\partial t} = \frac{v}{r},$$

and the symbols n_0 and n_1 have been introduced into the notation. We put $n_0 = 2, n_1 = 0$ for the case of spherical waves and $n_0 = 1, n_1 = 1$ for the case of radial cylindrical waves.

The second invariant of the deviatoric stress tensor, occurring in the relaxation function, takes the following form:

(20.2) $$J_2 = \frac{1}{3}\left[(\sigma_{rr}^2 + n_1\sigma_{zz}^2 + \sigma_{\varphi\varphi}^2) - (n_0\sigma_{rr}\sigma_{\varphi\varphi} + n_1\sigma_{\varphi\varphi}\sigma_{zz} + n_1\sigma_{zz}\sigma_{rr})\right].$$

The above equations, together with equation of motion (16.9), constitute the complete system of equations for the problem. This system is a system of first-order hyperbolic differential equations in the velocity v and the stress and strain tensor components. Taking into account (9.10) we deduce that the characteristics of the system take the form of straight lines with equations

(20.3) $$r = \text{const}, \qquad r = r_0 \mp at + \text{const}, \qquad \text{where} \qquad a = \sqrt{\frac{4\mu + 3K}{3\varrho}}.$$

The following relations are valid along the characteristics (on the basis of (9.17)):

$$n_0\left\{\pm\frac{3a}{r}(\sigma_{rr} - \sigma_{\varphi\varphi}) + 2\mu\gamma\left\langle\Phi\left[\frac{\sqrt{J_2}}{\varkappa} - 1\right]\right\rangle\frac{(1+n_1)\sigma_{rr} - \sigma_{\varphi\varphi} - n_1\sigma_{zz}}{\sqrt{J_2}} + \frac{v}{r}(2\mu - 3K)\right\}dr$$

$$-(4\mu + 3K)\,dv \pm 3a\,d\sigma_{rr} = 0, \quad \text{for} \quad r = r_0 \pm at + \text{const},$$

(20.4)
$$3K\left\{\frac{\sigma_{rr} - \sigma_{\varphi\varphi}}{\sqrt{J_2}}\gamma\left\langle\Phi\left[\frac{\sqrt{J_2}}{\varkappa} - 1\right]\right\rangle + (1+n_0)\frac{v}{r}\right\}dt + \left(\frac{3K}{2\mu} - 1\right)d\sigma_{rr}$$
$$-\left(\frac{3K}{2\mu} + n_0\right)d\sigma_{\varphi\varphi} - n_1\,d\sigma_{zz} = 0,$$

$$d\varepsilon_{\varphi\varphi} = \frac{v}{r}\,dt,$$

$$2\mu d\varepsilon_{rr} - d\sigma_{rr} + d\sigma_{\varphi\varphi} - 2\mu\left\{\frac{v}{r} + \frac{\sigma_{rr} - \sigma_{\varphi\varphi}}{\sqrt{J_2}}\gamma\left\langle\Phi\left[\frac{\sqrt{J_2}}{\varkappa} - 1\right]\right\rangle\right\}dt = 0,$$

$$d\varepsilon_{rr} - \frac{1}{3K}(d\sigma_{rr} + n_0\,d\sigma_{\varphi\varphi} + n_1\,d\sigma_{zz}) + \frac{v}{r}dt = 0 \qquad \text{for} \qquad r = \text{const}.$$

In the general case of the constitutive equations (3.5) the solution of the problem of propagation of spherical waves or cylindrical radial waves is not known in an exact form. The solution on the fronts of strong discontinuity waves reduces to integral equations while in the regions of viscoplastic strain the solution is constructed numerically using the characteristic net method or using the method of successive approximations (see section 9).

In the case of the boundary condition in the form shown in (17.1) (and for zero initial conditions) a wave of strong discontinuity which coincides with the characteristic of equation $r = r_0 + at$ will propagate into the undisturbed region $r > r_0$. The conditions of dynamic (7.18) and kinematic (8.9) continuity must hold on the front of this wave. The conditions assume the forms

(20.5) $$\sigma_{rr} = -\varrho a v, \qquad v = -a\varepsilon_{rr}.$$

Making use of the fact that the strain rate is infinite at the front of a strong discontinuity wave, we obtain from (20.1) and (20.5) the relation between the stress tensor components at the front of the wave $r = r_0 + at$. In the case of the spherical wave we have

(20.6) $$\sigma_{\varphi\varphi} = \sigma_{rr}\left(1 - \frac{2\mu}{\varrho a^2}\right),$$

while for the radial cylindrical wave we find that $\sigma_{zz} = \sigma_{\varphi\varphi}$. In order to obtain the solution on the front of the strong discontinuity wave $r = r_0 + at$ we make use of the relations on the positive characteristic $(20.4)_1$ and of (20.5) and (20.6). After some algebra and expressing all the quantities in terms of σ_{rr} we obtain the equation

$$\frac{\partial \sigma_{rr}}{\partial r} = -\frac{n_0 \sigma_{rr}}{2r} - \frac{n_0 \mu}{(3)^{1/n_0} a} \gamma \Phi \left[\frac{2\mu \sigma_{rr}}{\sqrt{3} \varrho a^2 \varkappa\left(\int_0^t \sigma_{ij} d\varepsilon_{ij}^p \right)} - 1 \right], \tag{20.7}$$

where the yield function takes the form (3.1). We shall be able to solve (20.7) provided the hardening parameter $\varkappa$, determined by formula (3.1) on the front of wave $r = r_0 + at$, depends only on variables σ_{rr} and r. In the case of spherical or cylindrical symmetry we have, using (3.1),

$$W^p \overset{\text{def}}{=} \int_0^t \sigma_{ij} d\varepsilon_{ij}^p = \int_0^t \sigma_{rr} d\varepsilon_{rr}^p + 2 \int_0^t \sigma_{\varphi\varphi} d\varepsilon_{\varphi\varphi}^p. \tag{20.8}$$

The second term on the right-hand side of (20.8) is equal to zero since

$$[\varepsilon_{\varphi\varphi}] = \left[\frac{u}{r} \right] = 0.$$

Dividing the strain tensor component into elastic $\overset{e}{\epsilon}_{rr}$ and plastic $\overset{p}{\epsilon}_{rr}$ parts and making use of conditions (20.5), we find that

$$\overset{p}{\varepsilon}_{rr} = \varepsilon_{rr} - \overset{e}{\varepsilon}_{rr} = \frac{\sigma_{rr}}{\varrho a^2} - \overset{e}{\varepsilon}_{rr}.$$

From Hooke's law and (20.6) we obtain the following expression for $\overset{e}{\epsilon}_{rr}$:

$$\overset{e}{\varepsilon}_{rr} = \frac{1}{\varrho a^2} \sigma_{rr},$$

and hence directly that $W^p = 0$.

Since the material strain energy is equal to zero on the front of the strong discontinuity wave, the problem for a material with hardening reduces to that for an ideally plastic medium. The hardening parameter $\varkappa$ is equivalent to the shear yield limit k.

Integrating both sides of (20.7) and making use of the initial condition,

$$\sigma_{rr}(r_0, 0) = -p_0, \tag{20.9}$$

we obtain the non-linear Volterra equation of the second kind,

$$\sigma_{rr} = p_0 - \int_{r_0}^{r} \Psi[\xi, \sigma_{rr}(\xi)] \, d\xi, \tag{20.10}$$

where function $\Psi[r, \sigma_{rr}(r)]$ takes the form

$$\Psi[r, \sigma_{rr}(r)] = \frac{\sigma_{rr}}{2n_0 r} + \frac{n_0 \mu \gamma}{(3)^{\frac{1}{n_0}} a} \Phi \left[\frac{2\mu \sigma_{rr}}{\sqrt{3} \varrho a^2 k} - 1 \right].$$

Equation (20.10) is solved by the method of successive approximations. It can be proved that the integrand $\Psi[r, \sigma_{rr}(r)]$ is bound and satisfies the Lipschitz condition. Thus the process of successive iterations is convergent.

The recurrence formula for the solution of (20.10) assumes the form

$$\sigma_{rr} = \lim_{n\to\infty} \sigma_{rr}^{(n)}, \tag{20.11}$$

where

$$\sigma_{rr}^{(n+1)} = p_0 - \int_{r_0}^{r} \Psi[\xi, \sigma_{rr}^{(n)}(\xi)]\, d\xi,$$

and

$$v^{(n)} = -\frac{1}{\varrho a}\sigma_{rr}^{(n)}, \qquad \varepsilon_{rr}^{(n)} = \frac{1}{\varrho a^2}\sigma_{rr}^{(n)}, \qquad \varepsilon_{\varphi\varphi} = 0,$$

$$\sigma_{\varphi\varphi}^{(n)} = \sigma_{\theta\theta}^{(n)} = \left(1 - \frac{2\mu}{\varrho a^2}\right)\sigma_{rr}^{(n)} \quad \text{for spherical waves,}$$

$$\sigma_{\varphi\varphi}^{(n)} = \sigma_{zz}^{(n)} = \left(1 - \frac{2\mu}{\varrho a^2}\right)\sigma_{rr}^{(n)} \quad \text{for radial cylindrical waves.}$$

The above solutions, in the case of spherical waves, are valid for $r < r^*$ (Fig. 67), where r^* is determined from condition $J_2(r^*) = k^2$. For $r \geqslant r^*$ the solution on the strong discontinuity

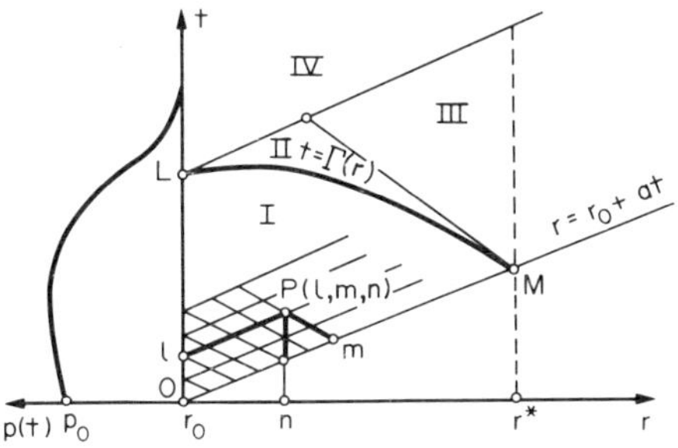

Fig. 67

wave $r = r_0 + at$ can be obtained in an exact form (for spherical waves), namely:

$$\sigma_{rr} = -\frac{\sqrt{3}\,\varrho a^2 k}{2\mu}\frac{r^*}{r}, \qquad \sigma_{\varphi\varphi} = \sigma_{\theta\theta} = -\left(1 - \frac{2\mu}{\varrho a^2}\right)\frac{\sqrt{3}\,\varrho a^2 k}{2\mu}\frac{r^*}{r},$$

$$v = \frac{\sqrt{3}\,ak}{2\mu}\frac{r^*}{r}, \qquad \varepsilon_{rr} = -\frac{\sqrt{3}\,k}{2\mu}\frac{r^*}{r}, \qquad \varepsilon_{\varphi\varphi} = 0. \tag{20.12}$$

In the case of radial cylindrical waves the solution of (20.10) is valid in the interval $[r_0, +\infty)$. In this connection the illustration of the solution in the phase plane (r, t) changes a little (Fig. 67). Namely, point M on the characteristic $r = r_0 + at$ tends to infinity, therefore the region of viscoplastic deformation I has a form similar to the case of a plane wave (see section 15, Fig. 53). The stress σ_{rr} on wave $r = r_0 + at$ decreases monotonically, reaching, at infinity, the value $\sigma_{rr} = \sigma^*_{rr} = -\dfrac{\sqrt{3}\,\varrho a^2 k}{2\mu}$.

In the case of the constitutive equations (3.13) when function $\gamma\Phi\left(\dfrac{\sigma_i}{f_1(\varepsilon_i)} - 1\right)$ takes the form of (3.30) or the equivalent one:

$$\gamma\Phi = \alpha_n(\varepsilon_i)\left(\frac{\sigma_i}{f_1(\varepsilon_i)} - 1\right) + \beta_n(\varepsilon_i), \tag{20.13}$$

where $\alpha_n(\epsilon_r)$ and $\beta_n(\epsilon_r)$ are functions of ϵ_i determined from the results of the experiments, we can give the following estimation of the error in (20.10) for the case of spherical waves. Assuming

$$\begin{gathered} |\alpha_n(\varepsilon_i^*)| \leqslant \alpha_{0n}, \quad |\alpha_n'(\varepsilon_i^*)| \leqslant \alpha_{0n}', \quad |\beta_n(\varepsilon_i^*)| \leqslant \beta_{0n}, \\ |\beta_n'(\varepsilon_i^*)| \leqslant \beta_{0n}', \quad f_1'(\varepsilon_i^*) \leqslant f_0', \end{gathered} \tag{20.14}$$

where $\varepsilon_i^* = \dfrac{2}{\sqrt{3}}\dfrac{\sigma_{rr}}{\varrho a^2}$, we can write [57]:

$$\overline{\Psi[r, \sigma_{rr}(r)]} \leqslant \frac{p_0}{r_0} + \frac{2\mu}{\sqrt{3}a}\left\{\alpha_{0n}\left[\frac{\frac{2}{\sqrt{3}}\frac{\mu}{\varrho a^2}p_0}{k} - 1\right] + \beta_{0n}\right\} = M,$$

$$\overline{\left|\frac{\partial\Psi}{\partial\sigma_{rr}}\right|} = \frac{1}{r_0} + \frac{2\mu}{\sqrt{3}a}\left\{\frac{2}{\sqrt{3}}\frac{1}{\varrho a^2}\alpha_{0n}'\left[\frac{\frac{2}{\sqrt{3}}\frac{\mu}{\varrho a^2}p_0}{k} - 1\right] + \alpha_{0n}\left[\frac{2}{\sqrt{3}}\frac{\mu}{\varrho a^2 k} + \left(\frac{2}{\sqrt{3}}\frac{\mu}{(\varrho a^2)^2}p_0 f_0'\right)\frac{1}{k^2}\right\} = R. \tag{20.15}$$

Then we have

$$|\sigma_{rr}| \leqslant p_0 + \frac{M}{R}(e^{Rh} - 1), \tag{20.16}$$

where h is equal to the smaller of the two following quantities:

$$\frac{r^* - r_0}{2}, \quad \left|\frac{\sqrt{3}\varrho a^2 k - 2\mu p_0}{4\mu M}\right|.$$

The solution in the region of viscoplastic deformation (region I) is constructed numerically by the method of characteristic nets, making use of the relations along the characteristics (20.4). The differentials occurring in these formulae are replaced by finite differences. The recurrence formulae for the approximate values of the parameters of the solution at point $P(l, m, n)$ (Fig. 67) are as follows [11], [57]:

$$3K\left\{\gamma\langle\Phi\left[\frac{\sqrt{J_2}}{\varkappa}-1\right]\rangle\frac{\sigma_{rr}-\sigma_{\varphi\varphi}}{\sqrt{J_2}}+(1+n_0)\frac{v}{r}\right\}_{(l-1,\,m-1,\,n)}\Delta t$$

$$+\left(\frac{3K}{2\mu}-1\right)[\sigma_{rr}(l,m,n)-\sigma_{rr}(l-1,m-1,n)]-\left(\frac{3K}{2\mu}+n_0\right)[\sigma_{\varphi\varphi}(l,m,n)$$

$$-\sigma_{\varphi\varphi}(l-1,m-1,n)]-n_1[\sigma_{zz}(l,m,n)-\sigma_{zz}(l-1,m-1,n)]=0,$$

$$n_0\left\{\frac{3a}{r}(\sigma_{rr}-\sigma_{\varphi\varphi})+2\mu\gamma\langle\Phi\left[\frac{\sqrt{J_2}}{\varkappa}-1\right]\rangle\frac{(1+n_1)\sigma_{rr}-\sigma_{\varphi\varphi}-n_1\sigma_{zz}}{\sqrt{J_2}}\right.$$

$$\left.+\frac{v}{r}(2\mu-3K)\right\}_{(l,m-1,n-1)}\Delta r-(4\mu+3K)[v(l,m,n)-v(l,m-1,n-1)]$$

(20.17)
$$+3a[\sigma_{rr}(l,m,n)-\sigma_{rr}(l,m-1,n-1)]=0,$$

$$n_0\left\{-\frac{3a}{r}(\sigma_{rr}-\sigma_{\varphi\varphi})+2\mu\gamma\langle\Phi\left[\frac{\sqrt{J_2}}{\varkappa}-1\right]\rangle\frac{(1+n_1)\sigma_{rr}-\sigma_{\varphi\varphi}-n_1\sigma_{zz}}{\sqrt{J_2}}\right.$$

$$\left.+\frac{v}{r}(2\mu-3K)\right\}_{(l-1,\,n,\,n+1)}\Delta r-(4\mu+3K)[v(l,m,n)-v(l-1,m,n+1)]$$

$$-3a[\sigma_{rr}(l,m,n)-\sigma_{rr}(l-1,m,n+1)]=0,$$

$$\varepsilon_{\varphi\varphi}(l,m,n)-\varepsilon_{\varphi\varphi}(l-1,m-1,n)-\frac{1}{r}v(l-1,m-1,n)\Delta t=0,$$

$$2\mu[\varepsilon_{rr}(l,m,n)-\varepsilon_{rr}(l-1,m-1,n)]-[\sigma_{rr}(l,m,n)-\sigma_{rr}(l-1,m-1,n)]$$

$$+[\sigma_{\varphi\varphi}(l,m,n)-\sigma_{\varphi\varphi}(l-1,m-1,n)]-2\mu\left\{\frac{v}{r}+\gamma\langle\Phi\left[\frac{\sqrt{J_2}}{\varkappa}-1\right]\rangle\frac{\sigma_{rr}-\sigma_{\varphi\varphi}}{\sqrt{J_2}}\right\}_{(l-1,m-1,n}\Delta t=0,$$

$$[\varepsilon_{rr}(l,m,n)-\varepsilon_{rr}(l-1,m-1,n)]-\frac{1}{3K}\{[\sigma_{rr}(l,m,n)-\sigma_{rr}(l-1,m-1,n)]$$

$$+n_0[\sigma_{\varphi\varphi}(l,m,n)-\sigma_{\varphi\varphi}(l-1,m-1,n)]+n_1[\sigma_{zz}(l,m,n)-\sigma_{zz}(l-1,m-1,n)]\}$$

$$+\frac{1}{r}v(l-1,m-1,n)\Delta t=0.$$

Symbols l, m, n denote the position of point $P(l, m, n)$ at which the characteristics intersect: lth positive, mth negative, and nth of equation $r = \text{const}$.

The required discrete values of the stress tensor components, strain tensor components and the particle velocity v are determined in an approximate way from the above system of difference equations. In the case when the effect of material hardening is disregarded in the constitutive equations (3.15), the above system of equations is reduced either to three equations in the case of spherical waves or to a system of four equations for radial cylindrical waves since the strain tensor components ε_{rr} and $\varepsilon_{\varphi\varphi}$ do not appear.

The region of viscoplastic strain I is bounded by a curve of equation $t = \Gamma(r)$, constituting an "unloading wave". This curve limits the validity of the solutions of the system of (20.17). Curve $\Gamma(r)$ is determined from the condition $F = 0$, where F denotes the yield function of material, defined, in general case, by formula (3.1). For example, in the case of constitutive equations (3.13) the condition for curve $\Gamma(r)$ takes the form

$$\sigma_i[r, \Gamma(r)] = f_1\{\varepsilon_i[r, \Gamma(r)]\}. \tag{20.18}$$

Here it has been also assumed that the negative characteristic, departing from point M (Fig. 67), does not intersect curve $\Gamma(r)$.

In region II, which is bounded by curve $\Gamma(r)$ and the following characteristics, the positive one departing from point L and the negative departing from point M (Fig. 67) (in the case of radial cylindrical waves point M is at infinity, therefore region II is no longer bounded, it links with region III), the motion of the medium is governed by the following system of equations:

$$\frac{\partial \sigma_{rr}}{\partial r} + \frac{n_0}{r}(\sigma_{rr} - \sigma_{\varphi\varphi}) = \varrho \frac{\partial v}{\partial t},$$

$$\varepsilon_i = \frac{1}{2\mu}(\sigma_i - \sigma_{io}) + \varepsilon_{io}, \tag{20.19}$$

$$3K(\varepsilon_{rr} + n_0 \varepsilon_{\varphi\varphi}) = \sigma_{rr} + n_0 \sigma_{\varphi\varphi} + n_1 \sigma_{zz},$$

where σ_{io} and ϵ_{io} denote the stress intensity and the strain intensity respectively on curve $\Gamma(r)$. In the case of spherical waves the above system of equations can be reduced to a second-order equation for the displacement $u(r, t)$:

$$\frac{\partial^2 u}{\partial t^2} - a^2\left(\frac{\partial^2 u}{\partial r^2} + \frac{2}{r}\frac{\partial u}{\partial r} - 2\frac{u}{r}\right) = \Psi(r), \tag{20.20}$$

where

$$\Psi(r) = \frac{1}{\varrho}\left[\frac{2}{\sqrt{3}}\mu\frac{\partial}{\partial t}\left(\frac{\sigma_{io}}{2\mu} - \varepsilon_{io}\right) + \frac{2\sqrt{3}}{r}\mu\left(\frac{\sigma_{io}}{2\mu} - \varepsilon_{io}\right)\right].$$

In the case of radial cylindrical waves the system of (20.19) cannot be reduced to a single differential equation of the second order as it can be for spherical waves.

In region II we have to solve the Cauchy problem for (20.20) in the case of spherical waves (or the system of (20.19) in the case of radial cylindrical waves), making use of the data on curve $\Gamma(r)$ determined from the solution in region I. Equation (20.20) is solved similarly as in region I by the method of finite differences along the net of characteristics. Here we utilize (20.17) assuming that $\gamma\Phi \equiv 0$.

In the case of spherical waves, in region III, which is bounded by characteristics $r = r_0 + at$ and by the negative characteristic departing from point M, we have to solve the Darboux problem for (20.20). In the case of radial cylindrical waves region III disappears. For $r \geqslant r^*$ in the case of spherical waves function $\Psi(r)$ occurring on the right-hand side of the equation is identically equal to zero. This is obvious from the physical standpoint since the medium for $r \geqslant r^*$ is not subjected to viscoplastic deformation.

Finally, in region IV we have to solve the Picard problem for the system of (20.19) with the following known initial conditions: boundary condition for $r = r_0$ and the displacements of the points on the positive characteristic departing from point L (the displacements are known from the solution in region II). In the case of spherical waves an exact form for the solution can be found in region IV. The above solutions for the case of spherical waves were given with full particulars in [57] and [11]. The detailed analysis of the numerical solutions of the posed problem is performed there. The effect of material hardening on the value of the stress and strain fields is also examined. It is asserted (for the numerical data for mild steel) that the changes in the stresses σ_{rr} and $\sigma_{\varphi\varphi}$ between the case when hardening is taken into account and the case when it is disregarded are not too high (of the order 15–20%), while they are considerable in the strains ε_{rr} and $\varepsilon_{\varphi\varphi}$ (of the order 40%).

The problem of the reflection of spherical stress wave from an undeformable obstacle, concentric with the spherical cavity, was also considered in [57]. The pressure on the boundary of the cavity of radius $r = r_0$ is chosen so that the wave reflected from the undeformable obstacle is also a plastic wave. The solution for the reflected wave is constructed in the same way as for the incident wave. The solutions in the remaining regions do not contribute a qualitatively new picture. However, it was discovered that the effect of material hardening in the neighbourhood of the undeformable obstacle is even more pronounced than it is in the neighbourhood of the cavity boundary of radius r_0.

The problem of the propagation of spherical waves in a non-homogeneous elastic/viscoplastic medium was solved in [112]. Freudenthal's equations were taken as the starting point: these equations can be obtained as a particular case of the constitutive equations (3.25) by setting $\Phi(F) = F$. It is assumed in these equations that the material constants vary with radius r. The problem of the generation of a weak discontinuity wave and a strong discontinuity wave is considered. The solutions in the regions of viscoplastic deformation are constructed numerically by means of the method of characteristic nets. At the front of a strong discontinuity wave the solution is reduced to that of an integral equation. Making further simplifications and assuming, as a starting point, the constitutive equations of Hohenemser and Prager (equations (3.25) in which the elastic strain is neglected as small compared with the inelastic strain, as well as a linear form for the relaxation function), the solution of the problem of the propagation of a strong discontinuity stress wave was obtained in an exact form in [167]. The form of the elastic unloading wave is also determined for a certain particular case.

21. Cylindrical shear waves

Now we consider the case when a shear stress $p(t)$ is uniformly distributed on the surface of a cylindrical cavity of radius r_0 in an unbounded elastic/viscoplastic space. In this case the constitutive equations (3.5), because of (16.10)–(16.12), assume the following form:

$$\frac{\partial v}{\partial r} - \frac{v}{r} = \frac{1}{\mu}\frac{\partial \tau_{r\varphi}}{\partial \tau} + 2\gamma \left\langle \Phi\left(\frac{\tau_{r\varphi}}{\varkappa} - 1\right)\right\rangle, \tag{21.1}$$

where $v = \partial u/\partial t$.

Constitutive equations (21.1), together with the equation of motion (16.3) and the relation for small strain

$$\frac{\partial v}{\partial r} - \frac{v}{r} = 2\frac{\partial \varepsilon_{r\varphi}}{\partial t}, \tag{21.2}$$

constitute the complete system of equations for the considered problem. The system of equations is of the hyperbolic type with the following families of characteristics:

$$r = r_0 \mp at + \text{const}, \qquad r = \text{const}, \tag{21.3}$$

where $a = \sqrt{\mu/\varrho}$.

The following relations,

$$(21.4)\qquad \begin{aligned} &dv \mp \frac{2}{\mu\gamma}\, a\tau_{r\varphi}\, dr - \left[2\gamma\langle\Phi\left(\frac{\tau_{r\varphi}}{\varkappa}-1\right)\rangle + \frac{v}{r}\right] dr \mp \frac{a}{\mu}\, d\tau_{r\varphi} = 0,\\ &\frac{1}{2\mu}\, d\tau_{\varphi r} - d\varepsilon_{r\varphi} + \left[\frac{v}{r} + 2\gamma\langle\Phi\left(\frac{\tau_{r\varphi}}{\varkappa}-1\right)\rangle\right] dt = 0, \end{aligned}$$

are valid along the characteristics (21.3).

If the stress $p(t)$ is suddenly applied at instant $t = 0$, then the solution, at the front of the strong discontinuity wave of equation $r = r_0 + at$, takes the form

$$(21.5)\qquad \tau_{r\varphi} = p_0 - \int_{r_0}^{r} \Psi[\xi, \tau_{r\varphi}(\xi)]\, d\xi,$$

where

$$(21.6)\qquad \Psi[r, \tau_{r\varphi}(r)] = \frac{1}{2r}\, \tau_{r\varphi} + \varrho a\, \gamma\langle\Phi\left(\frac{\tau_{r\varphi}}{k}-1\right)\rangle.$$

The above solution has been obtained by making use of $(21.4)_1$ on the positive characteristic and of the conditions of dynamic continuity and kinematic continuity across the front of the strong discontinuity

$$(21.7)\qquad \tau_{r\varphi} = -\varrho a v, \qquad v = -a\varepsilon_{r\varphi}$$

(when we assume homogeneous, zero initial conditions).

Similar to the cases of the spherical waves or the radial cylindrical waves, (21.5) is solved by the successive approximation method with a known error. Equation (21.5) is valid only for $r \leqslant r^*$ where r^* is determined from the condition $\tau_{r\varphi}(r^*) = k$. For $r \geqslant r^*$ at the front of the strong discontinuity wave $r = r_0 + at$ the solution can be obtained in an exact form, namely:

$$(21.8)\qquad \tau_{r\varphi} = k\left(\frac{r^*}{r}\right)^{1/2}.$$

The remaining parameters of the solution on wave $r = r_0 + at$ are determined by means of relations (21.7).

The solution in the region of viscoplastic deformation is constructed by means of the method of the characteristic nets by utilizing relations (21.4) written down in a form suitable for finite increments.

The solution of the problem of the reflection of a wave from a concentric obstacle does not present much difficulty. It can be readily obtained in a similar way to the solutions of the reflection of plane waves in a half-space or of spherical waves in an elastic/viscoplastic medium.

CHAPTER V

PLASTIC LONGITUDINAL–TRANSVERSE WAVES

In this chapter we shall discuss first of all the solution of problems involving the propagation of simple waves. We shall analyse the case of a two-parameter loading of the boundary of the body under consideration. We shall consider in turn bodies whose properties are governed by the equations of the theory of plastic flow as exemplified by the constitutive equations due to Grigorian for soil dynamics and by the equations of the bilinear theory of plasticity. Following this we shall present solutions of problems concerning the propagation of longitudinal–transverse waves in homogeneous elastic/viscoplastic bodies (both plane waves and radial cylindrical waves).

At the present time many solutions exist to problems of wave propagation involving complex stress states (for a single spatial variable and for two-parameter loading). Early papers in this field were confined to the solutions of self-similar problems [13]–[15], [127], [145], [146]. Thus the solutions were restricted to that class of the boundary conditions which are such that it was possible to make the stress state, the strain state, and the particle velocities of the medium dependent upon one independent variable only. This then reduced the system of partial differential equations describing the motion of the medium to ordinary differential equations. On account of the character of the external loading assumed in the papers quoted, the problem of the generation of plastic wave fronts, which appear due to the interaction of the longitudinal and transverse waves did not arise. Problems of unloading wave generation also did not occur. These non-self-similar problems were discussed in references [59]–[62], [163], [164] where the problem of the propagation of longitudinal–transverse waves in an elastic/viscoplastic medium was considered for arbitrary, time-dependent external loads.

22. Simple waves in an elastic–plastic half-space

We shall examine the wave motion in an elastic–viscoplastic half-space $x_1 > 0$. The normal and tangential stresses (Fig. 68),

$$\sigma_{11} = -p_0, \qquad \sigma_{12} = -q_0, \tag{22.1}$$

constant in time, appear suddenly at instant $t = 0$ on the bounding plane. In the case of the displacement boundary conditions the solution procedure is identical. The problem is solved using the constitutive equations (2.30) supplemented by the relations of small strain theory [14], [40], [92], [155], [2], [45].

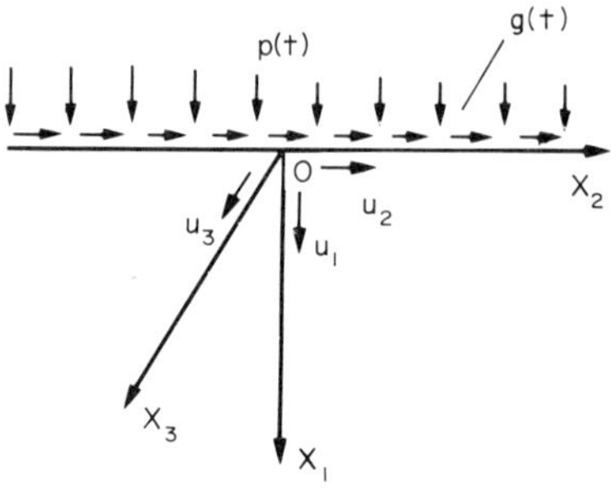

Fig. 68

In the case of small strains, body forces being neglected, the equations of dynamic equilibrium (5.5) take the following form:

(22.2)
$$\varrho \frac{\partial v_i}{\partial t} = \sigma_{ij,j}.$$

Shear stress σ_{13}, σ_{23} and the strain components in the direction of the x_3-axis vanish in the plane problem under consideration. The remaining stress tensor components and the displacement vector components do not depend on coordinate x_3. Since the loading on the boundary of the half-space (22.1) is independent of x_2, consequently the remaining quantities are functions of x_1 and t only. As a result we obtain

(22.3)
$$\begin{gathered} \varepsilon_{11} = u_{1,1} = \varepsilon, \quad \varepsilon_{12} = u_{2,1} = 2\gamma, \quad \varepsilon_{22} = \varepsilon_{33} = \varepsilon_{13} = \varepsilon_{23} = 0, \\ \sigma_{23} = \sigma_{13} = 0, \quad \sigma_{12} = \tau, \quad \sigma_{11} = \sigma_1, \\ \dot{u}_1 = v_1, \quad \dot{u}_2 = v_2, \quad \dot{u}_3 = u_3 = 0. \end{gathered}$$

Using conditions (22.3) we can reduce the equations of dynamic equilibrium to the forms (in the following we shall write $x_1 = x$):

(22.4)
$$\varrho \frac{\partial v_1}{\partial t} = \frac{\partial \sigma_1}{\partial x}, \quad \varrho \frac{\partial v_2}{\partial t} = \frac{\partial \tau}{\partial x},$$

and for small strain we have

(22.5)
$$\frac{\partial v_1}{\partial x} = \frac{\partial \varepsilon}{\partial t}, \quad \frac{\partial v_2}{\partial x} = 2\frac{\partial \gamma}{\partial t}.$$

The yield condition, in the case considered, takes the form

(22.6)
$$F = \frac{1}{3}(\sigma_1 - \sigma_2)^2 + \tau^2 = k^2,$$

where, in the case of the Huber–Mises condition, we have $\sigma_2 = \sigma_{22} = \sigma_{33}$, k is the yield limit.

From the constitutive equations (2.30), taking into account (22.3), (22.5), and (22.6), we obtain the following system of first-order differential equations:

$$\frac{\partial v_1}{\partial x} = \frac{1}{E}\frac{\partial \sigma_1}{\partial t} - \frac{2\nu}{E}\frac{\partial \sigma_2}{\partial t} + s\frac{\partial \Lambda}{\partial t},$$

(22.7)
$$-\frac{2\nu}{E}\frac{\partial \sigma_1}{\partial t} + \frac{2(1-\nu)}{E}\frac{\partial \sigma_2}{\partial t} - s\frac{\partial \Lambda}{\partial t} = 0,$$

$$\frac{\partial v_2}{\partial x} = \frac{1}{\mu}\frac{\partial \tau}{\partial t} + 2\tau\frac{\partial \Lambda}{\partial t},$$

where the following symbols have been introduced:

(22.8)
$$s = \frac{2}{3}(\sigma_1 - \sigma_2), \qquad \lambda = \frac{\partial \Lambda}{\partial t}.$$

Quantity $\partial \Lambda/\partial t$, occurring in (22.7), can be expressed by the function F in the following form:

(22.9)
$$\frac{\partial \Lambda}{\partial t} = \frac{3}{4k^2 E}\alpha(k)\frac{\partial F}{\partial \sigma_{kl}}\frac{\partial \sigma_{kl}}{\partial t},$$

where $\alpha(k)$ is a parameter describing the material hardening. This function can be determined from the material characteristic in the case of a uniaxial stress state. If $E_p(\sigma)$ denotes the slope of the σ–ϵ curve in the uniaxial stress state in terms of stress function σ, we obtain

(22.10)
$$\alpha(k) = \frac{E}{E_p(3k)} - 1.$$

In the limiting case when $E_p = E$ (region of elastic strain), then $\alpha = 0$; if $E_p = 0$ (the case of ideal plasticity) then $\alpha = \infty$. If we assume that E_p is a monotonically decreasing function of k then $\alpha(k)$ is a monotonically increasing function of k.

The system of (22.4), (22.7), and (22.9) constitutes the complete system of equations for the problem considered. This system can be represented in matrix form, namely

(22.11)
$$\mathbf{A}\mathbf{u}_{,x} + \mathbf{B}\mathbf{u}_{,t} = 0,$$

where $\mathbf{u}$ is defined as follows:

(22.12)
$$\mathbf{u} = \begin{vmatrix} v_1 \\ v_2 \\ \sigma_1 \\ \sigma_2 \\ \tau \\ \Lambda \end{vmatrix}$$

while the matrices **A** and **B** take the forms:

$$
\text{(22.13)} \qquad \mathbf{A} = \begin{bmatrix} 0 & 0 & -1 & 0 & 0 & 0 \\ 0 & 0 & 0 & 0 & -1 & 0 \\ -1 & 0 & 0 & 0 & 0 & 0 \\ 0 & 0 & 0 & 0 & 0 & 0 \\ 0 & -1 & 0 & 0 & 0 & 0 \\ 0 & 0 & 0 & 0 & 0 & 0 \end{bmatrix}, \quad \mathbf{B} = \begin{bmatrix} \varrho & 0 & 0 & 0 & 0 & 0 \\ 0 & \varrho & 0 & 0 & 0 & 0 \\ 0 & 0 & \dfrac{1}{E} & -\dfrac{2\nu}{E} & 0 & s \\ 0 & 0 & -\dfrac{2\nu}{E} & \dfrac{2(1-\nu)}{E} & 0 & -s \\ 0 & 0 & 0 & 0 & \dfrac{1}{\mu} & 2\tau \\ 0 & 0 & s & -s & 2\tau & \dfrac{4k^2E}{3\alpha(k)} \end{bmatrix}
$$

Matrices **A** and **B** are symmetric.

The characteristic curves of (22.11) are the solutions of equation $|\mathbf{A}-a\mathbf{B}| = 0$ (see section 9.2). Expanding the determinant $|\mathbf{A}-a\mathbf{B}|$ and equating it to zero we obtain the following equation:

$$
\text{(22.14)} \qquad a^2 D(a) = 0,
$$

where

$$
\text{(22.15)} \qquad D(a) = s^2\left(\frac{a^2}{a_2^2}-1\right)\left(\frac{3a^2}{\beta a_2^2}-1\right) + \frac{4\tau^2 a^2}{\beta a_2^2}\left(\frac{a^2}{a_2^2}-\frac{a_1^2}{a_2^2}\right) + \frac{12k^2}{3\alpha(k)(\beta+1)}\left(\frac{a^2}{a_1^2}-1\right)\left(\frac{a^2}{a_2^2}-\frac{a_1^2}{a_2^2}\right).
$$

Here a_1 is the speed of longitudinal elastic waves, a_2 is the speed of transverse elastic waves, and β is a parameter given by $\beta = \dfrac{2(1+\nu)}{1-2\nu}$. Since Poisson ratio varies in range (0, 1/2), the parameter β varies in the range $(2, \infty)$.

A trivial solution of (22.14) is $a = 0$. Analysing (22.15) we can assert that $D(0) \geqslant 0$, $D(a_1) \geqslant 0$, and $D(a_2) \leqslant 0$. Equation (22.15) is of the fourth degree with respect to a. If its roots are denoted by $\pm a_s$, $\pm a_f$, then we can write that

$$
\text{(22.16)} \qquad 0 \leqslant a_s \leqslant a_2 \leqslant a_f \leqslant a_1 .
$$

Magnitudes a_s and a_f are the speeds of the slow and fast waves respectively. In the limiting cases when $\alpha = 0$ (the region of elastic deformation or of elastic unloading), we deduce from equation $D(a) = 0$ that $a = a_1$ and $a = a_2$; if $\alpha \to \infty$ (an elastic-ideally-plastic body) we obtain the characteristic equation for the waves speeds exactly as in [14]:

$$
\text{(22.17)} \qquad s^2\left(\frac{a^2}{a_2^2}-1\right)\left(\frac{3a^2}{\beta a_2^2}-1\right) + \frac{4\tau^2 a^2}{\beta a_2^2}\left(\frac{a^2}{a_2^2}-\frac{a_1^2}{a_2^2}\right) = 0.
$$

We obtain the following special cases from (22.15):

$$\text{if} \quad s = 0, \quad \text{then} \quad a_f = a_1, \quad 0 \leqslant a_s \leqslant a_2,$$

(22.18)
$$\text{if} \quad \tau = 0 \quad \text{and} \quad \beta \geqslant 3, \quad \text{then} \quad \sqrt{\frac{\beta}{3}}\, a_2 \leqslant a_f \leqslant a_1, \quad a_s = a_2,$$

$$\text{if} \quad \tau = 0 \quad \text{and} \quad \beta < 3, \quad \text{then} \quad a_2 \leqslant a_f \leqslant a_1,\ a_s = a_2 \ \text{for} \ s \leqslant s^*,$$

$$a_f = a_2, \ \sqrt{\frac{\beta}{3}}\, a_2 \leqslant a_s \leqslant a_2 \quad \text{for} \quad s \geqslant s^*,$$

where s^* corresponds to the value for which the speed of propagation of a plane plastic longitudinal wave has the same value as the speed of an elastic shear wave a_2. This value is determined from

(22.19)
$$\alpha\left(\frac{2s^*}{\sqrt{3}}\right) = \frac{\beta}{3-\beta}.$$

The relations along the characteristics are obtained using (9.14) from

(22.20)
$$\mathbf{l}^T \mathbf{B}\left(\frac{d\mathbf{u}}{dt}\right) = 0,$$

where $d\mathbf{u}/dt$ is the derivative along the characteristic $d\mathbf{u}/dt = \mathbf{u}_{,x}\, a + \mathbf{u}_{,t}$; $\mathbf{l}^T$ denotes the transposed left eigenvector $\mathbf{l}$, which is determined from (9.12):

(22.21)
$$\mathbf{l}^T(\mathbf{A} - a\mathbf{B}) = 0.$$

On account of the symmetry of the matrices **A** and **B**, the above equation, in the case under consideration, can be written in the form

(22.21′)
$$(\mathbf{A} - a\mathbf{B})\,\mathbf{l} = 0.$$

Consequently, the left and the right eigenvectors are identical. For $a = \pm a_f$ or $a = \pm a_s$ the left eigenvector of (22.21′) takes the form

(22.22)
$$\mathbf{l} = \begin{vmatrix} \Psi \\ 1 \\ -\varrho a \Psi \\ \Phi \\ -\varrho a \\ \Theta \end{vmatrix},$$

where

(22.23)
$$\Psi = \frac{s}{\tau}\,\frac{a^2 - a_2^2}{a^2 - a_1^2},$$

$$\Phi = \frac{\varrho s (a^2 - a_2^2)(a^2 - \beta a_2^2)}{2\tau a\,(a^2 - a_1^2)},$$

$$\Theta = \frac{1}{2\tau a}\left(\frac{a^2}{a_2^2} - 1\right).$$

The particular solutions of (22.11) are the so-called *simple waves*. Simple waves are solutions of (22.11) which are such that vector **u** is a constant value along the characteristics. The waves propagating with speed a_f we shall term the *fast waves*, and those travelling with speed a_s will be termed the *slow waves.*

Since the characteristic directions are functions of s and τ, the characteristics of (22.11) are therefore straight lines for solutions in the form of simple waves. Thus $\mathbf{u}$ = const along the line the slope of which is $dx/dt = a$. Hence we obtain

(22.24) $$\mathbf{u}_{,x}a + \mathbf{u}_{,t} = 0.$$

Eliminating $\mathbf{u}_{,t}$ between (22.24) and (22.11) we find that

(22.25) $$(\mathbf{A} - a\mathbf{B})\mathbf{u}_{,x} = 0.$$

On the other hand, eliminating $\mathbf{u}_{,x}$ we obtain

(22.26) $$(\mathbf{A} - a\mathbf{B})\mathbf{u}_{,t} = 0.$$

Since $d\mathbf{u} = n\mathbf{u}_{,x} + m\mathbf{u}_{,t}$ is the total differential along each direction $dx/dt = n/m$, for arbitrary n and m, we deduce from (22.25) and (22.26) that

(22.27) $$(\mathbf{A} - a\mathbf{B})\,d\mathbf{u} = 0.$$

Comparing (22.21′) and (22.27) we see that $d\mathbf{u}$ is proportional to the left eigenvector $\mathbf{l}$. Making use of this fact we obtain from (22.22)

(22.28) $$\frac{dv_1}{\Psi} = \frac{dv_2}{1} = \frac{d\sigma_1}{-\varrho a\Psi} = \frac{d\sigma_2}{\Phi} = \frac{d\tau}{-\varrho a} = \frac{d\Lambda}{\Theta}.$$

Equations (22.28) constitute five ordinary differential equations of the first order. Since Ψ, Φ, and Θ are functions of stress τ and s, then all these equations are coupled. In particular, since we have

(22.29) $$\frac{d\sigma_1 - d\sigma_2}{\varrho a\Psi + \Phi} = \frac{d\tau}{\varrho a},$$

then, by means of (22.8) and (22.29), we obtain

(22.30) $$\frac{ds}{d\tau} = \frac{s(a^2 - a_2^2)(3a^2 - \beta a_2^2)}{3\tau a^2(a^2 - a_1^2)}.$$

In order to obtain the changes in the stress tensor components on the $(\sigma_1, \sigma_2, \tau)$ plane, for the case of simple waves, we have to add to (22.30) the equation

(22.31) $$\frac{d\sigma_1}{d\tau} = \frac{s}{\tau}\frac{a^2 - a_2^2}{a^2 - a_1^2}$$

which can be obtained from (22.28) and (22.23).

Integrating (22.30) and (22.31) we obtain a two-parameter family of spatial curves in the three-dimensional stress space $(\sigma_1, \sigma_2, \tau)$ for each a_f and a_s. This spatial curve can be constructed provided its projections are known in each of two planes (σ_1, τ) and (σ_2, τ) (these planes are not parallel). On account of the form of (22.30) and (22.31) we can make use of the (s, τ) and (σ_1, τ) planes. Integrating (22.30) we obtain a one-parameter family of curves in the (s, τ) plane. A one-parameter family of curves is sufficient to give a projection of curves onto the (σ_1, τ) plane. The second parametric family of curves can be obtained by means of a rotation of the first parametric family of curves in the σ_1 direction. This can be proved by analysis of (22.30) and (22.31). The stress lines for the fast simple waves and for the slow simple waves are mutually orthogonal in the (σ_1, τ) plane. However, one should note that they are orthogonal in the (σ_1, τ) plane provided that they intersect in the $(\sigma_1, \sigma_2, \tau)$ space. If the stress line for the fast simple waves and the stress line for the slow simple waves do not intersect in $(\sigma_1, \sigma_2, \tau)$ space, then their projections onto the (σ_1, τ) plane are not necessarily mutually orthogonal. The stress lines in the (σ_1, τ) and (s, τ) planes are discussed with full particulars in [156].

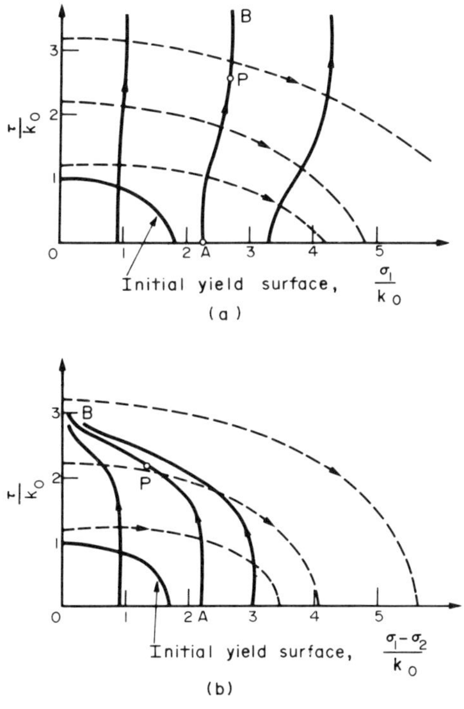

Fig. 69

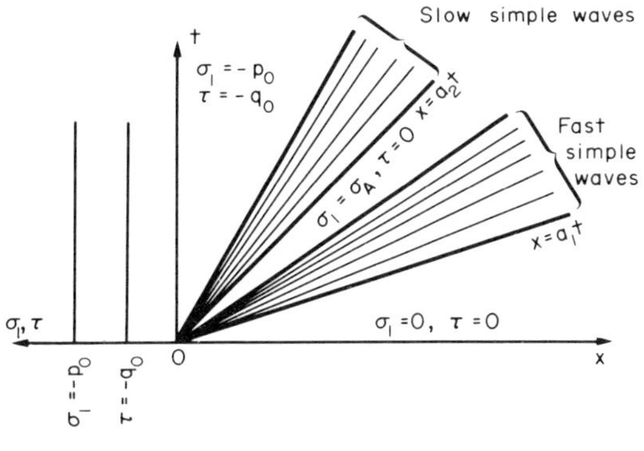

Fig. 70

As an example, the stress lines are shown in Fig. 69 on the (τ, σ_1) and (τ, s) planes respectively for the particular form of function $\alpha(k) = 3\left(\frac{k}{k_0} - 1\right)^4$, where k_0 denotes the initial yield surface, $\beta = 2.25$. The continuous lines represent the stress lines for slow simple waves, while the dashed lines signify fast simple waves. The curves in Fig. 69 correspond to the plastic strain range. In the regions of elastic strain the stress lines for the fast simple waves, propagating with speed $a = a_1$, are straight lines parallel to the σ_1 and s axes; the stress lines corresponding to the slow simple waves propagating with speed $a = a_2$ are parallel to the τ-axis. The stress lines in the region of elastic strain are not inclined, i.e. the state of stress can change in both directions.

The solution corresponding to the boundary conditions (22.1), i.e. to the case of normal and tangential tractions which are constant in time and suddenly applied to the boundary of the half-space, is presented in Fig. 70. This solution corresponds to the stress lines in Fig. 69 along the path OAB. A zero initial state of stress is assumed, i.e. $\sigma_1(0, x) = \tau(0, x) \equiv 0$. Point P in Figs. 69a and 69b corresponds to the state of stress on the boundary of the half-space, i.e. to stresses $\sigma_1 = -p_0$, $\tau = -q_0$.

Since the characteristic directions a are functions of the stresses σ_1, σ_2, and τ, and since the stresses are constant along the characteristics for simple waves, these characteristics are therefore straight lines. Each point on the stress line OAP corresponds to a characteristic in the phase plane (x, t). The position of the point on the stress line OAP determines the stresses along the characteristics as well as the characteristic direction a. Stresses σ_1 and τ can be determined from Fig. 69a and stress σ_2 from Fig. 69b.

In the case of an elastic-ideally-plastic body the above discussion is considerably simplified. The characteristic equation $D(a) = 0$ reduces to the form of (22.17). The characteristic directions a are independent of the stress component σ_1. The yield condition (22.6) is reduced to the form

$$\frac{1}{3}(\sigma_1 - \sigma_2)^2 + \tau^2 = k_0^2, \tag{22.32}$$

where k_0 is a constant. Stress σ_2 ceases to be an independent variable; it is expressed in terms of the stress tensor components σ_1 and τ. The problem of simple wave propagation in an elastic-ideally-plastic half-space, loaded on the boundary by stresses σ_1, τ, according to the boundary condition (22.1), was thoroughly examined by Bleich and Nelson [14].

The method presented at this point cannot be applied to the case of arbitrary boundary conditions. It can be applied only in the case of constant tractions on the boundary of the half-space. The method can be generalized to the case in which pressure arbitrarily increases in time from zero to a constant value; this was done in [156]. In the case when the time change of the tractions on the boundary of the half-space is arbitrary, the solution to the problem can be found numerically. To this end the relations along the characteristics obtained from (20.20) are used.

The problem of an unloading wave in the case of two-parameter loading of an elastic–plastic medium was considered by Clifton [28]. The problem of wave propagation in a semi-infinite thin-walled cylinder loaded on the boundary by a normal pressure and a torque was examined in that paper. Clifton and Lipkin [29], [81] demonstrated the existence of fast and slow simple waves experimentally also, and a comparison of the experimental and theoretical results is given in [29].

We now present the problem of two-parameter loading of the boundary of an elastic–plastic half-space based on the soil dynamics equations proposed by Grigorian [48]. On the bounding plane of the half-space $x_1 \geqslant 0$, Fig. 68, of the medium, described by Grigorian's equations (see section 4.1), the boundary conditions are prescribed in the form of (22.1).

The shear stresses σ_{13} and σ_{23} and the component of displacement in the direction of the x_2-axis vanish as in the previously considered plane problem. The remaining stress tensor components and displacement components are independent of coordinates x_2 and x_3, consequently

$$
(22.33) \quad
\begin{aligned}
&\sigma_{13} = \sigma_{23} = 0, \qquad u_3 = \dot{u}_3 = 0,\\
&\dot{e}_{11} = 2\frac{\partial v_1}{\partial x_1}, \qquad \dot{e}_{21} = \dot{e}_{12} = \frac{\partial v_2}{\partial x_1}, \qquad \dot{e}_{13} = \dot{e}_{22} = \dot{e}_{23} = \dot{e}_{31} = \dot{e}_{32} = \dot{e}_{33} = 0,\\
&\omega_{11} = \omega_{22} = \omega_{33} = \omega_{13} = \omega_{31} = \omega_{32} = \omega_{23} = 0,\\
&\omega_{12} = -\omega_{21} = -\frac{1}{2}\frac{\partial v_2}{\partial x_1}.
\end{aligned}
$$

By means of the above relations, equations (4.8) are reduced to the forms

$$
(22.34) \quad
\begin{aligned}
\frac{4}{3}\mu\frac{\partial v_1}{\partial x_1} &= \frac{ds_{11}}{dt} + s_{12}\frac{\partial v_2}{\partial x_1} + \lambda s_{11},\\
-\frac{2}{3}\mu\frac{\partial v_1}{\partial x_1} &= \frac{ds_{22}}{dt} - s_{21}\frac{\partial v_2}{\partial x_1} + \lambda s_{22},\\
-\frac{2}{3}\mu\frac{\partial v_1}{\partial x_1} &= \frac{ds_{33}}{dt} + \lambda s_{33},\\
\mu\frac{\partial v_2}{\partial x_1} &= \frac{ds_{12}}{dt} + \frac{1}{2}(s_{22} - s_{11})\frac{\partial v_2}{\partial x_1} + \lambda s_{12}.
\end{aligned}
$$

It should be noted that $(22.34)_{1-3}$ are linearly independent. Equations (22.34), on assuming that $s_{ij} = \sigma_{ij} + p\delta_{ij}$ and that $\frac{d}{dt} = \frac{\partial}{\partial t} + v_i\frac{\partial}{\partial x_i}$ and introducing the notation $\gamma = \sigma_{22} - \sigma_{33}$, take the following forms:

$$
(22.35) \quad
\begin{aligned}
&\frac{4}{3}\mu\frac{\partial v_1}{\partial x_1} = \frac{d(\sigma_{11}+p)}{dt} + v_1\frac{\partial(\sigma_{11}+p)}{\partial x_1} + \sigma_{12}\frac{\partial v_2}{\partial x_1} + \lambda(\sigma_{11}+p),\\
&\frac{d\gamma}{dt} - \sigma_{12}\frac{\partial v_2}{\partial x_1} + v_1\frac{\partial \gamma}{\partial x_1} + \lambda\gamma = 0,\\
&\mu\frac{\partial v_2}{\partial x_1} = \frac{d\sigma_{12}}{dt} + v_1\frac{\partial \sigma_{12}}{\partial x_1} + \frac{1}{4}[\gamma - 3(\sigma_{11}+p)]\frac{\partial v_2}{\partial x_1} + \lambda\sigma_{12},
\end{aligned}
$$

where $3p = -(\sigma_{11} + \sigma_{22} + \sigma_{33})$.

Yield condition (4.7) is expressed by the formula

$$
(22.36) \quad \frac{3}{4}(\sigma_{11}+p)^2 + \frac{1}{4}\gamma^2 + \sigma_{12}^2 = F(p).
$$

The equations of motion and the equation of mass conservation (5.21) take the following forms:

$$
(22.37) \quad
\begin{aligned}
&\varrho\frac{\partial v_1}{\partial t} + \varrho v_1\frac{\partial v_1}{\partial x_1} = \frac{\partial \sigma_{11}}{\partial x_1},\\
&\varrho\frac{\partial v_2}{\partial t} + \varrho v_1\frac{\partial v_2}{\partial x_1} = \frac{\partial \sigma_{12}}{\partial x_1}, \qquad \frac{d\varrho}{dt} + \varrho\frac{\partial v_1}{\partial x_1} = 0.
\end{aligned}
$$

Parameter λ entering into (22.35) is defined as follows:

$$\lambda = \frac{\mu}{F(p)}\left[(\sigma_{11}+p)\frac{\partial v_1}{\partial x_1}+\sigma_{12}\frac{\partial v_2}{\partial x_1}\right]-\frac{F'(p)\dfrac{dp}{dt}}{2F(p)}. \tag{22.38}$$

The system of (22.35)–(22.38) together with (4.1), in the case when $dp/dt>0$, or with (4.2) in the case when $dp/dt<0$, constitutes the complete system of equations for the problem considered (including the effect of finite strain).

For small strain gradients the systems of equations mentioned above can be reduced to the forms

$$\frac{4}{3}\mu\frac{\partial v_1}{\partial x_1}=\frac{\partial(\sigma_{11}+p)}{\partial t}+\lambda(\sigma_{11}+p),\qquad \mu\frac{\partial v_2}{\partial x_1}=\frac{\partial\sigma_{12}}{\partial t}+\lambda\sigma_{12},$$

$$\varrho_0\frac{\partial v_1}{\partial t}=\frac{\partial\sigma_{11}}{\partial x_1},\qquad \varrho_0\frac{\partial v_2}{\partial t}=\frac{\partial\sigma_{12}}{\partial x_1},$$

$$\frac{3}{4}(\sigma_{11}+p)^2+\sigma_{12}^2=F(p),\qquad \gamma=0, \tag{22.39}$$

$$\frac{\partial p}{\partial t}=-K\frac{\partial v_1}{\partial x_1}\quad\text{for}\quad\frac{dp}{dt}>0\quad\text{and}\quad\varrho=\varrho^*(x_1)\quad\text{for}\quad\frac{dp}{dt}<0,$$

$$\lambda=\mu\frac{\sigma_{11}+p}{F(p)}\frac{\partial v_1}{\partial x_1}+\mu\sigma_{12}\frac{1}{F(p)}\frac{\partial v_2}{\partial x_1}-\frac{F'(p)}{2F(p)}\frac{\partial p}{\partial t}.$$

In the region of elastic strain, since λ = 0, (22.39) can be considerably simplified. The system of (22.39) can be replaced by an equivalent system of equations along the characteristics:

$$\begin{aligned} d\sigma_{11}\mp\varrho_0 a_1\,dv_1&=0\quad\text{for}\quad x=\pm a_1 t+\text{const},\\ d\sigma_{12}\mp\varrho_0 a_2\,dv_2&=0\quad\text{for}\quad x=\pm a_2 t+\text{const},\end{aligned} \tag{22.40}$$

where $a_1^2=\dfrac{(3K+4\mu)}{\varrho_0}$, $a_2^2=\dfrac{\mu}{\varrho_0}$. Thus we have to deal with two non-interacting Riemann waves – one longitudinal and the other transverse.

In the region of plastic strain (λ > 0) the transverse waves are coupled with the longitudinal ones.

We present a particular solution of this problem, namely for the case of the propagation of longitudinal–transverse waves generated by the following boundary conditions:

$$\sigma_{11}=\sigma_1,\qquad \sigma_{12}=\tau_1\qquad\text{for}\qquad x_1=0,\qquad t\geqslant 0, \tag{22.41}$$

where the tractions σ_1 and τ_1 are constant and occur suddenly on the boundary of the half-space. We assume the boundary condition in the forms

$$\sigma_{11}=\sigma_0,\qquad \sigma_{12}=\tau_0,\qquad v_1=v_2=0\qquad\text{for}\qquad t=0,\qquad x\geqslant 0. \tag{22.42}$$

In the case of the boundary conditions in the form (22.41) the problem is considerably simpler since it is then a self-similar one.

Equating to zero the characteristic determinant of the system (22.30), we obtain the characteristic equation

(22.43)
$$a^4-\left\{[1+ks]K+\frac{4-s^2}{3}\mu\right\}a^2+\mu Ks(s+k)=0,$$

where

$$a^2=\varrho_0\left(\frac{dx}{dt}\right)^2,\qquad s^2=1-\frac{\sigma_{12}}{F(p)},\qquad k=\frac{F'(p)}{\sqrt{3F(p)}}.$$

Equation (22.43) has been obtained without imposing any restrictions on quantities k, μ, and k. We assume that these quantities are constant and that $\mu<(1+k)K$. We consider the case when the motion is such that s is constant along each characteristic. Since a depends on s, the characteristics therefore are straight lines.

Because of the assumed boundary conditions (22.41) all quantities depend on a single, spatial, dimensionless variable:

(22.44)
$$U=\frac{x}{t}=\frac{a}{\sqrt{\varrho_0}}.$$

Since a depends only on s, all the quantities appearing in (22.39) depend solely on s. By the use of (22.44) the system of partial differential equations (22.39) can be replaced by a system of ordinary differential equations. After some manipulations we deduce from these equations that

(22.45)
$$\frac{d\sqrt{F}}{ds}=\frac{kK\sqrt{F}}{a^2-(1+ks)K}.$$

Upon integration we obtain

(22.46)
$$\sqrt{F(p)}=\sqrt{F(p_0)}\exp\left(\int_{s_0}^{s}\frac{kK}{a^2(\xi)-(1+k\xi)K}d\xi\right),$$

where p_0 and s_0 correspond to the values of p and s in front of the first wave; they are determined from the initial conditions of the problem.

If $F(p)$ is known, all the required unknowns can be determined. Substituting into (22.46) the roots of (22.43), we obtain the family of solutions of the governing system of equations (22.39). Equation (22.43) possesses four roots: two of them are positive $a^{\mathrm{I}}_{1,2}$ and correspond to the simple waves both longitudinal and transverse ($a^{\mathrm{I}}_1>a^{\mathrm{I}}_2$). The negative roots $a^{\mathrm{II}}_{1,2}$ correspond to waves travelling in the reverse direction, again longitudinal and transverse ($a^{\mathrm{II}}_1>a^{\mathrm{II}}_2$).

If $\lambda>0$ the system of (22.39) describes the plastic shear strain. From the system of (22.39), making use of (22.43) and of the transformation

$$\frac{\partial}{\partial t}=-\frac{a}{\sqrt{\varrho_0}}\frac{\partial}{\partial x_1}$$

(resulting from the change to dimensionless variable (22.44)), we obtain the following expression for parameter λ:

(22.47)
$$\lambda=\frac{\sqrt{3}}{2s\sqrt{F}}\left[a^2-\left(K+\frac{4}{3}\mu\right)\right]\frac{\partial v_1}{\partial x_1}.$$

If the medium is compressed then the condition $\lambda > 0$ leads to the inequalities for loading and unloading processes:

$$a^2 < K + \frac{4}{3}\mu \quad \text{for} \quad \frac{\partial v_1}{\partial x_1} < 0,$$

(22.48)

$$a^2 > K + \frac{4}{3}\mu \quad \text{for} \quad \frac{\partial v_1}{\partial x_1} > 0.$$

For example, if condition $\frac{4}{3}\mu > kK$ is valid for a loading process and the condition $kK > \frac{4}{3}\left\{1 + \frac{4+k}{4(1+k)}\right\}$ holds for an unloading process then the picture of the wave solution in the phase plane is as in Fig. 71. Regions I, III, and V are constant stress regions. In region II

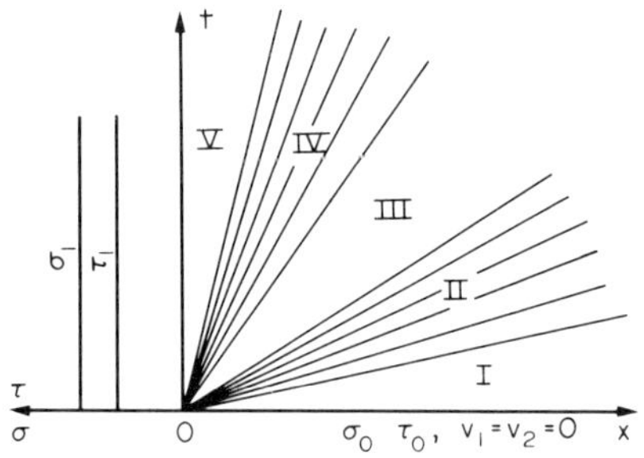

Fig. 71

fast simple waves propagate, while in region IV we have slow simple waves. In region I, on account of the initial conditions, we have $\sigma_{11} = \sigma_0$, $\sigma_{12} = \tau_0$, $v_1 = v_2 = 0$. In region V, for boundary conditions (22.9) we obtain $\sigma_{11} = \sigma_1$ and $\sigma_{22} = \tau_1$. In order to determine the solutions in regions II, III, and IV we have to make use of (22.46), condition $\sigma_{12} = \sqrt{1-s^2}F(p)$, and of the yield condition (22.39). The solutions are analysed in the (σ_{12}, z) plane, where $z = \sqrt{F(p)}$ (eq. (22.39)$_4$ uniquely determines the relations between σ_{11}, σ_{12}, and σ_{12}, z). Examining the change of stress with time for a fixed x, we make use of (22.46) in which, when passing across the fast simple waves, one has to assume $a^2 = a_1^2$ whilst across the slow simple waves $a^2 = a_2^2$. Depending on the position of the point (z_0, τ_0) (the initial conditions of the problem), we obtain different solutions of the problem.

The complete solution of the problem of the propagation of longitudinal–transverse waves in a medium governed by Grigorian's equations for soil dynamics so far has not been found for the cases of loads which vary arbitrarily with time. The construction of the plastic loading wave for the case when the loads σ_{11} and σ_{12} increase monotonically from zero on the boundary of the half-space does not present any difficulty. This wave is constructed in the same manner as in the case of an elastic/viscoplastic medium (see section 23), described by means of condition (4.7). If function $F(p)$ is taken in the form (4.14), then the local speed of propagation of the plastic loading wave is given by the formula

(22.49)

$$c^*(x^*,t^*) = a_1 \frac{\frac{3}{2}(\sigma_{11}+p)(\sigma_{11}+p)_{,t}+2\sigma_{12}\sigma_{12,t}-2\alpha(\alpha p+\beta)p_{,t}}{\frac{3}{2}(\sigma_{11}+p)(\sigma_{11}+p)_{,t}-2\alpha(\alpha p+\beta)p_{,t}+2\sigma_{12}[\sigma_{12,t}-\sigma_{12}(x^*+a_1(t-t^*),t^*)]},$$

where σ_{11}, σ_{12} and p are known values, determined on the front of the plastic loading wave from the side of the elastic region. It can also be proved that the speed of propagation of the plastic loading wave is limited to the interval

$$a_2 \leqslant c^* \leqslant a_1,$$

i.e. in the phase plane the wave of plastic loading lies between the characteristics corresponding to the elastic longitudinal and transverse waves or in the limit case coincides with one of the two. The solution in the region of plastic strain can be constructed in the same way as has been done in many papers devoted to problems of elastic–plastic wave propagation produced by two-parameter loading [92], [153]–[157].

To conclude this point we present some remarks referring to problems related to the propagation of longitudinal–transverse waves in a medium governed by the equations of the bilinear theory of plasticity.

The solution in the range of active loading in a half-space loaded on the boundary by stresses $p(t)$ and $q(t)$ arbitrarily increasing in time was examined in detail in [99]. The analysis of the generation of the loading wave fronts, for both weak and strong discontinuities, is given fully (for various ratios of speeds a_{1p}/a_{2s}, a_{1s}, and a_{2s} denote the speeds of the elastic longitudinal and transverse waves; a_{1p} and a_{2p} those for the corresponding plastic waves). It was shown in [99] that, in the case when $a_{2s} > a_{1p}$, the local speed of the plastic loading wave c^* is bounded by the inequalities

$$a_{1p} \leqslant c^* \leqslant a_{2s} \quad \text{or} \quad c^* \geqslant a_{1s}, \quad \text{or} \quad c^* \leqslant a_{2p}.$$

In the case $a_{1p} > a_{2s}$ we have

$$c^* = a_{1p} \quad \text{or} \quad c^* \geqslant a_{1s}, \quad \text{or} \quad c^* \leqslant a_{2p}.$$

Furthermore, $c^* = a_{1p}$ only occurs if, under the wave of plastic loading, a region exists in which the normal stress σ_{11} is constant.

Also certain remarks are given in [99] concerning the loading wave for the case when classical elastic unloading is assumed. It was shown, from the analysis of the loading of the half-space boundary by tractions $p(t)$ and $q(t)$, monotonically increasing from zero and then from instant $t = t_0$ monotonically decreasing, that two different cases may exist in which the unloading wave can be uniquely determined. These correspond to the cases when the initial speed of the unloading wave is bounded in the interval (when $a_{2s} > a_{1p}$) $a_{2s} \leqslant c_0 \leqslant a_{1s}$ or in the interval $a_{2p} \leqslant c_0 \leqslant a_{1p}$, while if $a_{1p} > a_{2s}$ in the interval $a_{2p} \leqslant c_0 \leqslant a_{2s}$ or $a_{1p} \leqslant c_0 \leqslant a_{1s}$. The first case occurs when the sign of the gradient of the normal loading on the boundary $dp(t)/dt$ changes while the gradient of the tangential loading $dq(t)/dt$ does not change in sign or in the case of the simultaneous change. Thus in this case the generation of the unloading wave is determined only by the change of the sign of the gradient of normal loading $p(t)$. The second case occurs when the normal stress increases monotonically, or when, at constant normal stress, the sign of the gradient of the tangential stress $dq(t)/dt$ changes. The conditions restricting the speed of the unloading wave agree with the results of Clifton [27], concerning simple fast and slow plastic waves in thin-walled elastic–plastic cylinders.

The problem of the propagation of longitudinal–transverse waves was also solved in [98]. The difference, as compared with [99], consists in the assumption that the motion of the medium for the process of loading is governed by (2.13) and the rigid unloading characteristic was assumed for the unloading process (the stress intensity during unloading is independent of time). Such a model is suitable for an approximate description of the deformation of sandy soils of small moisture content in the medium pressure range. Under these conditions we can assume that the voluminal strain is irreversible and practically constant during the unloading process. The introduction of rigid unloading, similarly as in the case of one-parameter loadings of the boundary, permitted the solution of the problem of the propagation of the longitudinal–transverse waves in a closed form.

23. Plane longitudinal–transverse waves in an isotropic, elastic/viscoplastic space

We consider the motion of an elastic/viscoplastic medium, filling a half-space $x_1 > 0$ (Fig. 68) with normal and tangential tractions

$$\sigma_{11} = -p(t), \qquad \sigma_{12} = q(t) \tag{23.1}$$

varying arbitrarily with time, applied to the bounding plane. The type of boundary conditions, as well as the geometry of the problem considered, lead to relations identical with those for the problems discussed in the preceding section, i.e. to (22.3). The equations of dynamic equilibrium and the relations resulting from the assumption of small strain take the form (22.4) and (22.5) respectively.

The constitutive equations (3.10) or (3.25), on taking into account condition (22.3) and equations (22.5), assume the forms

$$\left(1+\frac{3K}{4\mu}\right)\frac{\partial v_1}{\partial x} - \frac{3}{4\mu}\frac{\partial \sigma_1}{\partial t} = \langle \Phi_1 \rangle, \qquad \frac{1}{2}\frac{\partial v_2}{\partial x} - \frac{1}{2\mu}\frac{\partial \tau}{\partial t} = \langle \Phi_2 \rangle, \tag{23.2}$$

where we have introduced the notation

$$\begin{aligned}\langle \Phi_1 \rangle &= \frac{\gamma}{2}\langle \Phi(F)\rangle \frac{\sigma_1-\sigma_2}{\sqrt{J_2}},\\ \langle \Phi_2 \rangle &= \frac{\gamma}{2}\langle \Phi(F)\rangle \frac{\tau}{\sqrt{J_2}},\\ J_2 &= \frac{1}{\sqrt{3}}\sqrt{(\sigma_1-\sigma_2)^2+3\tau^2}.\end{aligned} \tag{23.3}$$

Equations (22.4) and (23.2) constitute a complete system of the equations for the problem posed. They are valid in the regions of elastic deformation, the regions of unloading (then $\Phi \equiv 0$), and in the regions of viscoplastic deformation as well.

The straight lines

$$x \pm a_1 t = \text{const}, \qquad x \pm a_2 t = \text{const}, \tag{23.4}$$

where

$$a_1 = \sqrt{\frac{3K+4\mu}{3\varrho}}, \qquad a_2 = \sqrt{\frac{\mu}{\varrho}},$$

constitute the real characteristics of the system of (22.4) and (23.2).

The relations along the characteristics have the forms:

$$(23.5)\qquad \begin{aligned} &d\sigma_1 \mp \varrho a_1\, dv_1 + \frac{4}{3}\mu\langle\Phi_1\rangle\, dt = 0 \qquad \text{for} \quad x \mp a_1 t = \text{const},\\ &d\tau \mp \varrho a_2\, dv_2 + 2\mu\langle\Phi_2\rangle\, dt = 0 \qquad \text{for} \quad x \mp a_2 t = \text{const}. \end{aligned}$$

Thus the system of the semi-linear partial differential equations (22.4) and (23.2) can be replaced by the equivalent system of (23.5) along the characteristics (23.4). In the elastic zone (and unloading zone) the system of (22.4) and (23.2) becomes uncoupled because $\Phi \equiv 0$, and its solution consists of two non-interacting systems of Riemann waves: the longitudinal one propagating with speed a_1 and transverse wave propagating with speed a_2.

Depending on the nature of the time change of the tractions $p(t)$ and $q(t)$ on the bounding plane $x_1 = 0$, waves of weak or strong discontinuity can propagate into the medium $x_1 > 0$. Since the assumed elastic/viscoplastic model refuses to admit the generation of shock waves, waves of strong discontinuity can be produced only by step changes in the $p(t)$ or $q(t)$.

We consider, in turn, the generation of weak and strong discontinuity waves.

23.1. WAVES OF WEAK DISCONTINUITY

In this case $p(t)$ and $q(t)$ are continuous functions, and $p(0) = q(0) = 0$. The wave picture in the phase plane is as follows. As tractions $p(t)$ and $q(t)$ increase, two non-interacting systems of Riemann waves propagate into the space $x_1 > 0$ with the speeds a_1 and a_2 (regions I and II, Fig. 72).

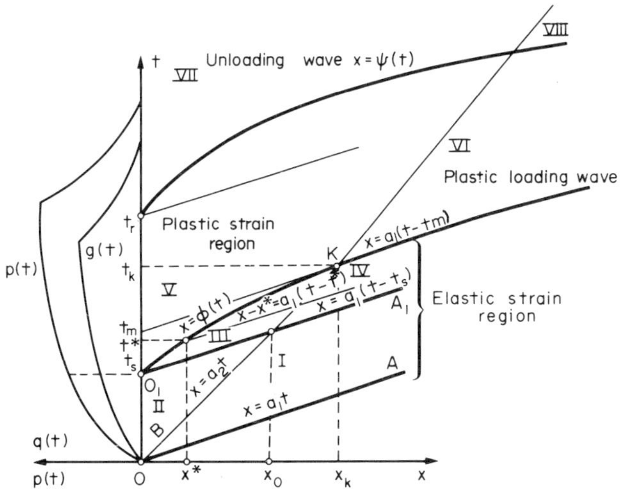

Fig. 72

The elastic longitudinal Riemann waves propagate in region I:

$$\sigma_1 = -p\left(t - \frac{x}{a_1}\right), \qquad \sigma_2 = \sigma_3 = -\nu p\left(t - \frac{x}{a_1}\right), \qquad \tau \equiv 0,$$

(23.6)
$$v_1 = \frac{1}{\varrho a}\, p\left(t - \frac{x}{a_1}\right), \qquad v_2 \equiv 0.$$

In region II the transverse waves superimpose themselves on the effects of the longitudinal elastic Riemann waves. The following solution is obtained in this region:

$$\sigma = -p\left(t - \frac{x}{a_1}\right), \qquad \sigma_2 = \sigma_3 = -\nu p\left(t - \frac{x}{a_1}\right), \qquad \tau = -q\left(t - \frac{x}{a_2}\right),$$

(23.7)
$$v_1 = \frac{1}{\varrho a_1}\, p\left(t - \frac{x}{a_1}\right), \qquad v_2 = \frac{1}{\varrho a_2}\, q\left(t - \frac{x}{a_2}\right),$$

where

$$\nu = \frac{3K - 2\mu}{3K + 4\mu}.$$

The fundamental difficulty in constructing the solution of the problem considered consists in finding the velocity of propagation of the plastic loading wave which separates the region of elastic deformation (in front of the plastic wave) from the region of viscoplastic strain (behind the plastic wave front). This problem does not exist in the case of a uniaxial state of stress or strain since there the front of the plastic loading wave coincides (in a homogeneous medium) with the positive characteristic of the equations which govern the problem independently of the boundary condition.

We determine the local velocity of propagation of the plastic loading wave.† Differentiating the stress tensor components along the characteristic $x = x^* + a_1(t-t^*)$ and along the front of the plastic wave $x = \varphi(t)$, we obtain

(23.8)
$$\begin{aligned} a_1\sigma_{1,1} + \sigma_{1,t} &= 0, \qquad a_1\sigma_{2,1} + \sigma_{2,t} = 0, \\ a_1\tau_{,1} + \tau_{,t} &= \tau^{*\prime}(t), \qquad \sigma_{i,1}\varphi'(t) + \sigma_{i,t} = 0, \end{aligned}$$

where $\tau^{*\prime}(t) = \tau[x^* + a_1(t-t^*), t]$.

Since wave $x = \varphi(t)$ is a wave of weak discontinuity, the condition $J_2 = \frac{1}{3}[(\sigma_1-\sigma_2)^2 + 3\tau^2] = k^2$ and the condition of stress continuity have to be satisfied across the wave of plastic loading. Also, since region I is a region of elastic strain, we deduce that $\sigma_2 = \nu\sigma_1$ from the constitutive equations (3.10) or (3.25). Taking into account these conditions we obtain, from the system of (23.8), the expression for the local velocity of the plastic wave

(23.9)
$$c^* = \varphi'(t)|_{t=t^*} = \lim_{\substack{t\to t^* \\ x\to x^*}} a_1 \frac{(1-\nu)^2\sigma_1\sigma_{1,t} + 3\tau\tau_{,t}}{(1-\nu)^2\sigma_1\sigma_{1,t} + 3\tau\tau_{,t} - 3\tau\tau^{*\prime}(t)}.$$

Passing, in the above expression, to the limit as $t \to t^*$ and $x \to x^*$, we find the required formula for the local velocity of the plastic loading wave

†In exactly the same way that we determine the local velocity of an unloading wave in the case of an elastic/viscoplastic medium without hardening.

$$c^* = a_1 \frac{(1-\nu)^2 \sigma_1(x^*, t^*)\sigma_{1,t}(x^*, t^*) + 3\tau(x^*, t^*)\tau_{,t}(x^*, t^*)}{(1-\nu)^2\sigma_1(x^*, t^*)\sigma_{1,t}(x^*, t^*) + 3\tau(x^*, t^*)\tau_{,t}(x^*, t^*) - 3\tau(x,^* t^*)\tau^{*\prime}(t^*)}. \tag{23.10}$$

In the limiting cases we have

$$\begin{aligned} &\text{if} \quad \tau(x,t) = -q(t) \equiv 0, \quad \text{then} \quad c^* = a_1, \\ &\text{if} \quad \sigma_1(x,t) = -p(t) \equiv 0, \quad \text{then} \quad c^*\tau_{,1} + \tau_{,t} = 0 \rightarrow c^* = a_2. \end{aligned} \tag{23.11}$$

In the particular case when $t = t_s$, we obtain the expression for the initial velocity of the plastic wave. The velocity of propagation of the plastic wave $x = \varphi(t)$ depends only on the values of the stresses and their time derivatives on its front. It has been provėd, in [48], that the speed c^* is contained in the interval

$$a_2 \leqslant c^* \leqslant a_1. \tag{23.12}$$

This means that the plastic wave lies between characteristics $x = a_1(t-t_s)$, $x = a_2(t-t_s)$, or, in a limiting case, coincides with one of them.

The shape of the plastic loading wave can be constructed by solving for regions III and V simultaneously, and taking into account the stress and the velocity continuity conditions across its front. Behind wave $x = \varphi(t)$ the region of viscoplastic deformation spreads out. The longitudinal waves are coupled with the transverse waves and they interact.

The solution in regions III and V and the front of wave $x = \varphi(t)$ were determined numerically in [60] using the finite difference method for characteristic nets, utilizing, for this purpose, the relations (23.5) on the characteristics.

In order to simplify the discussion the following dimensionless quantities are introduced:

$$\begin{aligned} &\xi = \frac{x}{x_0}, \quad \eta = \frac{a_1 t}{x_0}, \quad S = \frac{\sigma_1}{k}, \quad T = \frac{\tau}{k}, \quad Y = \frac{\sigma_2}{k}, \quad U = \frac{\varrho a_1}{k} v_1, \\ &V = \frac{\varrho a_2}{k} v_2, \quad S_i = \frac{\sigma_i}{k}, \quad \alpha_0 = \frac{a_2}{a_1}, \quad \varkappa_0 = \frac{\mu}{k}\frac{x_0}{a_1}\gamma, \quad E = \frac{\varrho a_1^2}{k}\varepsilon, \quad \Gamma = \frac{\varrho a_2^2}{k}\gamma. \end{aligned} \tag{23.13}$$

In regions III and V the relations on the characteristics (23.5) take the forms

$$\begin{aligned} dS \mp dU &= 0 \quad \text{for} \quad \xi \mp \eta = \text{const}, \\ dT \mp dV &= 0 \quad \text{for} \quad \xi \mp \alpha_0 \eta = \text{const}, \end{aligned} \tag{23.14}$$

whereas in region V we have the following relations:

$$\begin{aligned} dS \mp dU + \Psi_1\, d\eta &= 0 \quad \text{for} \quad \xi \mp \eta = \text{const}, \\ dT \mp dV + \Psi_2\, d\eta &= 0 \quad \text{for} \quad \xi \mp \alpha_0 \eta = \text{const}, \end{aligned} \tag{23.15}$$

where

$$\Psi_1 = \frac{2}{3}\varkappa_0\left(1 - \frac{\sqrt{3}}{\sqrt{(S-Y)^2 + 3T^2}}\right)(S-Y),$$

$$\Psi_2 = \varkappa_0\left(1 - \frac{\sqrt{3}}{\sqrt{(S-Y)^2 + 3T^2}}\right)T.$$

From the above relations and constitutive equations (3.10) or (3.18), taking into account conditions (22.3) and writing them down in the form of finite differences along the characteristics, we obtain the recurrence formulae for the discrete values of the parameters at point i of the element of the mesh of characteristics shown in Fig. 73.

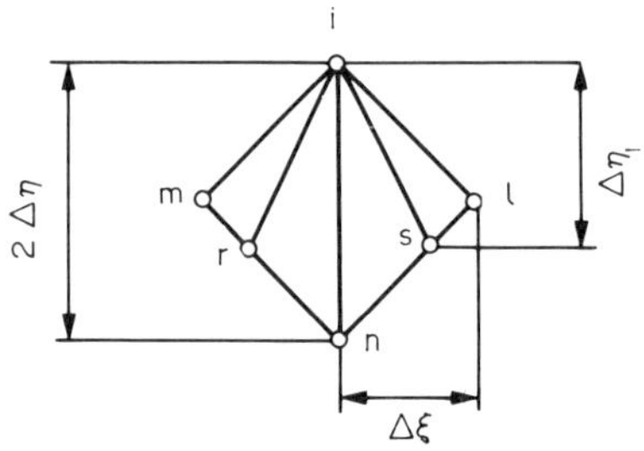

Fig. 73

For region III we have

$$
\begin{aligned}
S(i) &= \frac{1}{2}[S(m)+S(l)] - \frac{1}{2}[U(m)-U(l)],\\
Y(i) &= Y(n)+v[S(i)-S(n)],\\
T(i) &= \frac{1}{2}[T(r)+T(s)] + \frac{1}{2}[V(s)-V(r)],\\
U(i) &= \frac{1}{2}[U(m)+U(l)] + \frac{1}{2}[S(l)-S(m)],\\
V(i) &= \frac{1}{2}[V(r)+V(s)] + \frac{1}{2}[T(s)-T(r)].
\end{aligned}
\tag{23.16}
$$

For region V we obtain

$$
\begin{aligned}
S(i) &= \frac{1}{2}[S(m)+S(l)] - \frac{1}{2}[U(m)-U(l)] - \frac{1}{2}[\Psi_1(m)+\Psi_1(l)]\Delta\eta,\\
Y(i) &= Y(n)+v[S(i)-S(n)] + \frac{9K}{3K+4\mu}\Psi_1(n)\Delta\eta,\\
T(i) &= \frac{1}{2}[T(r)+T(s)] + \frac{1}{2}[V(s)-V(r)] - \frac{1}{2}[\Psi_2(r)+\Psi(s)]\Delta\eta_1,\\
U(i) &= \frac{1}{2}[U(m)+U(l)] + \frac{1}{2}[S(l)-S(m)] + \frac{1}{2}[\Psi_1(m)-\Psi_1(l)]\Delta\eta,\\
V(i) &= \frac{1}{2}[V(r)+V(s)] + \frac{1}{2}[T(s)-T(r)] + \frac{1}{2}[\Psi_2(r)-\Psi_2(s)]\Delta\eta_1.
\end{aligned}
\tag{23.17}
$$

At points lying on the boundary of regions III and V we find that

$$
\begin{aligned}
S(i) &= \frac{1}{2}[S(m)+S(l)] - \frac{1}{2}[U(m)-U(l)] - \frac{1}{2}\Psi_1(m)\Delta\eta,\\
Y(i) &= Y(n)+v[S(i)-S(n)],\\
T(i) &= \frac{1}{2}[T(r)+T(s)] + \frac{1}{2}[V(s)-V(r)],
\end{aligned}
\tag{23.18}
$$

(23.18) [cont.]

$$U(i) = \frac{1}{2}[U(m) + U(l)] + \frac{1}{2}[S(l) - S(m)] + \frac{1}{2}\Psi_1(m)\Delta\eta,$$

$$V(i) = \frac{1}{2}[V(r) + V(s)] + \frac{1}{2}[T(s) - T(r)].$$

At points lying on the boundary of the half-space $\xi = 0$ values $S(i)$, $T(i)$ are prescribed (boundary conditions (25.1)). The remaining quantities are determined by the formulae

(23.19)

$$Y(i) = Y(n) + \nu[p(n) - p(i)] - \frac{9K}{3K + 4\mu}\Psi_1(n)\Delta\eta,$$

$$U(i) = U(l) + p(i) + S(l) - \Psi_1(l)\Delta\eta,$$

$$V(i) = V(s) + q(i) + q(s) - \Psi_2(s)\Delta\eta_1.$$

The procedure for solving the equations is as follows. The discrete values of the parameters along characteristic $x = a_1(t + 2\Delta t - t_s)$ $(\xi = \eta + 2\Delta\eta - \eta_s)$ are computed starting from the data on characteristic O_1A_1 the equation of which is $x = a_1(t - t_s)$ (Fig. 72) (in the dimensionless variables $\xi = \eta - \eta_s$). On the boundary $\xi = 0$ we determine the solution from (23.10), and at the subsequent points from (23.17). From the moment when the stress intensity, $S_i(i) = \frac{1}{\sqrt{3}}[(S(i) - Y(i))^2 + 3T^2(i)]$, at the current point of characteristic $\xi = \eta + 2\Delta\eta - \eta_s$ decreases to the unity, then, instead of (23.17), we take (23.16). We perform the calculations up to the instant when characteristic $\xi = \eta + 2\Delta\eta - \eta_s$ intersects characteristic $\xi = \alpha_0\eta$, $(x = a_2 t)$. Next we pass to the adjacent strip $\xi = \eta + 4\Delta\eta - \eta_s$. At a point on the boundary $\xi = 0$ the parameters are calculated as before from formulae (23.17). If the stress intensity at the ith point of the characteristic $\xi = \eta + 4\Delta\eta - \eta_s$ reaches a value $S_i(i) \leqslant 1$, then we repeat, at this point, the calculation by means of (23.18); at the subsequent points we make use of (23.16). Similarly, as before, we complete the calculations to characteristic $\xi = \alpha_0\eta$. The calculations for the remaining strips are performed in the same way as for the second strip. The curve separating regions III and V is the wave of plastic loading $\xi = \varphi(\eta)$, on which $S_i = 1$.

If, beginning from a certain time instant, the stress intensity decreases and, for example, at instant $\eta = \eta_r (t = t_r)$ (Fig. 72) reaches value $S_i \leqslant 1$ (in the case of the constitutive equations (3.25), i.e. for a material without hardening) then an unloading wave, the equation of which is $\xi = \psi(\eta)$, $(x = \psi(t))$ starts propagating into the medium. Behind the unloading wave there exists the unloading region; equations (23.16) govern that region. The unloading wave is determined together with the solution in regions V and VII. Along characteristic $\xi = \eta - \eta_r$, $(x = a_1(t - t_r))$ the algorithm does not change. On characteristic $\xi = \eta + 2\Delta\eta - \eta_r$ at a point on the boundary $\xi = 0$ we compute the parameters from formulae (23.19) assuming there that $\varkappa_0 = 0$, i.e. $\Psi_1 = \Psi_2 = 0$. At the subsequent points of the characteristic we apply (29.16) until the instant when the stress intensity on this characteristic reaches a value $S_i(i) \geqslant 1$, then we repeat the calculations using the formulae:

(23.20)

$$S(i) = \frac{1}{2}[S(m) + S(l)] - \frac{1}{2}[U(m) - U(l)] - \frac{1}{2}\Psi_1(l)\Delta\eta,$$

$$Y(i) = Y(n) + \nu[S(i) - S(n)] + \frac{9K}{4\mu + 3K}\Psi_1(n)\Delta\eta,$$

$$T(i) = \frac{1}{2}[T(r) + T(s)] + \frac{1}{2}[V(s) - V(r)] - \frac{1}{2}[\Psi_2(r) + \Psi_2(s)]\Delta\eta_1,$$

(23.20) [cont.]

$$U(i) = \frac{1}{2}[U(m)+U(l)] + \frac{1}{2}[S(l)-S(n)] + \frac{1}{2}\Psi_1(l)\Delta\eta,$$

$$V(i) = \frac{1}{2}[V(r)+V(s)] + \frac{1}{2}[T(s)-T(r)] + \frac{1}{2}[\Psi_2(r)-\Psi_2(s)]\Delta\eta_1,$$

and at the subsequent points of the characteristic using (23.17). For the points further along the characteristic we repeat the calculations using the algorithm as for characteristic $\xi = \eta + 2\Delta\eta - \eta_r$. The unloading wave constitutes the geometrical locus of points in the phase plane such that $S_i = 1$.

In the case when the hardening of the material is taken into account, i.e. in the case when the constitutive equations (3.10) or (3.13) are applied, the solution procedure is the same as above; only the condition from which the front of the unloading wave is determined will change. Condition on the unloading wave $S(i) = 1$, in the case of the medium without hardening, has to be replaced by the condition

(23.21)
$$S_i[\psi(\eta), \eta] = \frac{\varkappa}{k}\left(\int_0^t \sigma_{kl}\, d\varepsilon_{kl}^p\right),$$

in the case of constitutive equations (3.10), or by the condition

(23.22)
$$S_i[\psi(\eta), \eta] = \frac{f(\varepsilon_i)}{k}$$

when the constitutive equations (3.13) are used. The introduction of material hardening does not make qualitative changes; only quantitative changes are observed.

The parameters on characteristic $\xi = \alpha_0 \eta$, $(x = a_2 t)$ are known from the solution in region III. Region IV is a region of elastic deformation; Riemann waves propagate there. Denoting by $S_0(\eta)$ the stress along characteristic $\xi = \alpha_0 \eta$ determined from region III, we can write down the solution in region IV in the forms:

(23.23)
$$S(\xi,\eta) = S_0(\xi-\eta), \qquad Y(\xi,\eta) = v\, S_0(\xi-\eta),$$
$$U(\xi,\eta) = -S(\xi,\eta), \qquad T(\xi,\eta) = V(\xi,\eta) = 0.$$

In region VI only longitudinal viscoplastic waves propagate. The solution in this region is constructed using (23.17) with $T = V = 0$ (likewise in functions Ψ_1, Ψ_2 we have to assume that $T = V = 0$). Region VIII is the unloading region, only elastic longitudinal waves propagate in it; the solution in this region is obtained by means of (23.16). On the boundary of regions VI and VIII we have to apply (23.20), taking into account that $T = V = 0$ and $\Psi_2 \equiv 0$.

23.2. WAVES OF STRONG DISCONTINUITY

Many cases of the propagation of strong discontinuity waves have been discussed in [60], depending on the character of the changes in the tractions $p(t)$ and $q(t)$ on the boundary of the half-space $x_1 = 0$. Only one of these cases is presented here, namely the case when both the tractions at instant $t = 0$ suddenly appear on the boundary and then change arbitrarily in time (e.g. as in Fig. 74). It is assumed that the initial values of the stresses p_0 and q_0 are sufficiently high that they produce viscoplastic deformation in the medium. Zero initial conditions have been assumed.

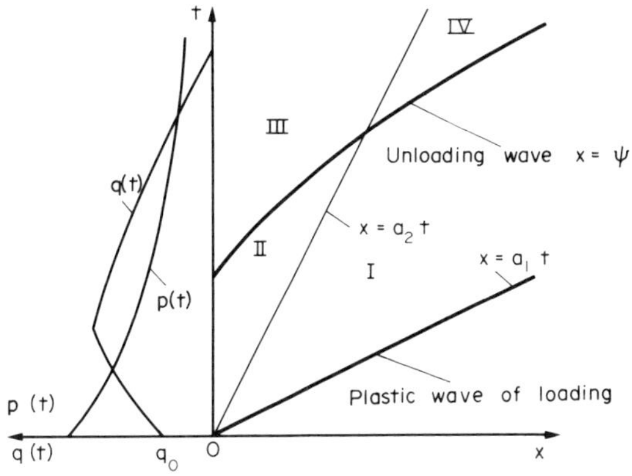

Fig. 74

Two strong discontinuity waves with the speeds a_1 and a_2, the fronts of which coincide with the characteristics $x = a_1 t$ and $x = a_2 t$, propagate from the origin of the coordinate system in the phase plane (x, t). Across the front of wave $x = a_1 t$, stresses S, Y, E and velocity U suffer jump changes. In the case of small strains the equations of the dynamic and kinematic continuity (7.18) and (8.9) take the forms

(23.24) $$S = -U, \qquad U = -E,$$

where $G = a_1$ and we have taken into account the fact that wave $x = a_1 t$ propagates into the undisturbed medium.

Making use of the above conditions of the law compressibility $(3.10)_2$

(23.25) $$S + 2Y = \frac{3K}{\varrho a_1^2} E$$

and of the condition on the positive characteristic (23.5)

(23.26) $$dS - dU + \Psi_1\, d\eta = 0,$$

we obtain the integral equation

(23.27) $$S(\eta) = -\frac{1}{2}\int\limits_0^{\eta} \Psi'\, d\xi - \frac{p_0}{k},$$

which in general can be solved by the successive approximation method with a known error. In the case of the constitutive equations (3.25), disregarding the hardening of the material and for a linear function Φ $(\Phi(F) = F)$, since

$$\Psi' = \frac{2}{3}\varkappa_0(S - Y - \sqrt{3})$$

we obtain the exact form solution of (23.27):

(23.28)
$$S(\eta) = -\frac{p_0}{k} e^{-ht} + \frac{\sqrt{3}}{3}\frac{\gamma}{h}(1-e^{ht}),$$

where

$$h = \left(1 - \frac{3K}{\varrho a_1^2}\right)\frac{\gamma}{2}.$$

The remaining parameters of the solution across the strong discontinuity wave are expressed as follows:

(23.29)
$$U(\eta) = -S(\eta), \quad Y(\eta) = -\left(1 - \frac{3K}{\varrho a_1^2}\right)\frac{S(\eta)}{2},$$
$$T = V = \Gamma \equiv 0.$$

It was proved in [60] that across wave $x = a_2 t$ only the magnitudes T, Γ, and V suffer jumps while S, Y, U and E change continuously across its front. The continuity conditions (23.24) across wave $x = a_2 t$ assume the form

(23.30)
$$T = -V, \quad V = -\Gamma.$$

The solution in the separate regions of phase space (Fig. 74) is constructed numerically using the finite difference method with the net of characteristics by making use of solution (23.27) on wave $x = a_1 t$ and of conditions (23.30) on wave $x = a_2 t$.

The front of the elastic unloading wave $x = \Psi(t)$ is constructed in the same way as in the case of the weak discontinuity waves.

23.3. ANALYSIS OF NUMERICAL RESULTS

By virtue of the numerical calculations which have been performed we can present certain interesting excerpts from the solutions of cases of the propagation of weak and strong discontinuity waves [61]. In Fig. 75 the solution is illustrated of four variants of a change of traction $q(t)$ with a stepwise applied stress $\sigma(t) = \sigma_S$ which then varies from the instant $t = t_S$. Stress $p_0 < \sigma_{is}$ produces only a longitudinal elastic wave of strong discontinuity; it does not have any effect on the configuration of plastic waves. We see from the figure that the plastic loading wave for a continuous change of $q(t)$ within $0 \leqslant t \leqslant t_s$ possesses the character of a curvilinear wave of weak discontinuity. With the increase of $q(t)$ the plastic wave undergoes refraction. In the limiting case of loading $q_4(t) = q_0$, the curvilinear loading wave becomes a characteristic of the longitudinal wave. This effect is not typical, nor is it encountered in other problems of this kind.

The influence of the initial gradient of the increase in the pressure $p(t)$, for $t > t_s$, on the position of the upper point of refraction of the plastic wave (Fig. 76) is examined. It has been found that the higher the gradient the smaller is the segment of wave refraction. In Fig. 76 the position of the unloading wave is presented for three kinds of change of pressure $p(t)$ on the boundary of the half-space. The calculations are performed using the constitutive equations for a medium without hardening and for a fixed value of the viscosity coefficient. The calculations can also be performed for various values of the viscosity coefficient. It can be shown that the higher the viscosity of the medium the smaller the region of viscoplastic deformation.

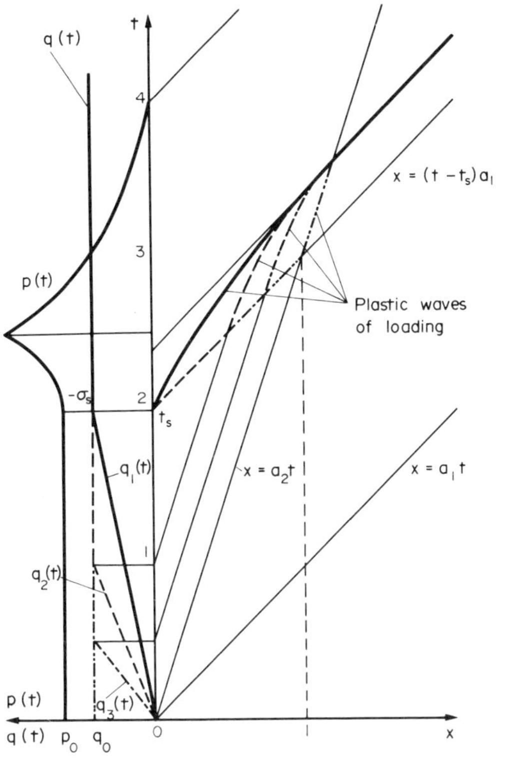
q (t)
t
4
3
2
1
p (t)
-σs
ts
q1(t)
q2(t)
q3(t)
x = (t − ts)a1
Plastic waves
of loading
x = a2t
x = a1t
p (t)
q (t)
p0
q0
0
x

Fig. 75

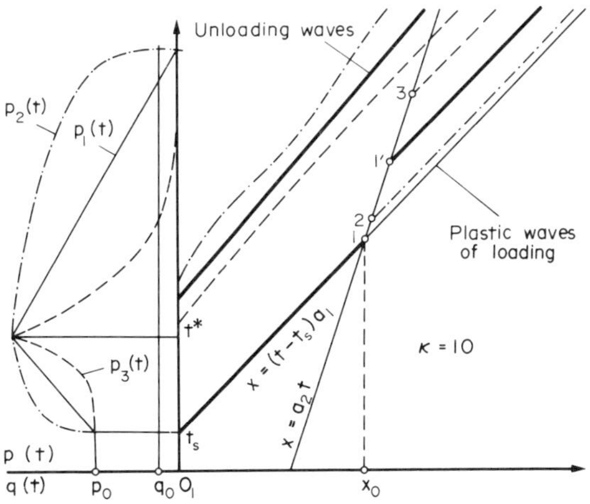
Unloading waves
p2(t)
p1(t)
p3(t)
t*
ts
3
1'
2
1
Plastic waves
of loading
κ = 10
x = (t − ts)a1
x = a2t
p (t)
q (t)
p0
q0
O1
x0

Fig. 76

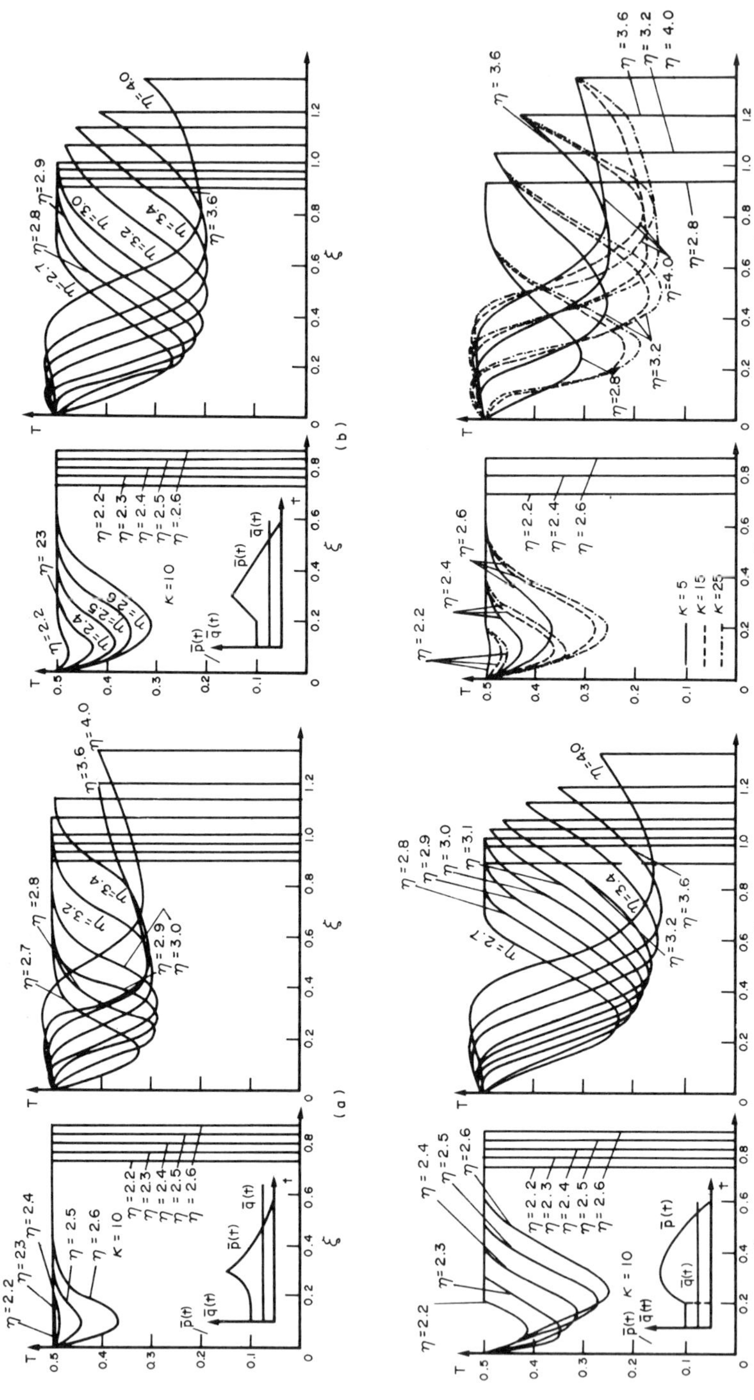
(a)
(b)
(c)
(d)
T
ξ
κ = 10
κ = 5
κ = 15
κ = 25
$\bar{p}(t)$
$\bar{q}(t)$
t
η = 2.2
η = 2.3
η = 2.4
η = 2.5
η = 2.6
η = 2.7
η = 2.8
η = 2.9
η = 3.0
η = 3.1
η = 3.2
η = 3.4
η = 3.6
η = 4.0

Fig. 77

The changes of shear stress T in terms of function ξ and time η for the three variations of loading $p(t)$ shown in Fig. 76 are presented in Fig. 77 a–c. The effect of the viscosity on the change of shear stress T is given in Fig. 77d. It is easy to note that the effect of the gradient of the stress $p(t)$ on the region of viscoplastic deformation is the same as the effect of viscosity. The characteristic feature of the figures is that the stress T decreases as a function of position and then it increases in spite of the constant tangential stress q_0 on the bounding plane. The maximum values in the half-space exceed the value q_0.

By virtue of the numerical calculations presented in [61] we can draw the following conclusions:

(1) Loading rates have an influence on the wave configuration.

(2) There is an essential dependence of the tangential stress state components on the tractions $p(t)$ and their gradients.

(3) We observe that the influence on the solution of increasing the viscosity is similar to that of increasing the pressure rate.

(4) A conclusion specific to a complex stress state: for constant q_0, and increasing and then decreasing $p(t)$ on the boundary, an inverse picture is obtained for T in the region of viscoplastic strain.

(5) Normal stress components S and Y depend little on the shear stress component T. Furthermore, the influence of T component on S and Y components vanishes with the distance from the boundary of the half-space. Thus in engineering calculations we can neglect the effect of shear components on the distribution of the normal stresses [100].

The calculations for an elastic/viscoplastic material when material hardening is taken into account (3.13) can also be performed. The influence of the hardening does not change the qualitative picture of the solutions. The change of stress intensity S_i against the strain intensity E_i is presented in Fig. 78: (a) for the case of the linear hardening, (b) for various values of the hardening parameter ($0 \leqslant \lambda \leqslant 0.6$; $\lambda = 0$ corresponds to a medium without hardening). It can

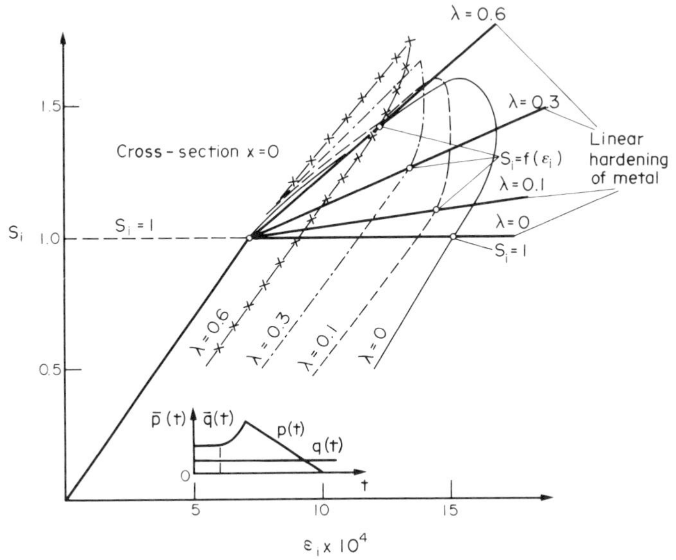

Fig. 78

be stated that the viscoplastic hysteresis loop is contracted with increase of the hardening parameter. This is obvious since increased hardening reduces the value of the permanent strain.

24. Longitudinal–transverse waves in a non-homogeneous elastic/viscoplastic medium

24.1. PLANE WAVES

The problem of the propagation of plane longitudinal–transverse waves in the case of a non-homogeneous, elastic/viscoplastic medium has been solved in [163]. The constitutive equations for soil deformation, (4.16) and (4.18) for a generally non-homogeneous medium, were taken as the starting point. For this medium the following physical quantities – shear strain modulus μ, bulk modulus K, the coefficients of viscosity γ, and of dilatation α, yield limit k, and density ϱ – are all functions of the spatial variable x_1. For these inhomogeneities, equations (4.16) and (4.18) take the form

(24.1)
$$\dot{\varepsilon}_{ij} = \frac{1}{2\mu(x_1)}\dot{s}_{ij} + \frac{1}{3K(x_1)}\dot{s}\,\delta_{ij} + \gamma(x_1)\langle\Phi\left[\frac{\alpha(x)J_1' + \sqrt{J_2}}{k(x_1)} - 1\right]\rangle\left[\alpha(x_1)\delta_{ij} + \frac{s_{ij}}{2\sqrt{J_2}}\right],$$
$$\dot{\varepsilon}_{ii} = \frac{1}{3K(x_1)}\dot{\sigma}_{ii} + 3\alpha(x_1)\gamma(x_1)\langle\Phi\left[\frac{\alpha(x)J_1' + \sqrt{J_2}}{k(x_1)} - 1\right]\rangle.$$

For small strains of a medium governed by the above constitutive equations, considering a half-space $x_1 > 0$ (Fig. 68) with the boundary conditions (22.1), taking account of conditions (22.3) and the equations of motion

(24.2)
$$\sigma_{ij,j} = \varrho(x_1)\,v_{i,t},$$

the problem is reduced to the solution of the following system of equations:

(24.3)
$$\left[1 + \frac{3}{4}\frac{K(x)}{\mu(x)}\right]v_{1,1} - \frac{3}{4\mu(x)}\sigma_{,t} = \gamma(x)\langle\Phi_1\rangle,$$
$$v_{2,1} - \frac{1}{\mu(x)}\tau_{,t} = \gamma(x)\langle\Phi_2\rangle, \qquad \sigma_{,1} - \varrho(x)\,v_{1,t} = 0, \qquad \tau_{,1} - \varrho(x)\,v_{2,t} = 0,$$

where

$$\langle\Phi_1\rangle = \langle\Phi\left[\frac{\alpha(x)J_1' + \sqrt{J_2}}{k(x)} - 1\right]\rangle\left[\frac{9}{4}\frac{K(x)}{\mu(x)}\alpha(x) + \frac{\sigma_1 - \sigma_2}{2\sqrt{J_2}}\right],$$
$$\langle\Phi_2\rangle = \langle\Phi\left[\frac{\alpha(x)J_1' + \sqrt{J_2}}{k(x)} - 1\right]\rangle\frac{\tau}{\sqrt{J_2}}.$$

The system of (24.3) is solved for zero initial conditions $u_1 = u_2 = v_1 = v_2 = 0$ for $t = 0$.

The differential equations of the characteristics of the system (24.3) have the forms

$$dx = \pm a_1(x)\,dt, \qquad dx = \pm a_2(x)\,dt. \tag{24.4}$$

The following relations are valid along the characteristics:

$$\begin{aligned} &d\sigma_1 \mp \varrho(x)\,a_1(x)\,dv_1 + \frac{4}{3}\mu(x)\gamma(x)\langle\Phi_1\rangle\,dt = 0 \quad \text{for} \quad dx = \pm a_1(x)\,dt, \\ &d\tau \mp \varrho(x)\,a_2(x)\,dv_2 + \mu(x)\gamma(x)\langle\Phi_2\rangle\,dt = 0 \quad \text{for} \quad dx = \pm a_2(x)\,dt. \end{aligned} \tag{24.5}$$

The additional relation is obtained from the constitutive equations (24.1) for $x = 0$:

$$d\sigma_2 - \frac{3K(x)-2\mu(x)}{3K(x)+4\mu(x)}\,d\sigma_1 + 6\mu(x)\gamma(x)\frac{K(x)}{3K(x)+4\mu(x)}\langle\Phi_3\rangle\,dt = 0, \tag{24.6}$$

where

$$\langle\Phi_3\rangle = \langle\Phi\left[\frac{\alpha(x)J_1' + \sqrt{J_2}}{k(x)} - 1\right]\rangle\left[-3\alpha(x) - \frac{\sigma_1-\sigma_2}{2\sqrt{J_2}}\right].$$

The shapes of the fronts of the loading and unloading waves are determined, in this case, not only by the boundary conditions but also by the inhomogeneous character of the medium.

In the case of a loading (or unloading) wave of weak discontinuity we can determine the local velocity of the wave front in a manner similar to that for the homogeneous medium (section 23.1). From the total differentials of the stress intensity $\sigma_i(x_1, t)$ in the direction of the tangent to the loading (or unloading) front and in the direction of the characteristic of the longitudinal or transverse wave we obtain the following formulae for the local velocity of propagation of the front of the loading (or unloading) wave:

$$\varphi'(t) = a_1(x)\frac{\sigma_{i,t}}{\sigma_{i,t} - a_1(x)\left[\sigma_{i1}'(x) + k'(x)\right]}\Bigg|_{x=\varphi(t)} \tag{24.7}$$

or

$$\varphi'(t) = a_2(x)\,\frac{\sigma_{i,t}}{\sigma_{i,t} - a_2(x)\left[\sigma_{i2}'(x) - k'(x)\right]}\Bigg|_{x=\varphi(t)}. \tag{24.7'}$$

Here $a_1(x)$ and $a_2(x)$ denote the propagation speeds of the longitudinal and transverse waves and $\sigma_{i1}'(x)$ and $\sigma_{i2}'(x)$ are the stress intensities along the directions determined by $a_1(x)$ and $a_2(x)$ respectively in the phase plane (x, t). It follows from an analysis of (24.7) that the front of the loading (or unloading) wave may assume all possible positions with respect to the characteristics of the longitudinal and transverse waves in contrast with the case of a homogeneous medium where the velocity of the plastic (or unloading) wave was bounded in the region $a_2 \leqslant c^* \leqslant a_1$ (see (23.12)). If $\sigma_{i1}' > k(x)$, then $\varphi'(t) > a_1[\varphi(t)]$ – the front of the plastic loading wave lies beneath the longitudinal wave characteristic. The front is determined from the condition $\sigma_{i1}'(x) \leqslant k'(x) \leqslant \sigma_{i2}'(x)$, where σ_i is the stress intensity in the region of elastic strain in front of the plastic wave front. If $\sigma_{i1}'(x) \geqslant k'(x) \geqslant \sigma_{i2}'(x)$ then $a_2(x) \leqslant \varphi'(t) \leqslant a_1(x)$, i.e. the front of the plastic loading waves lies between the characteristics of the longitudinal and transverse waves or, in a limiting case, coincides with one of them.

Also the case has been examined when $\sigma_{i2}' < k'(x)$, then $\varphi'(t) < a_2[\varphi(t)]$; this means that the front of the plastic loading wave lies above the characteristic of the transverse wave.

In general the system of (24.3) can be solved numerically by finite difference equations expanded along the characteristic net, using the relations on the characteristics (24.5) and equation (24.6) or by the reducing equations (24.3) to integro-differential equations and solving them by the successive approximation method. In this case the calculations are much more complicated as compared with the case of a homogeneous medium. A certain class of the inhomogeneous media can be defined for which an exact solution can be obtained in the regions of elastic deformation.

If we assume an exponential variation of the moduli $K(x)$ and $\mu(x)$ and of the medium density $\varrho(x)$, then the system of equations (24.3), for $\Phi_1 = \Phi_2 = 0$, reduces to the telegraph equation. Another way of solving the problem in the regions of elastic deformation was presented in [163], namely (24.3) were reduced (for $\Phi_1 = \Phi_2 \equiv 0$) to Euler–Darboux equations. As well as these, quite a wide class of inhomogeneities has been examined. The construction of the solution of the problem was presented in the above reference for the case of weak discontinuity waves (formulation analogous to that in section 23.1) and some remarks were given concerning solutions in the case of strong discontinuity waves.

In the second part of the cited paper, from the results of numerical calculations, the influences of the general inhomogeneity of the medium and of the coefficient of the voluminal dilatation on the dispersion process of plane longitudinal–transverse waves have been studied.

(1) It was ascertained that, as compared with the results for a homogeneous medium, the effect of the medium inhomogeneity which was examined introduces essential changes in the distribution of the stresses in both qualitative and quantitative respects. With increase of the moduli μ and K and density of medium, an increase of the stress tensor components takes place (and conversely in the case of the decrease). It results from the increase (or decrease) in the coefficient of continuous reflection of incident waves from the inhomogeneities of medium.

(2) The stress tensor components decrease with increase in the coefficient of voluminal dilatation; this effect is more pronounced with increase in the viscosity coefficient of the medium.

(3) The loop of hysteresis viscoplastic increases with increase of the coefficients of voluminal dilatation and of medium viscosity.

24.2. RADIAL CYLINDRICAL WAVES

The problem of the propagation of longitudinal–transverse loading and unloading waves of cylindrical symmetry in a generally inhomogeneous elastic/viscoplastic medium, with the permanent voluminal strain taken into account, constitutes a generalization of the problem posed in section 24.1 to the case of cylindrical symmetry. The waves of this type are generated by radial and tangential tractions uniformly distributed over a cylindrical surface of initial radius r_0.

We shall consider the motion of an unbounded medium (the physical parameters are functions of radius r), elastic/viscoplastic with a cylindrical cavity of initial radius r_0 on whose surface radial and tangential tractions in the transverse and axial directions (Fig. 79) are prescribed:

$$\sigma_r(r_0, t) = -|\sigma(t)|,$$
$$\tau_\varphi(r_0, t) = -|\tau_1(t)|, \qquad (24.8)$$
$$\tau_z(r_0, t) = -|\tau_2(t)|,$$

which vary arbitrarily with time. Zero initial conditions are assumed.

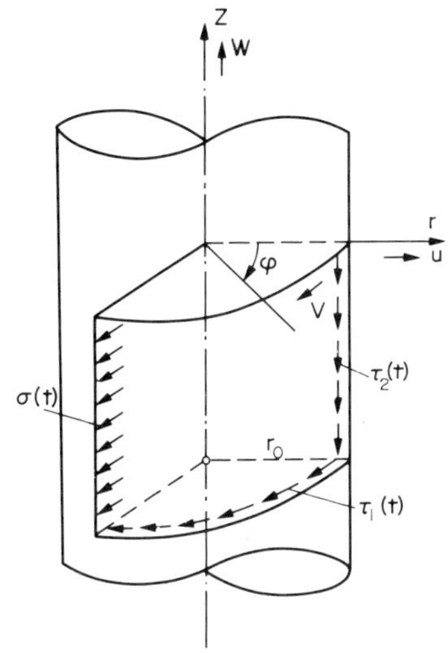

Fig. 79

In our treatment we use the constitutive equations (24.1) which, for the case of small strains, expressed in the cylindrical coordinates constitute, together with the equations of motion and the mass continuity condition, the system of equations for the problem considered [164]:

(24.9)
$$u_{,rt} = \frac{\mu(r)+3K(r)}{9K(r)\mu(r)}\dot{\sigma}_r + \frac{2\mu(r)-3K(r)}{18K(r)\mu(r)}(\dot{\sigma}_r+\dot{\sigma}_z)+\gamma(r)\langle\Phi_1\rangle,$$

$$\frac{1}{r}u_{,t} = \frac{\mu(r)+3K(r)}{9K(r)\mu(r)}\dot{\sigma}_\varphi + \frac{2\mu(r)-3K(r)}{18K(r)\mu(r)}(\dot{\sigma}_r+\dot{\sigma}_z)+\gamma(r)\langle\Phi_2\rangle,$$

$$0 = \frac{\mu(r)+3K(r)}{9K(r)\mu(r)}\dot{\sigma}_z + \frac{2\mu(r)-3K(r)}{18K(r)\mu(r)}(\dot{\sigma}_\varphi+\dot{\sigma}_r)+\gamma(r)\langle\Phi_2\rangle,$$

$$\frac{1}{2}\left(v_{,rt}-\frac{v_{,t}}{r}\right) = \frac{1}{2\mu(r)}\dot{\tau}_{r\varphi}+\gamma(r)\langle\Phi_4\rangle,$$

$$\frac{1}{2}w_{,rt} = \frac{1}{2\mu(r)}\dot{\tau}_{rz}+\gamma(r)\langle\Phi_5\rangle,$$

$$\varrho u_{,tt}-\sigma_{r,r} = \frac{\sigma_r-\sigma_\varphi}{r}, \quad \varrho v_{,tt}-\tau_{r\varphi,r} = 2\frac{\tau_{r\varphi}}{r}, \quad \varrho w_{,tt}-\tau_{rz,r} = \frac{\tau_{rz}}{r},$$

where

$$u_{,rt} = \varepsilon_{r,t}, \quad \frac{u_{,t}}{r} = \varepsilon_{\varphi,t}, \quad \frac{1}{2}\left(v_{,rt}-\frac{v_{,t}}{r}\right) = \gamma_{r\varphi,t}, \quad \frac{1}{2}w_{,rt} = \gamma_{rz,t},$$

(24.10)
$$\langle\Phi_1\rangle = \langle\Phi(F)\rangle\left[\alpha(r)+\frac{2\sigma_r-(\sigma_\varphi+\sigma_z)}{6\sqrt{J_2}}\right],$$

(24.10) [cont.]

$$\langle \Phi_2 \rangle = \langle \Phi(F) \rangle \left[\alpha(r) + \frac{2\sigma_\varphi - (\sigma_r + \sigma_z)}{6\sqrt{J_2}} \right],$$

$$\langle \Phi_3 \rangle = \langle \Phi(F) \rangle \left[\alpha(r) + \frac{2\sigma_z - (\sigma_r + \sigma_\varphi)}{6\sqrt{J_2}} \right],$$

$$\langle \Phi_4 \rangle = \langle \Phi(F) \rangle \frac{\tau_{r\varphi}}{2\sqrt{J_2}},$$

$$\langle \Phi_5 \rangle = \langle \Phi(F) \rangle \frac{\tau_{rz}}{2\sqrt{J_2}},$$

$$\langle \Phi(F) \rangle = \left\langle \Phi \left[\frac{\alpha(r) J_1' + \sqrt{J_2}}{k(r)} - 1 \right] \right\rangle.$$

The system of (24.9) is solved by the method of characteristic net, the differential relations along the characteristic directions being replaced by difference equations. Making use of the recurrence formula for the characteristic mesh, for the prescribed initial conditions, we determine the discrete fields of stress, velocity, and strain both in regions of elastic and viscoplastic deformation. The numerical calculations were performed on a digital computer. The solutions have a local character and give a good approximation in a sufficiently near vicinity of the cylindrical surface and for small times. For large values of r and time t the error resulting from the application of the finite difference method along the characteristic is considerable.

Similarly to the case of plane waves (section 24.1) we can determine a certain class of inhomogeneous media for which it is possible to construct the exact solution in the regions of elastic deformation, reducing the equations of the problem (24.9) to an equation of the Euler–Darboux type. In the second part of the paper [164] a discussion is given of the results obtained from the computations. The effects of the physical parameters on the distribution of the strain and velocity fields are examined.

A number of essential conclusions are drawn referring to the influence of the physical parameters and of the inhomogeneity of the medium on the extent of the range of viscoplastic deformation.

(1) The region of viscoplastic deformation decreases with increase of the absolute value of the coefficient α. The phenomenon is caused by the energy loss due to crushing the skeleton and to soil consolidation. In this way the permanent, irreversible, voluminal strain of the medium is produced.
(2) The depth of the viscoplastic region increases and the duration time of active viscoplastic strain decreases, at particular cross-sections, with increase of the viscosity coefficient.
(3) The depth of penetration of the viscoplastic waves decreases and, simultaneously, the duration time of the process of the viscoplastic deformation, in the vicinity of the cylindrical cavity in the medium, is extended with the increase of Poisson's ratio.
(4) Small increments of the radial stress $\sigma(t)$ lead to a considerable increase in the depth of penetration of the viscoplastic waves.
(5) The region of viscoplastic deformation decreases with increase of the yield limit $k(r)$, and the converse is also true.

Also a series of other conclusions, referring to the analysis of the stress and velocity fields, in the regions of the viscoplastic deformation, have been drawn. On the grounds of the results obtained and from experimental data, we can assert that the constitutive equations (24.1) well describe the physical properties of soils.

The problem of the propagation of longitudinal–transverse, radial, cylindrical waves in a homogeneous elastic/viscoplastic medium, e.g. in a thin-walled, semi-infinite cylinder with boundary conditions (Fig. 80),

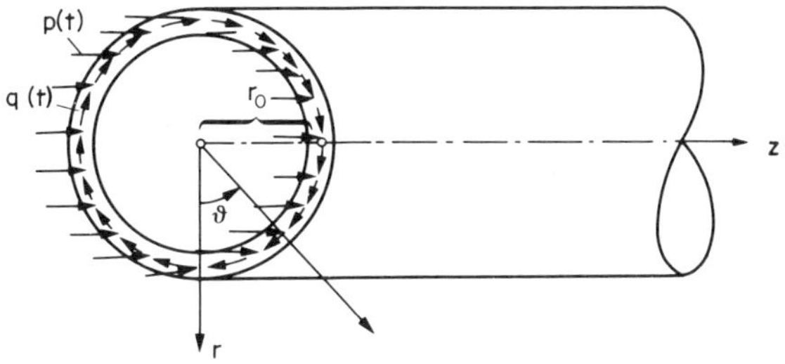

Fig. 80

(24.11) $$\sigma_z(r_0, t) = p(t), \qquad \tau_{r\varphi}(r_0, t) = q(t) \qquad \text{for} \qquad z = 0,$$

where $p(t)$ and $q(t)$ are arbitrary functions of time, is similar, from the mathematical standpoint, to the problem of the propagation of longitudinal–transverse waves in a half-space generated by the boundary conditions (22.1). In this case we have non-vanishing stress tensor components σ_z and $\tau_{r\varphi}$ and the strain tensor components ϵ_z and $\gamma_{r\varphi}$. The construction of the solution of this problem is the same as that of the problem discussed in section 23.

25. Stress waves in beams

The problems of the propagation in beams of transverse and bending waves are not so exhaustively treated as, for example, are the problems of longitudinal wave propagation in bars. Very little literature exists in this field at the moment. There is a lack of a general theory for solving problems in the case of transverse and bending wave propagation in elastic–plastic beams. Various authors have different ideas on the solution of these problems. The solution is difficult and labour consuming. Each time we have to resort to numerical computation. These computations are complicated because there are different propagation velocities for transverse and bending waves (elastic and plastic). Besides which we have to solve for unloading waves and, in the case of finite beams, for wave reflection from the beam boundary. Problems of the propagation of transverse and bending waves in elastic/viscoplastic beams are better worked out as compared to the case of elastic–plastic beams. On account of the existence of the medium viscosity plastic transverse and bending waves propagate with the same velocities as the corresponding elastic waves. This permits us to obtain a much simpler picture of the wave propagation. At this point we confine ourselves to a discussion of problems of transverse and bending wave propagation in beams whose material exhibits rheological effects.

We shall base the discussion upon the Timoshenko beam theory in which the effects of shear forces and of inertia forces of the rotation of beam elements on the transverse motion are taken into account.

First we consider a cantilever beam. We assume that a bending moment $M^0(t)$ and a shear force $N^0(t)$, both of which vary with time, act at the beam end. It is also assumed that the beam is homogeneous and of constant cross-section. The cartesian coordinate system (x, y, z) is oriented in such a way that the x-axis is directed along the axis of the beam and the y- and z-axes are directed along the axes of symmetry of the cross-section. The beam is bent with respect to the y-axis. In the selected coordinate system the stress state of the beam is determined by the normal stress σ_{xx} and the shear strain τ_{xz}, whilst the state of strain is determined by the longitudinal strain of the beam ϵ_{xx} and the shear strain γ_{xz}.

The normal stress σ_{xx} and shear stress τ_{xz} acting in the bent beam correspond to the bending moment M and the shear force N in the cross-section of the beam. These quantities are defined in the following way:

$$M = \iint_A \sigma_{xx} z \, dA, \qquad N = \iint_A \tau_{xz} \, dA, \tag{25.1}$$

where A denotes the area of the cross-section of the beam.

Now we shall define the displacements and the strains in the beam. We define by u and w the components of the displacement vector of an arbitrary point of the beam in the direction of the x- and z-axes respectively. The projections of the total displacement of an arbitrary point of the u and w on the z- and x-axes respectively (in the cross-section x at the distance z from the beam axis, $-\frac{h}{2} \leqslant z \leqslant \frac{h}{2}$, where h is the depth of the beam) are determined by the formulae

$$w = w(x, t), \qquad u = -\frac{\partial w}{\partial x} z + \gamma(x, t) z, \tag{25.2}$$

where $\gamma(x, t)$ is the function describing the change in the warping along the beam axis. The strain tensor components in the beam, on account of (25.2), take the forms

$$\begin{aligned} \varepsilon_{xx}(x, z, t) &= \frac{\partial u}{\partial x} = -\frac{\partial^2 w}{\partial x^2} z + \frac{\partial \gamma}{\partial x} z, \\ \gamma_{xz}(x, z, t) &= \frac{\partial u}{\partial x} + \frac{\partial w}{\partial x} = \gamma(x, t). \end{aligned} \tag{25.3}$$

The strain $\gamma(x, t)$ denotes a certain mean value of the shear angle $\gamma(x, z, t)$ in the middle surface of the beam.

If we take into account (25.2) then an element lying initially on the normal to the beam axis rotates by a certain angle α:

$$\alpha(x, t) = \frac{\partial w(x, t)}{\partial x} - \gamma(x, t). \tag{25.4}$$

Displacement w, deflection angle of the beam axis $\partial w/\partial x$, mean angle of rotation of an element α, strain ϵ_{xx}, mean shear strain $\gamma(x, t)$, and angular velocity ω in the case of small deflections and small rotation angles of the beam are coupled by the following equations, called the continuity relations:

$$\frac{\partial w}{\partial x} = \alpha + \gamma, \qquad v = \frac{\partial w}{\partial t}, \qquad \omega = \frac{\partial \alpha}{\partial t}, \qquad \varepsilon_{xx} = -z\frac{\partial \alpha}{\partial x}, \tag{25.5}$$

$$\frac{\partial \varepsilon}{\partial t} = -z\frac{\partial \omega}{\partial x}, \qquad \frac{\partial \gamma}{\partial t} = \frac{\partial v}{\partial x} - \omega. \tag{25.6}$$

The equations of motion of translation and rotation for an arbitrary cross-section of the beam have the forms

$$\frac{\partial N}{\partial x} = A\varrho\frac{\partial v}{\partial t}, \qquad \frac{\partial M}{\partial x} + I\varrho\frac{\partial \omega}{\partial t} = N, \tag{25.7}$$

where I is the moment of inertia of the cross-section with respect to the y-axis and ϱ is the material density.

For a complete description of the vibration of a beam we have to add the constitutive equations and the boundary and initial conditions.

We shall consider the vibrations of a beam of elastic/viscoplastic material. The constitutive equations are taken in the form of (3.10). In the case considered these equations reduce to the forms

$$\begin{aligned} \dot{\varepsilon} &= \frac{1}{E}\dot{\sigma} + \frac{2}{3}\bar{\gamma}\left\langle \Phi\left[\frac{\sqrt{J_2}}{\varkappa} - 1\right]\right\rangle \frac{\sigma}{\sqrt{J_2}}, \\ \dot{\gamma} &= \frac{1}{\mu}\dot{\tau} + \bar{\gamma}\left\langle \Phi\left[\frac{\sqrt{J_2}}{\varkappa} - 1\right]\right\rangle \frac{\tau}{\sqrt{J_2}} \end{aligned} \tag{25.8}$$

where we have written $\epsilon_{xx} = \epsilon$, $\sigma_{xx} = \sigma$, $\tau_{xz} = \tau$, and $\bar{\gamma}$ denotes the viscosity coefficient of the material. The second invariant of the stress deviator is expressed in the following manner:

$$J_2 = \frac{1}{3}\sigma^2 + \tau^2. \tag{25.9}$$

The system of (25.1) and (25.6)–(25.8) constitute the complete system of the equations for the problem. The system consists of eight equations in the first derivatives of the eight unknown functions $M, N, \sigma, \tau, v, \omega, \epsilon$, and γ. If we confine ourselves to the case of an elastic/visco-ideally-plastic beam material, i.e. when $\varkappa = k$ (k denotes the static yield limit for pure shear), the problem can be solved in terms of the six quantities: M, N, σ, τ, v, and ω. Thus making use of the continuity relations (25.5) and of equations (25.9) we obtain from the constitutive equations (25.8)

$$-z\frac{\partial \omega}{\partial x} = \frac{1}{E}\frac{\partial \sigma}{\partial t} + \frac{2}{3}\bar{\gamma}\left\langle \Phi\left[\frac{\sqrt{\frac{1}{3}\sigma^2 + \tau^2}}{k} - 1\right]\right\rangle \frac{\sigma}{\sqrt{\frac{1}{3}\sigma^2 + \tau^2}}, \tag{25.10}$$

(25.10) [cont.]

$$\frac{\partial v}{\partial x} - \omega = \frac{1}{\mu}\frac{\partial \tau}{\partial t} + \bar{\gamma}\left\langle \Phi\left[\frac{\sqrt{\frac{1}{3}\sigma^2 + \tau^2}}{k} - 1\right]\right\rangle \frac{\tau}{\sqrt{\frac{1}{3}\sigma^2 + \tau^2}}.$$

Accordingly, for the case of an elastic/visco-ideally-plastic beam we obtain the system of six equations ((25.1), (25.7), and (25.10)) with respect to the first derivatives of the six unknown functions M, N, σ, τ, v, and ω. The system of (25.10) will be presented in a different form. To do this, we multiply the first of the equations by zdA and the second one by dA. Next we integrate over the cross-section A. Making use of definitions (25.1) we obtain

(25.11)

$$-I\frac{\partial \omega}{\partial x} = \frac{1}{E}\frac{\partial M}{\partial t} + \frac{2}{3}\bar{\gamma}\iint_A \left\langle \Phi\left[\frac{\sqrt{\frac{1}{3}\sigma^2 + \tau^2}}{k} - 1\right]\right\rangle \frac{\sigma}{\sqrt{\frac{1}{3}\sigma^2 + \tau^2}}\, zdA,$$

$$A_s\left(\frac{\partial v}{\partial x} - \omega\right) = \frac{1}{\mu}\frac{\partial N}{\partial t} + \bar{\gamma}\iint_A \left\langle \Phi\left[\frac{\sqrt{\frac{1}{3}\sigma^2 + \tau^2}}{k} - 1\right]\right\rangle \frac{\tau}{\sqrt{\frac{1}{3}\sigma^2 + \tau^2}}\, dA,$$

where, as before, I denotes the moment of inertia of the cross-section and A_s is the effective shear surface:

(25.12)
$$A_s = \frac{1}{\gamma(x, t)}\iint_A \gamma(x, z, t)\, dA.$$

The shear surface A_s is different from the entire cross-section surface of the beam since, in reality, the true shear angle $\gamma(x, z, t)$ changes with depth in the beam. In strength of materials the effective shear surface A_s is determined by the coefficient k':

(25.13)
$$A_s = k'A,$$

where

$$k' = \frac{1}{\iint_A \frac{S^2(z)A}{I^2h^2}\, dA}.$$

Here S denotes the first moment of area of the beam cross-section and h is the depth of the beam. In the case of a rectangular beam of depth h and thickness b, we have $A_s = 2hb$.

Now, replacing stresses σ and τ by bending moment M and shear force N, we transform the statistical yield condition

$$J_2 = \frac{1}{3}\sigma^2 + \tau^2 = k^2.$$

We assume the following form for the yield condition (in the case of an elastic/visco-ideally-plastic body):

$$F = \sqrt{\left(\frac{M}{M_0}\right)^2 + \left(\frac{N}{N_0}\right)^2} - 1, \tag{25.14}$$

where M_0 and N_0 denote the corresponding integrals

$$M_0 = \sqrt{3}\iint_A kz\,dA, \qquad N_0 = \iint_A k\,dA. \tag{25.15}$$

The second invariant of the stress deviator obviously takes the form

$$J_2 = k^2\left[\left(\frac{M}{M_0}\right)^2 + \left(\frac{N}{N_0}\right)^2\right]. \tag{25.16}$$

If we take into account definition (25.1), then the system of (25.11) takes the following forms:

$$-I\frac{\partial \omega}{\partial x} = \frac{1}{E}\frac{\partial M}{\partial t} + \frac{2}{3}\bar{\gamma} \left\langle \Phi\left[\sqrt{\left(\frac{M}{M_0}\right)^2 + \left(\frac{N}{N_0}\right)^2} - 1\right]\right\rangle \frac{M}{k\sqrt{\left(\frac{M}{M_0}\right)^2 + \left(\frac{N}{N_0}\right)^2}},$$

(25.17)

$$A_s\left(\frac{\partial v}{\partial x} - \omega\right) = \frac{1}{\mu}\frac{\partial N}{\partial t} + \bar{\gamma} \left\langle \Phi\left[\sqrt{\left(\frac{M}{M_0}\right)^2 + \left(\frac{N}{N_0}\right)^2} - 1\right]\right\rangle \frac{N}{k\sqrt{\left(\frac{M}{M_0}\right)^2 + \left(\frac{N}{N_0}\right)^2}}.$$

As a result we have obtained a system of four equations ((25.7) and (25.17)) in the first derivatives of the four required functions M, N, ω, and v. The non-linear function Φ entering (25.17) can be determined, similarly to the case of uniaxial tension, from experimental data (see section 3.2). In the case of an elastic loading process or unloading process we have to assume in (25.17) that $\Phi \equiv 0$.

The system of (25.7) and (25.17) constitutes a system of first-order equations with respect to the derivatives of functions M, N, ω, and v, which is of hyperbolic type. The four families of straight lines with equations

$$x = \pm a_1 t + \text{const}, \qquad x = \pm a_2 t + \text{const}, \tag{25.18}$$

are the real characteristics of the system of equations. The characteristic directions have values

$$a_1 = \sqrt{\frac{E}{\varrho}}, \qquad a_2 = \sqrt{\frac{\mu A_s}{\varrho A}} \tag{25.19}$$

which constitute the speed of propagation of bending-longitudinal waves and the speed of shear wave propagation respectively.

The following relations are valid, on account of (9.17), along the characteristics:

for $\quad x = \pm a_1 t + \text{const},$

(25.20)

$$\pm\frac{I}{a_1}d\omega + \frac{dM}{E} - \left\{\frac{N}{E} \mp \frac{2}{3a_1}\bar{\gamma} \left\langle \Phi\left[\sqrt{\left(\frac{M}{M_0}\right)^2 + \left(\frac{N}{N_0}\right)^2} - 1\right]\right\rangle \frac{M}{\sqrt{\left(\frac{M}{M_0}\right)^2 - \left(\frac{N}{N_0}\right)^2}}\right\} dx = 0$$

and for $x = \pm a_2 t + \text{const}$,

(25.20) [cont.]
$$A_s\, dv \mp \frac{a_2}{\mu} dN - \left\{ A_s \omega + \overline{\gamma} \left\langle \Phi \left[\sqrt{\left(\frac{M}{M_0}\right)^2 + \left(\frac{N}{N_0}\right)^2} - 1 \right] \right\rangle \frac{N}{\sqrt{\left(\frac{M}{M_0}\right)^2 + \left(\frac{N}{N_0}\right)^2}} \right\} dx = 0.$$

The solution of the system of (25.20) for prescribed boundary and initial conditions enables us to determine the stress waves propagating in the beam. In the case of strong discontinuity waves the system of (25.20) has to be supplemented by the continuity relations across the discontinuity fronts. Waves of strong discontinuity can appear in the beam only if the external loading on the beam is described by discontinuous functions. Evidently, the continuity conditions of zero order must be satisfied across the fronts of strong discontinuity waves, i.e. the condition of displacement continuity and of the continuity of the rotation angle of the beam element hold:

(25.21)
$$[w] = 0, \quad [\alpha] = 0.$$

The conditions of dynamic continuity across the fronts of waves of strong discontinuity assume the following form:

(25.22)
$$\begin{aligned} [M] &= \pm a_1 I\varrho [\omega] \quad &&\text{for} \quad dx = \pm a_1 dt, \\ [N] &= \mp a_2 A\varrho [v] \quad &&\text{for} \quad dx = \pm a_2 dt. \end{aligned}$$

These conditions can easily be derived from the general form of the dynamic continuity conditions (7.18) by means of definitions (25.1).

Making use of the fact that across the fronts of strong discontinuity waves the elastic/viscoplastic material behaves like an elastic material, we can obtain from (25.8) expressions for the jump of stress and the jump of strain across the wave front, namely

(25.23)
$$[\varepsilon] = \frac{1}{E}[\sigma], \quad [\gamma] = \frac{1}{\mu}[\tau].$$

Integrating these relations through the depth of the beam we obtain

(25.24)
$$[M] = -EI\left[\frac{\partial \alpha}{\partial x}\right], \quad [N] = \mu A_s [\gamma].$$

By analysing (25.21)–(25.24) we can state that functions M, ω, σ, and $\partial\alpha/\partial x$ are subjected to jump changes across the fronts of the strong discontinuity waves $x = \pm a_1 t + \text{const}$. On the other hand the functions N, v, τ, and γ are continuous; their derivatives are subjected to jumps. Across the fronts of the strong discontinuity waves $x = a_2 t + \text{const}$ the functions N, v, τ, and γ are discontinuous whereas functions M, ω, σ, and $\partial\alpha/\partial x$ are continuous; only their derivatives are discontinuous.

Let us now consider the simplest case of the propagation of bending and shear waves in a cantilever beam. We assume that the loadings on the boundary of the beam, the shear force $N^0(t)$ and the bending moment $M^0(t)$ are suddenly applied at the initial instant (their values are such that at the initial instant the beam is brought into the plastic state) and then they decrease arbitrarily with time (Fig. 81). We assume homogeneous, zero, initial conditions.

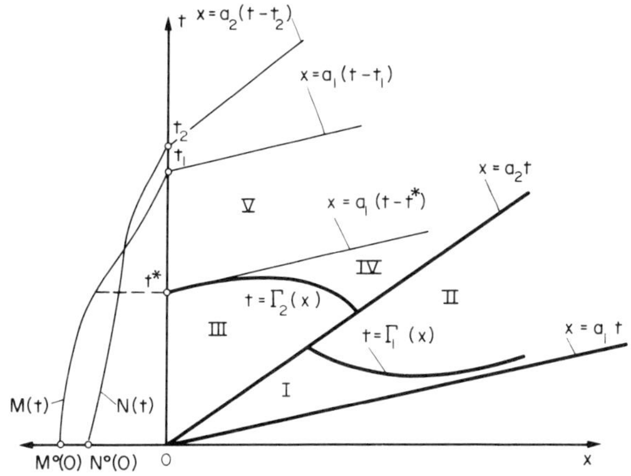

Fig. 81

In the case of such initial conditions a wave of strong discontinuity, of equation $x = a_1 t$, on which the quantities M and ω are subjected to jumps while the quantities N and v are continuous, first propagate into the undisturbed beam from its boundary $x = 0$. Since in front of the strong discontinuity wave the medium is undisturbed, $N = v = 0$ across the front of the wave and $[M] = M$, $[\omega] = \omega$. By means of these conditions, furthermore making use of the continuity condition $(25.22)_1$, we can integrate $(25.20)_1$ across the front of the strong discontinuity wave $x = a_1 t$. We obtain

(25.25)
$$\frac{dM}{dx} = -\frac{EM_0\bar{\gamma}}{3a_1 k}\left\langle \Phi\left(\frac{M}{M_0}-1\right)\right\rangle .$$

This equation can be reduced to the Volterra integral equation

(25.26)
$$M = M^0(0) - B\int_0^x \Phi\left(\frac{M}{M_0}-1\right)dx,$$

where we denoted $B = \dfrac{EM_0\bar{\gamma}}{3a_1 k}$.

Assuming a definite form for the function Φ, for example a function of the form $\Phi(F) = F^n$, we can obtain the solution of (25.25) in an explicit form. Examining the solution on the wave $x = a_1 t$ we can state that the value of the moment M along the entire span of the beam is greater in value than M_0 and asymptotically tends to value M_0 as $x \to \infty$, provided a bending moment $M^0(0) > M_0$ has been applied to the cross-section $x = 0$ at instant $t = 0$. The value M_0 represents the static yield moment of the cross-section in pure bending.

Behind the bending wave $x = a_1 t$ a slower shear wave proceeds with speed a_2. This wave is one of strong discontinuity; the shear force N and particle velocity v are subjected to jump changes across its front. The solution in the phase plane is illustrated in Fig. 81. Regions II and III denote the viscoplastic regions bounded by "unloading waves" of equations $t = \Gamma_1(x)$ and $t = \Gamma_2(x)$ respectively. Regions II and IV are unloading ones.

The solutions in the viscoplastic regions are constructed numerically, for example, by the method of characteristic nets using the relations along the characteristics (25.20). Writing them down in the form of finite differences along the characteristics, we obtain recurrence formulae for the determination of discrete values of the parameters in the solution at an arbitrary point of the element of the characteristic mesh, similarly as shown in section 23.

In the case considered, in spite of the far-reaching similarity in the wave picture to the problem of the propagation of longitudinal–transverse, strong discontinuity waves in a half-space (see section 23.2), there exists a fundamental difference in the construction of the solution in the region of plastic strain I. It might seem that, since the shear wave carrying the action of the shear force propagates slower than the bending wave, the shear force and the particles velocity v for the zero initial conditions should be equal to zero in region I. However, this is not the case because of the coupling of the bending moments and the shear forces (25.20). The existence of a bending moment in region I also generates a shear force. From the mathematical standpoint the assumption in region I, $N = v = 0$, means that these quantities are disregarded in (25.7) and (25.17). Thus they are reduced to the equations of engineering beam theory which are of parabolic type and for which all disturbances propagate with infinite velocity.

The construction of the solution in the remaining regions of the phase plane and the method of determining the unloading wave do not differ from the method presented previously in section 23.2. Also, for the sake of unifying the solution and adapting the calculations to digital computers, we make use of the same characteristic nets in the regions of elastic (or unloading) deformation (where it is assumed that $\langle \Phi \rangle = 0$) which is used in the regions of the viscoplastic deformation. This is of enormous importance, particularly in the case of finite beams when we have to deal with the reflections of waves from the boundaries of the beam with the mutual penetration of incident waves and waves reflected from the boundary.

26. Stress waves in plates

At this point we shall consider the problem of the propagation of transverse axially symmetric waves in unbounded plates. We assume that the material of the plate is strain-rate sensitive, and that its properties are described by the equations of viscoplasticity (3.10).

A system of cylindrical coordinates (r, φ, z) is associated with a plate of thickness $2h$. The z-axis is directed along the normal to the plate and r and φ denote the polar coordinates in the middle plane of the plate. Under conditions of axial symmetry the stress state is determined by the stress tensor components σ_{rr}, $\sigma_{\varphi\varphi}$, and τ_{rz}, and the strain state by the components ε_{rr}, $\varepsilon_{\varphi\varphi}$, and γ_{rz}. The remaining components of the stress tensor and strain tensor vanish. The constitutive equations (3.10) are reduced to the following three equations:

$$
\begin{aligned}
\dot{\varepsilon}_{rr} &= \frac{1}{E}(\dot{\sigma}_{rr} - \nu\,\dot{\sigma}_{\varphi\varphi}) + \frac{1}{3}\gamma\langle\Phi\,[F]\rangle\frac{2\sigma_{rr}-\sigma_{\varphi\varphi}}{\sqrt{J_2}},\\
\dot{\varepsilon}_{\varphi\varphi} &= \frac{1}{E}(\dot{\sigma}_{\varphi\varphi} - \nu\dot{\sigma}_{rr}) + \frac{1}{3}\gamma\langle\Phi\,[F]\rangle\frac{2\sigma_{\varphi\varphi}-\sigma_{rr}}{\sqrt{J_2}},\\
\dot{\gamma}_{rz} &= \frac{1}{\mu}\dot{\tau}_{rz} + \gamma\langle\Phi\,[F]\rangle\frac{\tau_{rz}}{\sqrt{J_2}},
\end{aligned}
\tag{26.1}
$$

where the intensity of the stress deviator is expressed in terms of the components of the stress tensor by the formula

$$(26.2)\qquad J_2 = \frac{1}{3}\left[\sigma_{rr}^2 - \sigma_{rr}\,\sigma_{\varphi\varphi} + \sigma_{\varphi\varphi}^2 + \tau_{rz}^2\right] \quad \text{and} \quad F = \frac{\sqrt{J_2}}{\varkappa} - 1 .$$

Now we introduce the following generalized quantities: shear force N, the radial and the tangential bending moments M_r and M_φ respectively, all of which are referred to unit lengths of the lines r = const or φ = const of the middle surface. These quantities are defined as follows:

$$(26.3)\qquad M_r = 2\int_0^h \sigma_{rr}\, z\, dz, \qquad M_\varphi = 2\int_0^h \sigma_{\varphi\varphi}\, z\, dz, \qquad N = 2\int_0^h \tau_{rz}\, dz.$$

In order to obtain equations for the problem which are of the hyperbolic type, we should, exactly as was done in the previous section for the Timoshenko beam, take into account the effects of shear strain and of rotary inertia. Thus we assume that the projections w and u on the z and r axes respectively of the total displacement of an arbitrary point of the plate in the cross-section r at the distance z from the middle surface of the plate are defined by formulae

$$(26.4)\qquad w = w(r, t), \qquad u = -\frac{\partial w}{\partial r} z + \gamma(r, t)\, z.$$

Accordingly the strain components ϵ_{rr} and γ_{rz} take the forms

$$(26.5)\qquad \begin{aligned} \varepsilon_{rr} &= \frac{\partial u}{\partial r} = -\frac{\partial^2 w}{\partial r^2} z + \frac{\partial \gamma}{\partial r} z, \\ \gamma_{rz} &= \frac{\partial u}{\partial z} + \frac{\partial w}{\partial r} = \gamma(r, t), \end{aligned}$$

where $\gamma(r, t)$ denotes the mean shear strain. Introducing, similarly as in the case of beams, the rotation angle α designating the rotation of an element initially situated on the normal to the middle surface of the plate,

$$(26.6)\qquad \alpha = \frac{\partial w}{\partial r} - \gamma,$$

we can write down the relations between the displacement w, the angle α, the component of normal velocity v, the angular velocity of an element ω and the components of strain ε_{rr}, $\varepsilon_{\varphi\varphi}$, and γ for the case of small deflections and rotations:

$$(26.7)\qquad \frac{\partial w}{\partial r} = \alpha + \gamma, \qquad v = \frac{\partial w}{\partial t}, \qquad \omega = \frac{\partial \alpha}{\partial t}, \qquad \varepsilon_{rr} = -z\frac{\partial \alpha}{\partial r}, \qquad \varepsilon_{\varphi\varphi} = -z\frac{\alpha}{r},$$

$$(26.8)\qquad \frac{\partial \varepsilon_{rr}}{\partial t} = -z\frac{\partial \omega}{\partial r}, \qquad \frac{\partial \varepsilon_{\varphi\varphi}}{\partial t} = -z\frac{\omega}{r}, \qquad \frac{\partial \gamma}{\partial t} = \frac{\partial v}{\partial r} - \omega .$$

Taking into account the continuity equations (26.8) we deduce from the constitutive equations (26.1) the following system of equations:

$$(26.9)\qquad -z\frac{\partial \omega}{\partial r} = \frac{1}{E}(\dot{\sigma}_{rr} - \nu\,\dot{\sigma}_{\varphi\varphi}) + \frac{1}{3}\bar{\gamma}\langle \Phi[F]\rangle \frac{2\sigma_{rr} - \sigma_{\varphi\varphi}}{\sqrt{J_2}},$$

(26.9) [cont.]

$$-z\frac{\omega}{r}=\frac{1}{E}(\dot{\sigma}_{\varphi\varphi}-\nu\,\dot{\sigma}_{rr})+\frac{1}{3}\bar{\gamma}\langle\Phi\,[F]\rangle\frac{2\sigma_{\varphi\varphi}-\sigma_{rr}}{\sqrt{J_2}},$$

$$\frac{\partial v}{\partial r}-\omega=\frac{1}{\mu}\dot{\tau}_{rz}+\bar{\gamma}\langle\Phi\,[F]\rangle\frac{\tau_{rz}}{\sqrt{J_2}}.$$

Now we transform the static yield condition $(26.2)_2$, replacing the stresses σ_{rr} and $\sigma_{\varphi\varphi}$ by the bending moments M_r and M_φ and the stress τ_{rz} by the shear force N in the same way as was done for beams. To simplify the problem we assume that $\varkappa = k$ (the body does not exhibit material hardening). We obtain the following static yield condition:

$$F=\sqrt{\left(\frac{M_r}{M_0}\right)^2-\frac{M_rM_\varphi}{M_0^2}+\left(\frac{M_\varphi}{M_0}\right)^2+\left(\frac{N}{N_0}\right)^2}-1, \tag{26.10}$$

where M_0 and N_0 are defined by the integrals

$$M_0=2\int_0^h kz\,dz,\qquad N_0=2\int_0^h k\,dz.$$

The second invariant of the stress deviator has the form

$$J_2=k^2\left[\left(\frac{M_r}{M_0}\right)^2-\frac{M_rM_\varphi}{M_0^2}+\left(\frac{M_\varphi}{M_0}\right)^2+\left(\frac{N}{N_0}\right)^2\right]. \tag{26.11}$$

Taking into account (26.10) and (26.11) in (26.9), next multiplying the first of these equations by zdz and the third one by dz, integrating through the depth of the plate from 0 to h, then making use of definitions (26.3), we obtain the following system of equations:

$$\begin{aligned}
-I\frac{\partial\omega}{\partial r}&=\frac{1}{E}(\dot{M}_r-\nu\,\dot{M}_\varphi)+\frac{1}{3}\bar{\gamma}\langle\Phi\,[F]\rangle\frac{2M_r-M_\varphi}{\sqrt{J_2}},\\
-I\frac{\omega}{r}&=\frac{1}{E}(\dot{M}_\varphi-\nu\,\dot{M}_r)+\frac{1}{3}\bar{\gamma}\langle\Phi\,[F]\rangle\frac{2M_\varphi-M_r}{\sqrt{J_2}},\\
A_s\left(\frac{\partial v}{\partial r}-\omega\right)&=\frac{1}{\mu}\dot{N}+\bar{\gamma}\langle\Phi\,[F]\rangle\frac{N}{\sqrt{J_2}},
\end{aligned} \tag{26.12}$$

where $I = 2h^3/3$, $A_s = 2h$, and F and J_2 are determined by formulae (26.10) and (26.11) respectively.

The equations of the translatory and rotary motion of the plate element take the forms

$$\begin{aligned}
\frac{\partial N}{\partial r}+\frac{N}{r}-2\varrho h\frac{\partial v}{\partial t}&=0,\\
\frac{\partial M_r}{\partial r}+\frac{M_r-M_\varphi}{r}+\varrho I\frac{\partial\omega}{\partial t}&=N.
\end{aligned} \tag{26.13}$$

The system of (26.12) and (26.13) constitutes the complete system for vibrations of an elastic/visco-ideally-plastic plate. The system consists of five equations in the first derivatives of the five required functions M_r, M_φ, N, ω, and v which are of hyperbolic type. It possesses the following families of real characteristics:

$$r = \pm a_1 t + \text{const},$$
$$r = \pm a_2 t + \text{const}, \tag{26.14}$$
$$r = \text{const},$$

where

$$a_1 = \sqrt{\frac{E}{\varrho(1-\nu^2)}}, \qquad a_2 = \sqrt{\frac{\mu}{\varrho}}$$

are the speeds of wave propagation in an elastic medium.

The following relations hold along the characteristics (26.14):

$$dM_r + \frac{M_r - M_\varphi}{r} dr - N\,dr \pm \frac{I}{a_1} d\omega + \left\{\nu I \frac{\omega}{r} + \frac{1}{3}\bar{\gamma}\langle\Phi[F]\rangle \frac{2M_r - M_\varphi}{\sqrt{J_2}}\right\} dt = 0 \quad \text{for} \quad r = \pm a_1 t + \text{const},$$

(26.15)

$$dN + \frac{N}{r} dr + \left\{\bar{\gamma}\langle\Phi[F]\rangle \frac{N}{\sqrt{J_2}} + A_s\omega \mp \frac{A_s}{a_2} dv\right\} dt = 0 \quad \text{for} \quad r = \pm a_2 t + \text{const},$$

$$\nu\,dM_r - dM_\varphi - \left\{\frac{1}{3}\bar{\gamma}\langle\Phi[F]\rangle \frac{2M_\varphi - M_r}{\sqrt{J_2}} + (1-\nu^2) I \frac{\omega}{r}\right\} dt = 0 \quad \text{for} \quad r = \text{const}.$$

The solution of the initial value problem can be constructed provided the boundary conditions and the initial conditions are prescribed, for example, in the forms

$$M_r = M_r(t), \qquad N = N(t) \qquad \text{for} \qquad r = r_0 \tag{26.16}$$

and provided we make use of the relations on the characteristics (26.15). If, however, the loads applied to the plate have discontinuities (e.g. they appear in a step-wise fashion), then waves of strong discontinuity occur. The following conditions have to be satisfied on waves of strong discontinuity:

(1) The continuity conditions of displacement and rotation angle:

$$[w] = 0, \qquad [\alpha] = 0. \tag{26.17}$$

(2) The dynamic continuity conditions (they can be obtained from (7.18) by means of definitions (26.3)):

$$[M_r]\,dt - \frac{I}{\varrho}[\omega]\,dr = 0, \qquad [N]\,dt + A_s\varrho[v]\,dr = 0. \tag{26.18}$$

(3) The conditions following from the fact that the material crossing the front of the strong discontinuity wave behaves as an elastic one; these conditions are a consequence of the constitutive equations (26.12):

(26.19) $$[M_r] = -I\left[\frac{\partial\alpha}{\partial r} + \frac{\alpha}{r}\right], \quad [M_\varphi] = -I\left[\frac{\alpha}{r} + \nu\frac{\partial\alpha}{\partial r}\right], \quad [N] = A_s[\gamma].$$

We assume that the loads M_r and N appear suddenly at the instant $t = 0$, at the surface $r = r_0$. Then a wave of strong discontinuity whose equations have the forms $r = r_0 \pm a_1 t$ and $r = r_0 \pm a_2 t$ start propagating in the plate. The quantities M_r, M_φ, ω, $\partial\alpha/\partial r$, σ_{rr} and $\sigma_{\varphi\varphi}$ are subjected to jumps across the fronts of waves $r = r_0 \pm a_1 t$, while functions N, v, γ, and τ_{rz} are continuous. On the other hand, the quantities N, v, γ, and τ_{rz} are discontinuous across the fronts of waves $r = r_0 \pm a_2 t$, while quantities M_r, M_φ, ω, $\partial\alpha/\partial r$, σ_{rr}, and $\sigma_{\varphi\varphi}$ are continuous.

The construction of the solution of the problem of the propagation of bending and transverse waves in an unbounded plate is identical to that in the case of beams (see section 25). On the fronts of the strong discontinuity waves the solution can be reduced to the solution of an integral equation. In the particular cases of a polynomial or linear relaxation function the solution can be obtained in an exact form. The solutions in the regions of the viscoplastic deformation and in the unloading regions can be constructed numerically by means of the relations along the characteristics (26.15).

27. Propagation of plane, two-dimensional stress waves

Many problems concerning the propagation of the stress waves in inelastic media rely upon the assumptions of symmetry of the medium and symmetry of the boundary conditions (e.g. spherical and cylindrical symmetry, see Chapter IV), or on the assumption that the tractions on the boundary are uniformly distributed and depend only on time (see the problems studied in the preceding sections of this chapter). However, uniform stress distribution over the boundary of the medium is very seldom realized in practice. As a rule, an explosion on the surface of the medium generates a pressure concentration over a small area which decreases outside the immediate vicinity of the area. The solution of this type of problem is complicated, and there exists only a very small number of papers solving problems of this type which are important in engineering applications. The solutions require comprehensive theoretical study and very complicated and labour-consuming numerical calculations. The theory of two-dimensional stress waves in elastic/viscoplastic media is reasonably well developed [3], [5]–[10], [107], [108]. This section of the book is mainly devoted to two-dimensional waves in such media.

Let us consider the plane strain deformation of a compressible elastic/viscoplastic medium. We assume that the medium is isotropic and homogeneous. We disregard the effects of external body forces and restrict the treatment to small strains in the medium. We take an orthogonal system of cartesian coordinates x_i. Consider the plane (x_1, x_2). For the case of a plane strain state the components of the displacement vector **u** take the form:

(27.1) $$u_1 = u_1(x_1, x_2, t), \qquad u_2 = u_2(x_1, x_2, t), \qquad u_3 \equiv 0.$$

We therefore have the following strain tensor components:

(27.2) $$\varepsilon_{11} = u_{1,1}, \qquad \varepsilon_{22} = u_{2,2}, \qquad \gamma_{12} = \frac{1}{2}(u_{1,2} + u_{2,1}), \qquad \varepsilon_{33} = \gamma_{13} = \gamma_{23} = 0.$$

Denoting the components of the velocity vector by $v_i = \partial u_i/\partial t$ we can write down the components of the strain rate tensor in the forms

(27.3) $\dot{\varepsilon}_{11} = v_{1,1}, \qquad \dot{\varepsilon}_{22} = v_{2,2}, \qquad \dot{\gamma}_{12} = \frac{1}{2}(v_{1,2} + v_{2,1}), \qquad \dot{\varepsilon}_{33} = \dot{\gamma}_{13} = \dot{\gamma}_{23} = 0.$

The stress state satisfies the condition $\tau_{13} = \tau_{23} \equiv 0$. The remaining components of the stress tensor as well as the components of the strain tensor of thè displacement vector and the velocity vector are functions of x_1, x_2, and t only; accordingly, we have

$$\begin{gathered}\sigma_{11} = \sigma_{11}(x_1, x_2, t), \qquad \sigma_{22} = \sigma_{22}(x_1, x_2, t), \qquad \sigma_{33} = \sigma_{33}(x_1, x_2, t), \\ \tau_{12} = \tau_{12}(x_1, x_2, t), \qquad \tau_{13} = \tau_{23} \equiv 0. \end{gathered} \tag{27.4}$$

Equations of motion (5.5) take the forms

$$\frac{\partial \sigma_{11}}{\partial x_1} + \frac{\partial \tau_{12}}{\partial x_2} = \varrho \frac{\partial v_1}{\partial t}, \qquad \frac{\partial \sigma_{22}}{\partial x_2} + \frac{\partial \tau_{12}}{\partial x_1} = \varrho \frac{\partial v_2}{\partial t}. \tag{27.5}$$

The constitutive equations, describing the dynamic behaviour of an elastic/visco-ideally-plastic medium, will be assumed in form (3.25). These equations, in the case of the strain state (27.3) and the stress state (27.4), take the following forms:

$$\begin{aligned} \dot{e}_{11} &= \frac{1}{2\mu}\dot{s}_{11} + \langle D\rangle s_{11}, \\ \dot{e}_{22} &= \frac{1}{2\mu}\dot{s}_{22} + \langle D\rangle s_{22}, \\ \dot{e}_{33} &= \frac{1}{2\mu}\dot{s}_{33} + \langle D\rangle s_{33} = 0, \\ \dot{\gamma}_{12} &= \frac{1}{\mu}\dot{\tau}_{12} + 2\langle D\rangle \tau_{12}, \\ \dot{\varepsilon}_{11} + \dot{\varepsilon}_{22} &= \frac{1}{3K}(\dot{\sigma}_{11} + \dot{\sigma}_{22} + \dot{\sigma}_{33}), \end{aligned} \tag{27.6}$$

where

$$\langle D\rangle = \gamma\langle \Phi(F)\rangle \frac{1}{\sqrt{J_2}}, \qquad F = \frac{\sqrt{J_2}}{k} - 1,$$

$$\langle D\rangle \doteq \begin{cases} D & \text{for} \quad \sqrt{J_2} > k, \\ 0 & \text{for} \quad \sqrt{J_2} \leqslant k, \end{cases}$$

$$\sqrt{J_2} = \sqrt{\sigma_{11}^2 + \sigma_{22}^2 + \sigma_{33}^2 - 3\sigma_{11}\sigma_{22} - 3\sigma_{22}\sigma_{33} - 3\sigma_{11}\sigma_{33} + \tau_{12}^2}, \tag{27.7}$$

$$s_{11} = \frac{1}{3}(2\sigma_{11} - \sigma_{22} - \sigma_{33}), \qquad s_{22} = \frac{1}{3}(2\sigma_{22} - \sigma_{11} - \sigma_{33}),$$

$$s_{33} = \frac{1}{3}(2\sigma_{33} - \sigma_{11} - \sigma_{22}).$$

The treatment presented here can easily be generalized to the case of a constitutive equation in which the effect of hardening of the material is taken into account, i.e. when the function F is in form (3.8) and the Huber–Mises yield condition takes the forms

(27.8) $$F = \frac{f(\sigma_{ij})}{\varkappa} - 1, \quad f(\sigma_{ij}) = \sqrt{J_2},$$

or, to the case of the dynamic deformation of an elastic/visco-ideally-plastic soil, governed by (4.16), where

(27.9) $$F = \frac{f(\sigma_{ij})}{k} - 1 \quad \text{and} \quad f(\sigma_{ij}) = \alpha J_1' + \sqrt{J_2}.$$

Following Clifton [30], let us introduce the following dimensionless quantities: velocities v_1 and v_2, time t, cartesian coordinates x_1 and x_2, the yield limit for pure shear k, and the coefficient of viscosity γ:

(27.10) $$v_1 = \frac{\bar{v}_1}{a_1}, \quad v_2 = \frac{\bar{v}_2}{a_1}, \quad t = \frac{\bar{t}a_1}{b}, \quad x_1 = \frac{\bar{x}_1}{b}, \quad x_2 = \frac{\bar{x}_2}{b},$$
$$k = \frac{\bar{k}}{\varrho a_1^2}, \quad \gamma = \frac{b}{a_1}\bar{\gamma}, \quad \Gamma = \frac{a_1}{a_2}.$$

A bar over the symbols indicates the dimensional quantities, a_1 and a_2 are the speeds of propagation of elastic longitudinal and transverse waves

(27.11) $$a_1 = \left(\frac{3K + 4\mu}{3\varrho}\right)^{1/2}, \quad a_2 = \left(\frac{\mu}{\varrho}\right)^{1/2},$$

and b denotes a characteristic quantity of the problem considered. Besides these we introduce the dimensionless stresses defined as follows:

(27.12) $$p = \frac{1}{2}\frac{(\sigma_{11} + \sigma_{22})}{\varrho a_1^2}, \quad q = \frac{1}{2}\frac{(\sigma_{11} - \sigma_{22})}{\varrho a_1^2}, \quad \tau = \frac{\tau_{12}}{\varrho a_1^2}, \quad r = \frac{\sigma_{33}}{\varrho a_1^2}.$$

Replacing the first two equations of (27.6) by their sum and difference, making use of relations (27.3), and taking into account the symbols (27.10) and (27.12), we obtain the following system of four constitutive equations:

(27.13)
$$-v_{1,1} - v_{2,2} + \frac{\Gamma^2}{3\Gamma^2 - 4}p_{,t} + \frac{\Gamma^2(2-\Gamma^2)}{3\Gamma^2 - 4}r_{,t} = -\frac{2}{3}\langle D\rangle(p - r),$$
$$-v_{1,1} + v_{2,2} + \Gamma^2 q_{,t} = -2q\langle D\rangle,$$
$$\frac{\Gamma^2(2-\Gamma^2)}{3\Gamma^2 - 4}p_{,t} + \frac{\Gamma^2 - 1}{3\Gamma^2 - 4}r_{,t} = -\frac{2}{3}\langle D\rangle(r - p),$$
$$-v_{1,2} - v_{2,1} + \Gamma^2 \tau_{,t} = -2\tau\langle D\rangle,$$

where

$$\langle D\rangle = \gamma\left\langle \omega\left(\frac{\sqrt{J_2}}{k} - 1\right)\right\rangle \frac{1}{\sqrt{J_2}}, \quad \sqrt{J_2} = \frac{\varrho a_1^2}{\sqrt{3}}\sqrt{(p-r)^2 + 3(q^2 - \tau^2)}.$$

Taking into account (27.10) and (27.12) we reduce the equations of motion (27.5) to the forms

(27.14) $$v_{1,t} - p_{,1} - q_{,1} - \tau_{,2} = 0, \quad v_{2,t} - p_{,2} + q_{,2} - \tau_{,1} = 0.$$

The system of (27.13) and (27.14) constitutes the complete system of the equations for the problem. The above system of equations can be presented in matrix form, namely:

$$L[\mathbf{w}] = \mathbf{A}^t\mathbf{w}_{,t} + \mathbf{A}^1\mathbf{w}_{,1} + \mathbf{A}^2\mathbf{w}_{,2} - \mathbf{B} = 0, \tag{27.15}$$

where vectors **w** and **B** can be represented as follows:

$$\mathbf{w} = \begin{bmatrix} v_1 \\ v_2 \\ p \\ q \\ r \\ \tau \end{bmatrix}, \quad \mathbf{B} = \begin{bmatrix} -\frac{2}{3}\langle D\rangle(p-r) \\ -2\langle D\rangle q \\ -\frac{2}{3}\langle D\rangle(r-p) \\ -2\langle D\rangle\tau \\ 0 \\ 0 \end{bmatrix}, \tag{27.16}$$

the matrices $\mathbf{A}^t$, $\mathbf{A}^1$, $\mathbf{A}^2$ take the forms

$$\mathbf{A}^t = \begin{bmatrix} 0 & 0 & M & 0 & Q & 0 \\ 0 & 0 & 0 & \Gamma^2 & 0 & 0 \\ 0 & 0 & Q & 0 & N & 0 \\ 0 & 0 & 0 & 0 & 0 & \Gamma^2 \\ 1 & 0 & 0 & 0 & 0 & 0 \\ 0 & 1 & 0 & 0 & 0 & 0 \end{bmatrix}, \quad \mathbf{A}^1 = \begin{bmatrix} -1 & 0 & 0 & 0 & 0 & 0 \\ -1 & 0 & 0 & 0 & 0 & 0 \\ 0 & 0 & 0 & 0 & 0 & 0 \\ 0 & -1 & 0 & 0 & 0 & 0 \\ 0 & 0 & -1 & -1 & 0 & 0 \\ 0 & 0 & 0 & 0 & 0 & -1 \end{bmatrix},$$

$$\mathbf{A}^2 = \begin{bmatrix} 0 & -1 & 0 & 0 & 0 & 0 \\ 0 & 1 & 0 & 0 & 0 & 0 \\ 0 & 0 & 0 & 0 & 0 & 0 \\ -1 & 0 & 0 & 0 & 0 & 0 \\ 0 & 0 & 0 & 0 & 0 & -1 \\ 0 & 0 & -1 & 1 & 0 & 0 \end{bmatrix}, \tag{27.17}$$

where we have introduced the symbols

$$M = \frac{\Gamma^4}{3\Gamma^2-4}, \quad N = \frac{\Gamma^2(\Gamma^2-1)}{3\Gamma^2-4}, \quad Q = \frac{\Gamma^2(2-\Gamma^2)}{3\Gamma^2-4}.$$

Matrices $\mathbf{A}^t$, $\mathbf{A}^1$, and $\mathbf{A}^2$ are symmetric; $\mathbf{A}^t$ is symmetric and positive definite. We can easily verify that (27.15) represents a semi-linear, symmetric, hyperbolic system of partial differential equations in which the derivatives have constant coefficients.

If, following Southwell and Allen, we make a simplifying assumption (see [6]) we obtain a simpler form of vectors (27.16) and matrices (27.17). The assumption of Southwell and Allen consists in the supposition that both the elastic and viscoplastic parts of the strain rate tensor $\dot{\varepsilon}_{33}$ are simultaneously zero.

Expressing the total strain rate tensor $\dot{\varepsilon}^c_{ij}$ as the sum of elastic and plastic parts,

$$\dot{\varepsilon}^c_{ij} = \dot{\varepsilon}^e_{ij} + \dot{\varepsilon}^p_{ij}, \tag{27.18}$$

where

$$\dot{\varepsilon}^e_{ij} = \frac{1}{2\mu}\dot{s}_{ij} + \frac{1}{3K}\dot{s}\delta_{ij}, \qquad \dot{\varepsilon}^p_{ij} = \langle D\rangle s_{ij}, \tag{27.19}$$

and equating to zero both the parts of the component $\dot{\varepsilon}^c_{33}$ of the strain rate tensor, we obtain

$$\begin{gathered} \frac{3K+\mu}{3K\mu}(\dot{\sigma}_{11}+\dot{\sigma}_{22}) + \frac{2\mu-3K}{6K\mu}\dot{\sigma}_{33} = 0, \\ \langle D\rangle(2\sigma_{33}-\sigma_{11}-\sigma_{22}) = 0. \end{gathered} \tag{27.20}$$

Hence, in the elastic range, we have

$$\dot{\sigma}_{33} = \nu(\dot{\sigma}_{11}+\dot{\sigma}_{22}), \tag{27.21}$$

where ν is Poisson's ratio. In the viscoplastic strain range we have

$$\sigma_{33} = \frac{1}{2}(\sigma_{11}+\sigma_{22}). \tag{27.22}$$

Thus the component of the stress tensor σ_{33} is not uniquely determined. It is uniquely determined only in the case of an incompressible medium. In such a case $\sqrt{J_2} = \varrho a_1^2\sqrt{p^2-q^2}$ and we have:

$$\mathbf{w} = \begin{bmatrix} v_1 \\ v_2 \\ p \\ q \\ \tau \end{bmatrix}, \qquad \mathbf{B} = \begin{bmatrix} 0 \\ -2\langle D\rangle q \\ -2\langle D\rangle \tau \\ 0 \\ 0 \end{bmatrix},$$

$$\mathbf{A}^t = \begin{bmatrix} 0 & 0 & \dfrac{\Gamma^2}{\Gamma^2-1} & 0 & 0 \\ 0 & 0 & 0 & \Gamma^2 & 0 \\ 0 & 0 & 0 & 0 & \Gamma^2 \\ 1 & 0 & 0 & 0 & 0 \\ 0 & 1 & 0 & 0 & 0 \end{bmatrix}, \qquad \mathbf{A}^1 = \begin{bmatrix} -1 & 0 & 0 & 0 & 0 \\ -1 & 0 & 0 & 0 & 0 \\ 0 & -1 & 0 & 0 & 0 \\ 0 & -1 & 0 & 0 & 0 \\ 0 & 0 & -1 & -1 & 0 \\ 0 & 0 & 0 & 0 & -1 \end{bmatrix},$$

$$\mathbf{A}^2 = \begin{bmatrix} 0 & -1 & 0 & 0 & 0 \\ 0 & 1 & 0 & 0 & 0 \\ -1 & 0 & 0 & 0 & 0 \\ 0 & 0 & 0 & 0 & -1 \\ 0 & 0 & -1 & 1 & 0 \end{bmatrix}.$$

The system of (27.15) constitutes a system of six partial differential equations, which are semi-linear, of the first order and of hyperbolic type, in the three independent variables x_1, x_2, and t, in which the derivatives have constant coefficients. The theory for these equations is given by Courant and Hilbert [33], it constitutes a generalization of the discussion presented briefly in section 9, to the case of systems of equations with a higher number of independent variables than two. This method was applied in spatial problems of gas dynamics by Butler [20], Burnat, Kielbasinski, and Wakulicz [19], Richardson [136], and Rusanow [138]. The method was also applied in dynamic problems in the theory of elasticity by Chou and Karp [26], Clifton [30], Recker [135], Sabodash and Cherednichenko [139],

Kukudzhanov [73], and Ziv [175]–[177]. In the dynamic problems in the theory of plasticity the method was applied by Estrin [38], Clifton [31], Bertholf [12], and Sauerwein [140]. In wave problems in the theory of viscoplasticity the method was adapted by Baltov [2], [3], Kolarov and Baltov [69], Bejda [5]–[10], and by Murakami and Bejda [90].

We now proceed to a discussion of the characteristic properties of the system of equations for the problem being considered. First of all we shall analyse the geometry of the characteristic surfaces connected with (27.15).

The condition for surface $\Phi(x_1, x_2, t) = \text{const}$ to be a characteristic surface of (27.15) is the disappearance of the determinant of the characteristic matrix **A**,

$$\det \mathbf{A} = 0, \tag{27.23}$$

where

$$\mathbf{A} = \mathbf{A}^t \Phi_{,t} + \mathbf{A}^1 \Phi_{,1} + \mathbf{A}^2 \Phi_{,2}. \tag{27.24}$$

Equation (27.23) is equivalent to

$$\{(\Phi_{,t})^2 - [(\Phi_{,1})^2 + (\Phi_{,2})^2]\}\left\{(\Phi_{,t})^2 - \frac{1}{\Gamma^2}[(\Phi_{,1})^2 + (\Phi_{,2})^2]\right\}(\Phi_{,t}) = 0. \tag{27.25}$$

We deduce from (27.25) that

$$(\Phi_{,t})^2 - a_1^2[(\Phi_{,1})^2 + (\Phi_{,2})^2] = 0$$

constitutes the equation of a wave surface propagating with the speed of longitudinal waves $\mp a_1$ while

$$(\Phi_{,t})^2 - a_2^2[(\Phi_{,1})^2 + (\Phi_{,2})^2] = 0$$

is the equation of a wave surface propagating with speed $\mp a_2$. The last term in (27.25) $\Phi_{,t}^2$ can be interpreted as the degenerated characteristic cones, reduced to the t-axis (their equations $x_1 = \text{const}, x_2 = \text{const}$). This being so it is as if the characteristic surface propagates with zero velocity.

The characteristic cones, being the envelope of all solutions of (27.25), passing through the point (x_1^0, x_2^0, t) have the form

$$(x_1 - x_1^0)^2 + (x_2 - x_2^0)^2 = a^2(t - t_0)^2, \tag{27.26}$$

where $a = 1$ or $a = 1/\Gamma$.

Introducing spherical coordinates we can represent the equations of the above characteristic cones in parametric form

$$x_1 - x_1^0 = a(t - t_0)\cos\alpha, \qquad x_2 - x_2^0 = a(t - t_0)\sin\alpha, \tag{27.27}$$

where α denotes an angle in the plane $t_0 = \text{const}$.

Equations (27.27) constitute the equations of the bicharacteristics. These are the generators of the characteristic cones (Fig. 82). The characteristic band, connected with the bicharacteristics (27.27), passing through the point (x_1^0, x_2^0, t_0), has the following form:

(27.28)
$$\Phi_{,t} = a,$$
$$\Phi_{,1} = -\cos\alpha, \qquad a = 1, \ a = \frac{1}{\Gamma},$$
$$\Phi_{,2} = -\sin\alpha.$$

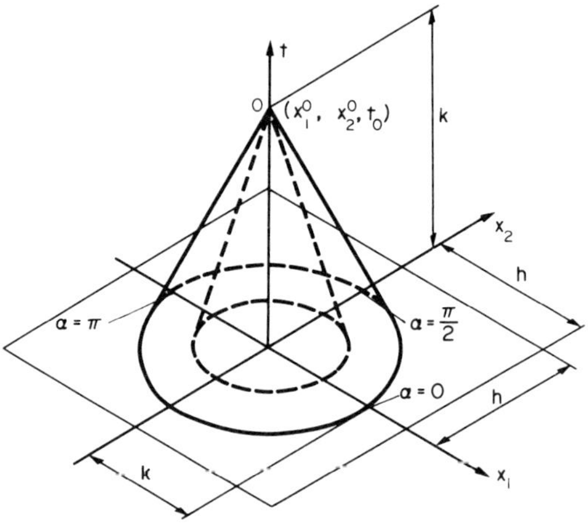

Fig. 82

If Φ is a characteristic surface of (27.15), then there exists an eigenvector **l** (see Courant and Hilbert [33]) such that

(27.29)
$$\mathbf{lA} = 0,$$

where **A** is the characteristic matrix determined by formula (27.24). The vector **l**, corresponding to the characteristic band defined by (27.27) and (27.28), takes the following form:

(27.30)
$$\mathbf{l} = \begin{vmatrix} -\Gamma^2 \cos\alpha \\ -\Gamma^2 \sin\alpha \\ \Gamma^2 - 1 \\ \cos 2\alpha \\ \sin 2\alpha \end{vmatrix} \quad \text{for} \quad a = 1$$

and

(27.30′)
$$\mathbf{l} = \begin{vmatrix} \Gamma \sin\alpha \\ -\Gamma \cos\alpha \\ 0 \\ -\sin 2\alpha \\ -\cos 2\alpha \end{vmatrix} \quad \text{for} \quad a = \frac{1}{\Gamma}.$$

If **l** is the eigenvector for the characteristic surface Φ then the differential equation for Φ is obtained from the condition

$$\mathbf{l}\cdot L[\mathbf{w}] = 0, \tag{27.31}$$

where the dot denotes the inner product. The derivatives in the direction normal to the surface Φ do not occur in this equation, consequently it can be expressed at each point by means of the derivatives in two directions on surface Φ. For example, we can choose the bicharacteristic directions and the direction for which time t is constant. The derivative of function f in the bicharacteristic direction can be expressed in the form

$$\frac{df}{dt} = f_{,t} + f_{,1}\frac{dx_1}{dt} + f_{,2}\frac{dx_2}{dt}, \tag{27.32}$$

where $dx_1/dt, dx_2/dt$ along the bicharacteristics can be determined by differentiating (27.27). Making use of formulae (27.15) and (27.30) and replacing, for the sake of simplicity, angle α by $\alpha + \pi$, we can write (27.31) in the following forms:

$$\cos\alpha\, dv_1 + \sin\alpha\, dv_2 + dp + \cos 2\alpha\, dq + \sin 2\alpha\, d\tau = -S_1(\alpha)\, dt \quad \text{for} \quad a = 1$$

(27.33) and

$$-\Gamma\sin\alpha\, dv_1 + \Gamma\cos\alpha\, dv_2 - \Gamma^2\sin 2\alpha\, dq + \Gamma^2\cos 2\alpha\, d\tau = -S_2(\alpha)\, dt \quad \text{for} \quad a = \frac{1}{\Gamma},$$

where

$$\begin{aligned} S_1(\alpha) = {} & \left[-\sin^2\alpha + \frac{1}{\Gamma^2}(1-\cos 2\alpha)\right] v_{1,1} + \left(\frac{1}{2}\sin 2\alpha - \frac{1}{\Gamma^2}\sin 2\alpha\right) v_{1,2} \\ & + (1+\cos 2\alpha)\, q_{,1}\cos\alpha + (1+\cos 2\alpha)\, q_{,2}\sin\alpha + \left(\frac{1}{2} - \frac{1}{\Gamma^2}\right) v_{2,1}\sin 2\alpha \\ & + \left[-\cos^2\alpha + \frac{1}{\Gamma^2}(1+\cos 2\alpha)\right] v_{2,2} + (\sin 2\alpha\cos\alpha - \sin\alpha)\,\tau_{,1} \\ & + (\sin 2\alpha\sin\alpha - \cos\alpha)\,\tau_{,2} + \frac{2D}{\Gamma^2}\left[\frac{1}{3}(p-r) + \tau\sin 2\alpha + q\cos 2\alpha\right], \end{aligned} \tag{27.34}$$

$$\begin{aligned} S_2(\alpha) = {} & \frac{1}{2}\sin 2\alpha\,(v_{1,1} - v_{2,2}) - \cos^2\alpha\, v_{1,2} + \Gamma\sin\alpha\, p_{,1} + \Gamma(\sin\alpha - \sin 2\alpha\cos\alpha)\, q_{,1} \\ & + \Gamma\sin\alpha\,(1+\cos 2\alpha)\,\tau_{,2} + \sin^2\alpha\, v_{2,1} - \frac{1}{2}\sin 2\alpha\, v_{2,2} - \Gamma\cos\alpha\, p_{,2} \\ & + (1 - 2\sin 2\alpha)\, q_{,2} - \Gamma\cos\alpha\,(1-\cos 2\alpha)\,\tau_{,1} - 2D(\sin 2\alpha\, q - \cos 2\alpha\,\tau). \end{aligned}$$

The spatial derivatives in (27.34) correspond to the derivatives in the tangential directions of the characteristic cone.

The solution of (27.15) by the finite difference method, for $B \equiv 0$, i.e. in the case of the equations of the theory of elasticity, is presented in [30]. This method was then generalized to semi-linear equations, i.e. the equations in the form (27.15), in [3], [5]–[10]. The finite difference method in the case of three independent variables x_1, x_2, t (and in the case of a greater number of independent variables) varies fundamentally from the method of characteristic net discussed in the preceding chapters. The characteristic method for the case of two independent variables consists in the reduction of a system of partial differential equations to a system of ordinary differential equations along the chosen characteristic curves. Next the total differentials of the function are replaced by the finite differences along the characteristics.

This method cannot be directly carried over to the case when the number of spatial variables is larger than two, i.e. to spatial problems. This is due to the fact that the differential equations along the bicharacteristics contain differentiation in more than one direction. In the expressions for $S_1(\alpha)$, $S_2(\alpha)$, determined by formulae (27.34), there appear differentiations with respect to spatial variables x_1 and x_2. In the papers cited ([3], [5]–[10] and [30]), the difference methods applied to mixed boundary problems of gas dynamics are transferred to the equations of the theory of elasticity and viscoplasticity. The method consists in the elimination of the derivatives of all the required functions v_1, v_2, p, q, r, and τ at point (x_1^0, x_2^0, t_0) (Fig. 82) where the discrete solution to the problem is sought.

In these papers cited ([3], [5]–[10], and [30]) only the problem of weak discontinuity waves was considered. It was assumed that the loading on the boundary of the half-space changed monotonically from zero to a certain definite value and then monotonically decreased. On surface $x_2 = 0$ either boundary conditions for stress or velocity or mixed conditions can be prescribed. A discussion of loading which is discontinuous in time entails the complicated analysis of the propagation of strong discontinuity waves [6]. The determination of the fronts of plastic loading waves and unloading waves also present difficulties. The fronts of the waves are determined only approximately.

The accuracy of the method depends on the value of the integration step. Clifton [30] demonstrated, in an example of elastic waves, the influence of the choice of the size of the integration step on the solution error. If the integration step is chosen such that the condition for the stability of the solution is satisfied, then the error increases linearly. Otherwise the error increases incommensurably. In the case of a system of semi-linear equations the problem of convergence of the solution and the stability of the method were not investigated.

The straightforward solution of the system of (27.13) and (27.14) by the method of finite elements (e.g. [135]) or by the method of finite differences for an implicit scheme seems to be more expedient. In this case the solution of the system of difference equations is stable independently of the chosen integration step. This method enables us to avoid numerous transforms in relations (27.34) and many computational complications, particularly in the case of complicated boundary conditions.

Estrin [38] analysed the propagation of weak discontinuity waves in an elastic–plastic half-space. He has derived an expression for the speed of the waves in the medium and has described two curves corresponding to the two types of waves. The dependence of the wave speeds on the direction of propagation of the disturbance and on the principal directions of the normal stress constitutes the characteristic features of these curves. For $\varphi - \alpha = 0$ (where α denotes an angle between the direction of the normal to the projection of the characteristic surface $\Phi =$ const on plane $t =$ const and x_1-axis) the waves propagate with the dilatational wave speed $a_1 = \sqrt{K/\varrho}$ and the shear wave speed $a_2 = \sqrt{\mu/\varrho}$ respectively. The speeds of propagation of the dilatational and shear waves are illustrated in Fig. 83 in the $\sigma_{11}-\sigma_{22}$ plane. We can easily see from the figure that the dilatational wave is almost independent of the direction of propagation of the disturbances (curve a). This curve almost does not differ from a circle. The speed of the shear wave (curve b) depends considerably on the angle $\varphi - \alpha$. It is equal to zero in the direction of principal shear stress. Curve c represents the elastic shear waves, i.e. a circle of radius $a_2 = \sqrt{\mu/\varrho}$.

In an elastic/viscoplastic medium circles of radii $a_1 = \sqrt{(3K+4\mu)/3\varrho}$, $a_2 = \sqrt{\mu/\varrho}$ [6], correspond to the speeds of longitudinal and shear waves respectively. They are identical to the case of an elastic medium. This results from the nature of the constitutive equations for elastic/viscoplastic media, discussed in Chapter I.

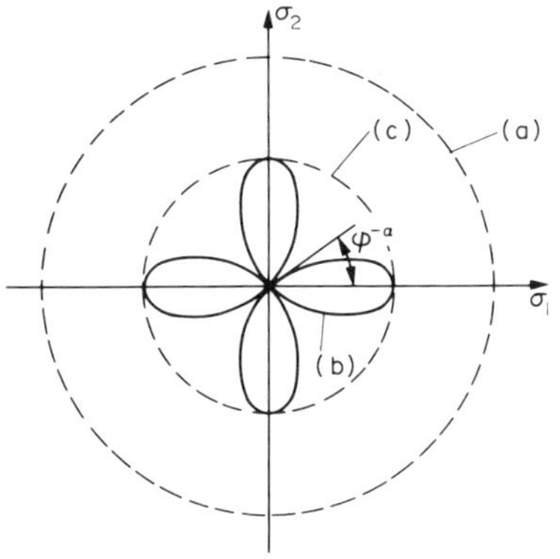

Fig. 83

We should also mention an interesting paper by Sauerwein [140] in which the speed of propagation of waves is investigated in an anisotropic elastic–plastic medium governed by Grigorian's soil dynamics equations [48]. The influence of the inhomogeneity of the medium on the velocity of propagation of dilatational and shear waves and the dependence on the medium anisotropy was examined by an extensive application of numerical techniques.

We move on to discuss the solution of the problem of wave propagation in an elastic/viscoplastic half-space containing a cylindrical cavity. The waves are generated by loads on the boundary which vary in time and space, a state of plane strain and small deformations of the medium being assumed.

We introduce the bipolar system of coordinates, [54], [86], [89], which enables us to simplify the geometry of the problem considered. The wave propagation problem will be solved by discretizing with respect to the introduced bipolar coordinates and integrating the equations of the problem with respect to time by the use of Treanor's method [158] of varying the integration step.

We shall examine the motion of an elastic/viscoplastic medium filling the half-space $x^1 > 0$ whose bounding plane is subjected to the normal stress (Fig. 84)

$$\sigma^{11} = -p_0(x^2, t) \tag{27.35}$$

dependent on time and variable x^2 and independent of x^3. It is assumed that these tractions constitute a continuous function of time which increases from zero at the instant $t = 0$ and whose intensity, in a certain region of x^2, exceeds the yield limit of the material. We can also assume that a tangential stress σ^{12}, dependent on time and the variable x^2, acts on the bounding plane.

Within the half-space $x^1 \geqslant 0$ there is, at a distance h from the boundary, a cylindrical cavity of radius $r = r_0$, the axis of which is parallel to the x^3-axis.

In this plane strain problem the motion of the medium is independent of the component of the displacement vector in the direction of the x^3-axis, i.e. $u^3 = 0$. Consequently, the following components of the strain tensor vanish:

$$\epsilon^{33} = \epsilon^{13} = \epsilon^{23} = 0,$$

as well as the stress tensor components

$$\sigma^{13} = \sigma^{23} = 0.$$

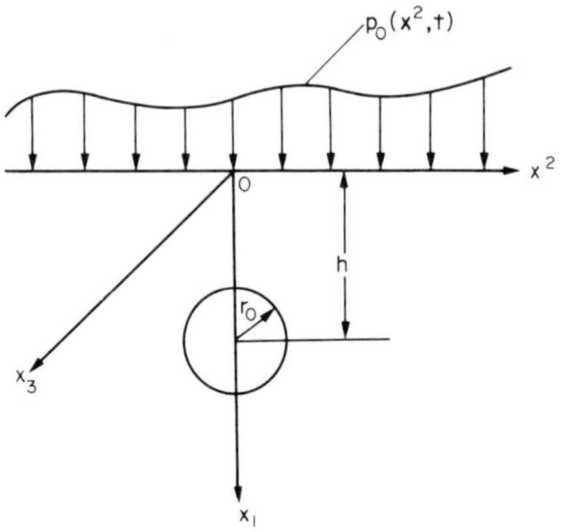

Fig. 84

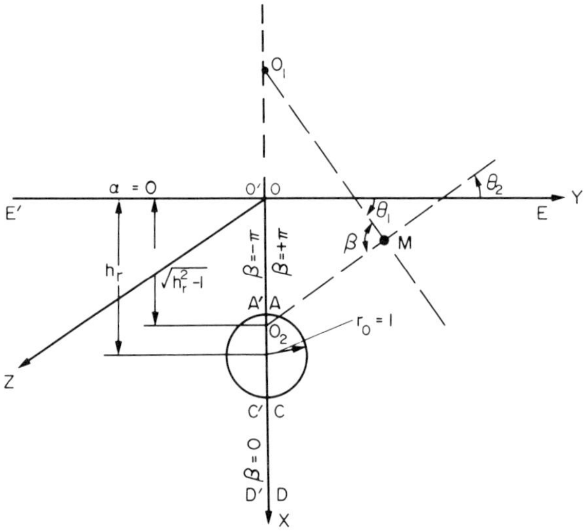

Fig. 85

In the case considered it is expedient to introduce curvilinear coordinates. In the $(x^1 - x^2)$ plane we choose the bipolar, curvilinear coordinate system (α, β) (Fig. 85) [54], [89], [86]. In Fig. 85 x^1 and x^2 denote the coordinates of a point M within the half-space, r_0 denotes the radius of the cylindrical cavity, and X and Y are the scaled coordinates of point M:

(27.36) $$X = \frac{x^1}{r_0}, \quad Y = \frac{x^2}{r_0},$$

h_r is the depth of submersion of the cylindrical cavity in the half-space:

$$h_r = \frac{h}{r_0}. \tag{27.37}$$

In the bipolar coordinate system (α, β, z) which has been introduced, points O_1 and O_2 (Fig. 85), of abscissae $\pm a = \sqrt{h_r^2 - 1}$, situated on the OX axis, constitute the poles of the transformation from the cartesian coordinate system X, Y, Z.

We denote

$$r_1 = |\overrightarrow{\mathbf{O_1 M}}|, \quad r_2 = |\overrightarrow{\mathbf{O_2 M}}|,$$

$$\theta_1 = (\overrightarrow{\mathbf{OY}}, \overrightarrow{\mathbf{O_1 M}}), \quad \text{and} \quad \theta_2 = (\overrightarrow{\mathbf{OY}}, \overrightarrow{\mathbf{O_2 M}}), \tag{27.38}$$

where r_1 and r_2 are the distances of point M from the poles of the transformation O_1 and O_2 respectively, in the X, Y plane, and θ_1 and θ_2 denote the angles between the radii and Y-axis. Hence we obtain

$$\alpha = \ln \frac{r_1}{r_2}, \quad \beta = \theta_1 - \theta_2, \tag{27.39}$$

and the inverse relations

$$X = a \frac{\sinh \alpha}{\cosh \alpha - \cos \beta}, \quad Y = a \frac{\sin \beta}{\cosh \alpha - \cos \beta}, \quad z = \frac{x^3}{r_0}. \tag{27.40}$$

Curves $\alpha = \text{const}$ constitute a family of circles of the radii $a \cosh \alpha$, possessing poles at $(0, -a)$ and $(0, \pm a)$. The x^2-axis corresponds to the curve $\alpha = 0$ and the circle coinciding with the contour of the cylindrical cavity corresponds to the value $\alpha = \alpha_1$ (positive), where $\alpha_1 = \text{arg sinh}\, a$. Curves $\beta = \text{const}$ represent the arcs of the circles passing through the poles $(0, \pm a)$, β is the angle between the radii r_1 and r_2. Curves $\beta = \text{const}$ have a discontinuity of value 2π between the poles O_1 and O_2 on the x^1-axis.

In the (α, β) plane the half-space $X > 0$, outside the cylindrical cavity, transforms into the rectangle $OO'A'A$ (Fig. 86). The additional advantage of the transformation is that the infinite region half-space in the rectangular coordinates X, Y, Z transforms into a bounded region. If we furthermore assume that the tractions on the bounding plane are symmetrical with respect to the OX axis then, considering the motion of the medium in bipolar coordinates (α, β) it is sufficient to restrict the discussion, for example, to region $O'A'C'D'$ only, i.e. to $\beta \leqslant 0$ (Fig. 86).

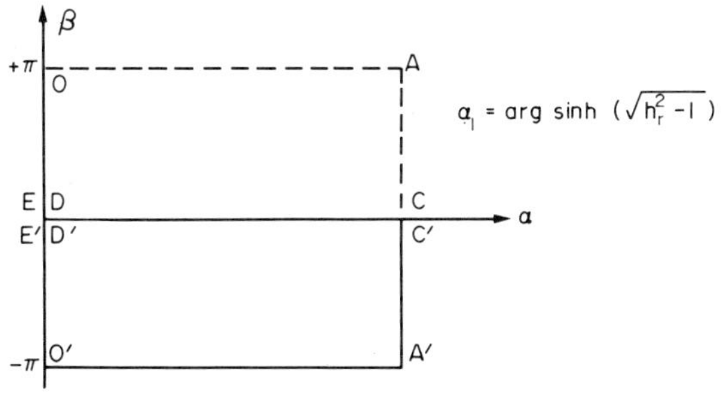

Fig. 86

In order to write down the dynamic equations of the problem in the bipolar, curvilinear system of coordinates we first introduce the metric of the space. The metric tensor g_{ij}, from (5.7), in bipolar coordinates (α, β, z) has the following components:

$$g_{11} = \frac{a^2}{(\cosh\alpha - \cos\beta)^2}, \quad g_{22} \frac{a^2}{(\cosh\alpha - \cos\beta)^2}, \quad g_{33} = 1, \tag{27.41}$$

while the components of the inverse matrix g^{ij} take the following forms (in the orthogonal curvilinear coordinates):

$$g^{11} = \frac{(\cosh\alpha - \cos\beta)^2}{a^2}, \quad g^{22} = g^{11}, \quad g^{33} = 1. \tag{27.42}$$

The determinant g of matrix g_{ij} is equal to

$$g = g_{11}g_{22}g_{33} = (g_{11})^2. \tag{27.43}$$

Next we determine the Christoffel symbols of the second kind. They will be expressed in terms of the coordinates of the metric tensor g_{ij}. In the case of orthogonal curvilinear coordinates the Christoffel symbols are determined from formulae (5.11), where the q^i denote the bipolar coordinates (α, β, z). In the case considered, the Christoffel symbols of the second kind take the forms:

$$\begin{aligned} &\Gamma_{11}^{1} = -\frac{\sinh\alpha}{\cosh\alpha - \cos\beta}, && \Gamma_{11}^{2} = \frac{\sin\beta}{\cosh\alpha - \cos\beta}, \\ &\Gamma_{12}^{1} = -\Gamma_{11}^{2}, && \Gamma_{12}^{2} = \Gamma_{11}^{1}, \\ &\Gamma_{22}^{1} = -\Gamma_{11}^{1}, && \Gamma_{22}^{2} = -\Gamma_{11}^{2}, \\ &\Gamma_{21}^{1} = -\Gamma_{11}^{2}, && \Gamma_{21}^{2} = \Gamma_{11}^{1}. \end{aligned} \tag{27.44}$$

The remaining components vanish.

In the bipolar coordinates (α, β, z) which have been introduced we can solve some other spatial dynamic problems for the case of a plane strain state, e.g. the following problems.

A. Half-space loaded on the boundary by varying tractions

If we assume that h is very large compared with the radius of the cylindrical cavity r_0 then in the system of bipolar coordinates (α, β, z) we can construct the solution of the problem of wave propagation in the half-space with the stresses σ^{11} and σ^{12} applied to the boundary, which are dependent on time and the coordinate x^2 but independent of x^3 (state of plane strain).

In the limiting case we can solve Lamb's dynamic problem, assuming that on the bounding plane $x^1 = 0$ the following tractions σ^{11} are applied along the x^3-axis:

$$\sigma^{11}|_{x^1 = 0} = \mathbf{1}\Delta(x)H(t),$$

where $\mathbf{1}$ denotes a concentrated force referred to a unit length, $\Delta(x)$ is the Dirac delta function and $H(t)$ is the Heaviside unit function. Because of the symmetry of the loading this problem also can be solved in the (α, β) plane with the calculations restricted to the region $OACD$, i.e. for $\beta \geqslant 0$ (Fig. 86).

B. Explosion inside a cylindrical cavity placed in a half-space

We can assume the time varying pressure

$$\sigma^{(\alpha\alpha)} = -p_0(t)g_{11} \quad \text{for} \quad \alpha = \alpha_1 \tag{27.45}$$

on the boundary of the cylindrical cavity placed at a distance h from the half-space boundary for $\alpha = \alpha_1$, where $\alpha_1 = \text{arg sinh}\, a$ (Fig. 85). We can examine, in bipolar coordinates, the problem of the propagation of cylindrical waves and their reflection from the bounding plane $\alpha = 0$. Again, due to the symmetry of the loading (27.45), we can confine the discussion to the region, e.g. $OACD$, i.e. for $\beta \geqslant 0$ (Fig. 86).

C. Diffraction of cylindrical waves by a cylindrical contour in an unbounded half-space

Suppose we have two cylindrical cavities of different radii $r = r_1$ and $r = r_2$ (Fig. 87), with mutually parallel axes, placed in a half-space. Let d be the distance between the cavities. Let pressure $p(t)$, boundary condition (27.45), be applied inside the cavity of radius r_1. The cavity of the radius r_2 is free from stress or has a constant pressure p_0 acting inside it. This is also a case of a state of plane strain.

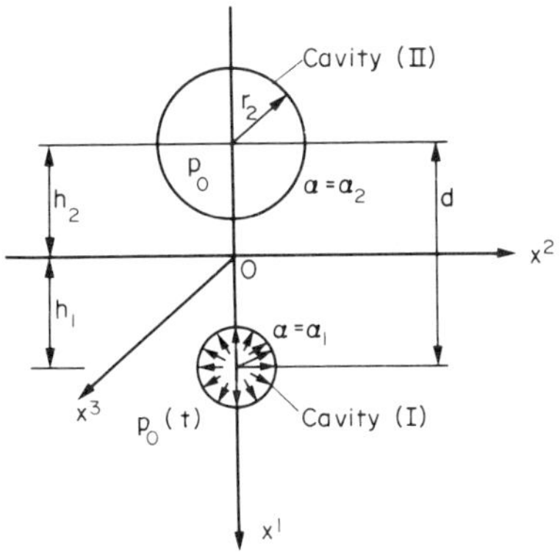

Fig. 87

The system of bipolar coordinates (α, β, z) is chosen such that the circles of radii $\alpha = \alpha_1$ and $\alpha = \alpha_2$, with $\alpha_1 > 0$ and $\alpha_2 < 0$, in the x^i coordinate system in the (x^1, x^2) plane, coincide with the contours of the cylindrical cavities.

The infinite region in the x^i space, containing the cylindrical cavities, is transformed in the bipolar coordinates (α, β, z) into the bounded region $AA'GG'$ (Fig. 88). Due to the symmetry of the problem with respect to x^1 it is sufficient to restrict the discussion, in (α, β, z) space, to the region in which $\beta \geqslant 0$, i.e. to region $ACHG$ (see Fig. 88).

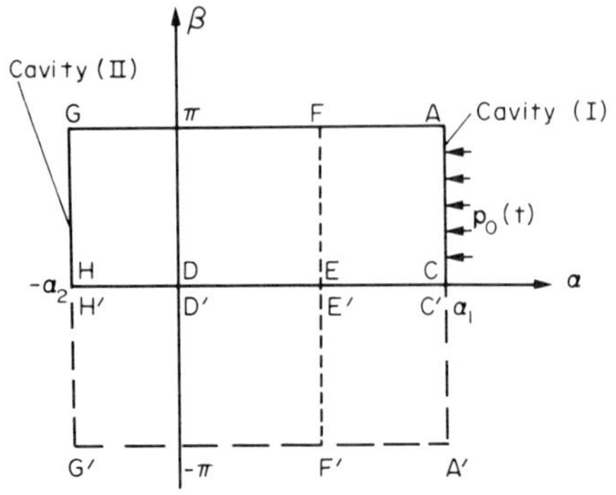

Fig. 88

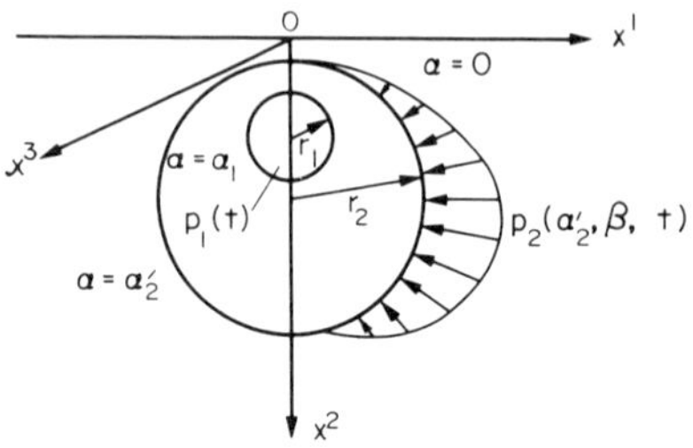

Fig. 89

D. A cylinder with an eccentric hole subjected to radial and tangential dynamic loads

Let a cylinder, the cross-section of which is bounded by two eccentric circles (Fig. 89), be subjected to a uniformly distributed normal pressure $-p_1(t)$ inside the inner surface and to an external pressure $p_2(\alpha_2', \beta, t)$ independent of x^3. We can simultaneously assume that the tangential stress $\sigma^{\alpha\beta}$ on both the surfaces is a function of position and time or we can assume that one of the cylindrical surfaces is free from normal stresses.

The above problem also constitutes a special case of plane strain since, from the assumptions, the motion of the medium does not depend on the variable x^3. In bipolar coordinates (α, β, z) we define the (α, β) plane, the outer surface of the cylinder, by $\alpha = \alpha_2'$, while the inner boundary is $\alpha = \alpha_1$. We assume that α_1 and α_2' are positive and that $\alpha_1 > \alpha_2$ (Fig. 89).

In the bipolar system of coordinates the problem reduces to a solution in rectangle $AA'FF'$ (Fig. 88). We take pressure $p_1(t)$ on the inner cylinder $\alpha = \alpha_1$ (boundary AA') and pressure $p_2(\alpha_2', \beta, t)$ outside the cylinder, on $\alpha = \alpha_2'$ (on boundary FF'). If $p_2(\alpha_2', \beta, t)$ is symmetric with respect to β, i.e. when the pressure is symmetric with respect to x^1, then the solution of the problem can be restricted to region $ACEF$ (Fig. 88).

Now we introduce a slightly different bipolar coordinate system (α, β, φ) from which we can formulate the dynamic problem of a spherical cavity submerged in a half-space in a manner similar to the case of a cylindrical cavity.

Suppose that in a half-space $x^1 > 0$ (Fig. 84), at a distance h from the boundary, there exists a spherical cavity of radius $r = r_0$. A pressure $p_0(t)$, varying in time, is applied inside the spherical cavity. The bounding plane $x^1 = 0$ is free from stresses. Thus we have to deal with an axially symmetric problem with respect to the x^1-axis.

In this case let us introduce the bipolar system of coordinates (α, β, φ), where φ denotes the angle between the vector **OM** and the positive axis Ox^2. Relations (27.36) and (27.39) now take the following forms:

$$x = \frac{x^1}{r_0}, \quad Y = \frac{x^2}{r_0}, \quad Z = \frac{x^3}{r_0}, \tag{27.46}$$

and

$$\alpha = \ln \frac{r_1}{r_2}, \quad \beta = \theta_1 - \theta_2, \quad \varphi = (\mathbf{OY}, \mathbf{OM}). \tag{27.47}$$

The inverse relations to (27.47) take the forms

$$X = a\frac{\sinh\alpha}{\cosh\alpha - \cos\beta}, \quad Y = a\frac{\sin\beta\cos\varphi}{\cosh\alpha - \cos\beta}, \quad Z = a\frac{\sin\beta\sin\varphi}{\cosh\alpha - \cos\beta}. \tag{27.48}$$

The metric tensor g_{ij} has, in the bipolar coordinates (α, β, φ), the following components:

$$\begin{aligned} g_{11} &= \frac{a^2}{(\cosh\alpha - \cos\beta)^2}, \quad g_{22} = \frac{a^2}{(\cosh\alpha - \cos\beta)^2}, \\ g_{33} &= \frac{a^2\sin^2\beta}{(\cosh\alpha - \cos\beta)^2}. \end{aligned} \tag{27.49}$$

We are dealing with an orthogonal system of coordinates so that consequently the Christoffel symbols of the second kind are determined from formulae (5.11). We obtain:

$$\begin{aligned} &\Gamma_{11}^1 = -\frac{\sinh\alpha}{\cosh\alpha - \cos\beta}, \quad \Gamma_{11}^2 = \frac{\sin\beta}{\cosh\alpha - \cos\beta}, \\ &\Gamma_{12}^1 = -\Gamma_{12}^2 = -\Gamma_{21}^2 = \Gamma_{11}^1, \\ &\Gamma_{33}^1 = -\sin^2\beta\, F_{11}^1, \quad \Gamma_{33}^2 = -\sin\beta\,(\cos\beta - \sin\beta\,\Gamma_{11}^2), \\ &\Gamma_{32}^3 = \frac{1}{\sin\beta}(\cos\beta - \Gamma_{11}^2). \end{aligned} \tag{27.50}$$

Similar to the case of the cylindrical cavity we can formulate, in an elastic/viscoplastic half-space, a number of different boundary value problems which can be formulated in the system of bipolar coordinates (α, β, φ).

Namely:

(1) diffraction of plane waves by the contour of a sphere;
(2) diffraction of the spherical waves by another spherical contour;
(3) propagation of waves in a sphere with an eccentric spherical hole, generated by a dynamic load on the inner or outer surface.

Let us now return to the solution of problems of dynamics of a cylindrical cavity in an elastic/viscoplastic half-space. We assume that the medium (half-space $x^1 \geqslant 0$, Fig. 84) is an elastic/viscoplastic one. We shall write down the dynamic equations of the problem in terms of the contravariant components of the stress tensor σ^{ij} using a natural basis, the strain rate tensor d^{ij}, and the contravariant components of the velocity vector v^i. The constitutive equations for the elastic/viscoplastic medium are thus taken in the form (3.14). If we confine ourselves to the Huber–Mises condition (3.8), then the equations assume the following form:

$$d^{ij} = \frac{1}{2\mu}\dot{s}^{ij} + \frac{1-2\nu}{3E}\dot{\sigma}^k{}_k g^{ij} + \gamma\left\langle \Phi\left(\frac{\sqrt{J_2}}{\varkappa} - 1\right)\right\rangle \frac{s^{ij}}{\sqrt{J_2}}, \tag{27.51}$$

where

$$\sigma^k{}_k = g_{km}\sigma^{km}, \quad J_2 = \frac{1}{2}s^{ij}s_{ij} = \frac{1}{2}s^{ij}s^{kl}g_{ki}g_{lj}.$$

The stress tensor σ^{ij}, in bipolar coordinates (α, β, z), has the following non-vanishing components:

$$(27.52) \qquad \begin{aligned} \sigma^{\alpha\alpha} &= \sigma^{\alpha\alpha}(\alpha, \beta, t), \quad & \sigma^{\beta\beta} &= \sigma^{\beta\beta}(\alpha, \beta, t), \\ \sigma^{\alpha\beta} &= \sigma^{\alpha\beta}(\alpha, \beta, t), \quad & \sigma^{zz} &= \sigma^{zz}(\alpha, \beta, t); \end{aligned}$$

the strain tensor has the following non-zero components:

$$(27.53) \qquad \epsilon^{\alpha\alpha} = \epsilon^{\alpha\alpha}(\alpha, \beta, t), \quad \epsilon^{\beta\beta} = \epsilon^{\beta\beta}(\alpha, \beta, t), \quad \epsilon^{\alpha\beta} = \epsilon^{\alpha\beta}(\alpha, \beta, t).$$

We determine the contravariant components of the strain rate tensor d^{ij}. By virtue of (3.17) and taking into account the values of the Christoffel symbols (27.44), we obtain the covariant derivatives v^i of the velocity vector $v^\alpha = v^\alpha(\alpha, \beta, t)$, $v^\beta = v^\beta(\alpha, \beta, t)$ in the forms

$$(27.54) \qquad \begin{aligned} \frac{Dv^\alpha}{\partial\alpha} &= \frac{\partial v^\alpha}{\partial\alpha} - \frac{1}{\cosh\alpha - \cos\beta}(v^\alpha \sinh\alpha + v^\beta \sin\beta), \\ \frac{Dv^\beta}{\partial\beta} &= \frac{\partial v^\beta}{\partial\beta} - \frac{1}{\cosh\alpha - \cos\beta}(v^\alpha \sinh\alpha + v^\beta \sin\beta), \\ \frac{Dv^\alpha}{\partial\beta} &= \frac{\partial v^\alpha}{\partial\beta} - \frac{1}{\cosh\alpha - \cos\beta}(v^\alpha \sin\beta - v^\beta \sinh\alpha), \\ \frac{Dv^\beta}{\partial\alpha} &= \frac{\partial v^\beta}{\partial\alpha} - \frac{1}{\cosh\alpha - \cos\beta}(v^\alpha \sin\beta - v^\beta \sinh\alpha). \end{aligned}$$

Using formula (3.16) we obtain the components of the strain rate tensor d^{ij}:

$$(27.55) \qquad \begin{aligned} d^{\alpha\alpha} &= \frac{\partial v^\alpha}{\partial\alpha} - \frac{1}{\cosh\alpha - \cos\beta}(v^\alpha \sinh\alpha + v^\beta \sin\beta)\, g^{11}, \\ d^{\beta\beta} &= \frac{\partial v^\alpha}{\partial\beta} - \frac{1}{\cosh\alpha - \cos\beta}(v^\alpha \sinh\alpha + v^\beta \sin\beta)\, g^{11}, \\ d^{\alpha\beta} &= \frac{1}{2}\left(\frac{\partial v^\alpha}{\partial\beta} + \frac{\partial v^\beta}{\partial\alpha}\right) g^{11}, \\ d^{\beta\alpha} &= d^{\alpha\beta}, \quad d^{\alpha z} = d^{z\alpha} = 0, \quad d^{\beta z} = d^{z\beta} = 0, \quad d^{zz} = 0. \end{aligned}$$

Taking into account the non-vanishing components of the strain rate tensor (27.55) and those of the stress tensor (27.52) as well as expressions (27.42), we can reduce the constitutive equations (27.51) to the following forms:

$$(27.56) \qquad \begin{aligned} d^{\alpha\alpha} &= \frac{1}{2\mu}\left(\dot{\sigma}^{\alpha\alpha} - \frac{1}{3}\dot{\sigma} g^{11}\right) + \frac{1}{9K}\dot{\sigma} g^{11} + \gamma\langle\Phi(F)\rangle\left(\sigma^{\alpha\alpha} - \frac{1}{3}\sigma g^{11}\right), \\ d^{\beta\beta} &= \frac{1}{2\mu}\left(\dot{\sigma}^{\beta\beta} - \frac{1}{3}\dot{\sigma} g^{11}\right) + \frac{1}{9K}\dot{\sigma} g^{11} + \gamma\langle\Phi(F)\rangle\left(\sigma^{\beta\beta} - \frac{1}{3}\sigma g^{11}\right), \\ 0 &= \frac{1}{2\mu}\left(\dot{\sigma}^{zz} - \frac{1}{3}\dot{\sigma}\right) + \frac{1}{9K}\dot{\sigma} + \gamma\langle\Phi(F)\rangle\left(\sigma^{zz} - \frac{1}{3}\sigma\right), \end{aligned}$$

(27.56) [*cont.*]

$$d^{\alpha\beta} = \frac{1}{2\mu}\,\dot{\sigma}^{\alpha\beta} + \gamma\langle\Phi(F)\rangle\,\sigma^{\alpha\beta},$$

where

$$\langle\Phi(F)\rangle = \left\langle\Phi\left(\frac{\sqrt{J_2}}{\varkappa} - 1\right)\right\rangle \frac{1}{\sqrt{J_2}}, \tag{27.57}$$

and

$$\sigma = g_{11}(\sigma^{\alpha\alpha} + \sigma^{\beta\beta}) + \sigma^{zz},$$

$$J_2 = \frac{1}{2}\Big\{ g_{11}^2\,[(\sigma^{\alpha\alpha})^2 + (\sigma^{\beta\beta})^2 + 2(\sigma^{\alpha\beta})^2] + (\sigma^{zz})^2 - \frac{1}{9}\,[\sigma^{\alpha\alpha} + \sigma^{\beta\beta} + \sigma^{zz}g^{11}]^2 \times [2g_{11}(3 - g_{11}) - 1]\Big\}.$$

In order to simplify the solutions we confine ourselves to an elastic/visco-ideally-plastic medium. We take function F in the form of (3.24). Then expression (27.57) takes the following form:

$$\langle\Phi\;(F)\rangle = \left\langle\frac{\sqrt{J_2}}{k} - 1\right\rangle \frac{1}{\sqrt{J_2}}. \tag{27.58}$$

The equations of motion of the medium will be derived in bipolar coordinates (α, β, z). In the case of orthogonal curvilinear coordinates (5.19), disregarding the vector of body forces $(\mathbf{X} \equiv 0)$, and taking into account (27.52), (27.41), and (27.43), we deduce from the equations of motion the following system of two equations:

$$\begin{aligned} &\frac{1}{g_{11}^2}\left\{\frac{\partial}{\partial\alpha}(g_{11}^2\,\sigma^{\alpha\alpha}) + \frac{\partial}{\partial\beta}(g_{11}^2\,\sigma^{\alpha\beta})\right\} + \frac{\sinh\alpha}{\cosh\alpha - \cos\beta}(\sigma^{\alpha\alpha} + \sigma^{\beta\beta}) = \varrho a^{\alpha}.\\ &\frac{1}{g_{11}^2}\left\{\frac{\partial}{\partial\alpha}(g_{11}^2\,\sigma^{\alpha\beta}) + \frac{\partial}{\partial\beta}(g_{11}^2\,\sigma^{\beta\beta})\right\} + \frac{\sin\beta}{\cosh\alpha - \cos\beta}(\sigma^{\alpha\alpha} + \sigma^{\beta\beta}) = \varrho a^{\beta}. \end{aligned} \tag{27.59}$$

The contravariant components of the acceleration vector **a**, occurring in equations of motion (27.59), will be determined from (5.16) with the values of Christoffel symbols given in (27.36). Thus we obtain

$$\begin{aligned} a^{\alpha} &= \frac{\partial v^{\alpha}}{\partial t} + \frac{\partial v^{\alpha}}{\partial\alpha}\,v^{\alpha} + \frac{\partial v^{\alpha}}{\partial\beta}\,v^{\beta} - \frac{1}{\cosh\alpha - \cos\beta}\left\{\sinh\alpha(v^{\alpha}v^{\alpha} - v^{\beta}v^{\beta}) + 2\sin\beta\,v^{\alpha}v^{\beta}\right\},\\ a^{\beta} &= \frac{\partial v^{\beta}}{\partial t} + \frac{\partial v^{\beta}}{\partial\alpha}\,v^{\alpha} + \frac{\partial v^{\beta}}{\partial\beta}\,v^{\beta} + \frac{1}{\cosh\alpha - \cos\beta}\left\{\sin\beta(v^{\alpha}v^{\alpha} - v^{\beta}v^{\beta}) - 2\sinh\alpha v^{\alpha}v^{\beta}\right\}. \end{aligned} \tag{27.60}$$

Eventually we obtain a system of six equations, namely (27.55) and (27.60), using (27.56) and (27.59). This is a system of six partial differential equations of the first order in the six required quantities: the four stress tensor components $\sigma^{\alpha\alpha}(\alpha, \beta, t)$, $\sigma^{\alpha\beta}(\alpha, \beta, t)$, $\sigma^{\beta\beta}(\alpha, \beta, t)$, and $\sigma^{zz}(\alpha, \beta, t)$, and the two velocity vector components $v^{\alpha}(\alpha, \beta, t)$ and $v^{\beta}(\alpha, \beta, t)$.

We solve the above system of equations assuming the zero initial conditions:

$$\begin{aligned} &\sigma^{\alpha\alpha}(\alpha, \beta, 0) = \sigma^{\beta\beta}(\alpha, \beta, 0) = \sigma^{zz}(\alpha, \beta, 0) = \sigma^{\alpha\beta}(\alpha, \beta, 0) \equiv 0,\\ &v^{\alpha}(\alpha, \beta, 0) = v^{\beta}(\alpha, \beta, 0) \equiv 0, \end{aligned} \tag{27.61}$$

and taking into account the following boundary conditions:

Case I: For the dynamic problem of a cylindrical cavity in an elastic/viscoplastic half-space, loaded on the boundary by a pressure which is dependent on time and the coordinate x^2, we assume:

for $\alpha = 0$ (half-space boundary) that

$$\sigma^{\alpha\alpha}(0, \beta, t) = (1 - \cos\beta)^2\, \sigma_0^{\alpha\alpha}(\beta, t),$$

(27.62)

and $$\sigma^{\alpha\beta}(0, \beta, t) = 0,$$

where

$$\sigma_0^{\alpha\alpha}(\beta, t) = -p_0(\beta, t)\frac{1}{a^2}$$

is the prescribed component of the stress tensor normal to the bounding plane, expressed in physical components;

for $\alpha = \alpha_1$, where $\alpha_1 = \operatorname{arg\,sinh} a$ (boundary of the cylindrical cavity),

$$\sigma^{\alpha\alpha}(\alpha_1, \beta, t) = 0, \quad \sigma^{\alpha\beta}(\alpha_1, \beta, t) = 0. \tag{27.63}$$

Case II: For the problem of cylindrical waves propagation in a half-space, generated by the dynamic loading of a cylindrical cavity placed inside the half-space $\left(\text{pressure } p_0(t)\frac{1}{a^2} \text{ is applied inside the cavity}\right)$ we assume:

for $\alpha = \alpha_1$ (boundary of the cylindrical cavity) that

$$\sigma^{\alpha\alpha}(\alpha_1, \beta, t) = (h_r - \cos\beta)^2 p_0(t),$$

(27.64)

and $$\sigma^{\alpha\beta}(\alpha_1, \beta, t) = 0,$$

while for $\alpha = 0$ (boundary of the half-space)

$$\sigma^{\alpha\alpha}(0, \beta, t) = 0, \quad \text{and} \quad \sigma^{\alpha\beta}(0, \beta, t) = 0. \tag{27.65}$$

We introduce the following dimensionless magnitudes:

$$U = \frac{v^\alpha}{a_1}, \quad V = \frac{v^\beta}{a_1}, \quad P = \frac{\sigma^{\alpha\alpha}}{\varrho a_1^2}, \quad Q = \frac{\sigma^{\beta\beta}}{\varrho a_1^2}, \quad S = \frac{\sigma^{zz}}{\varrho a_1^2},$$

(27.66)

$$T = \frac{\sigma^{\alpha\beta}}{\varrho a_1^2}, \quad \Gamma = \frac{a_1}{a_2}, \quad W = \frac{k_0}{\varrho a_1^2}, \quad \tau = \frac{a_1 t}{r_0}, \quad P_0 = \frac{p_0}{\varrho a_1^2},$$

where a_1 denotes the speed of propagation of the longitudinal wave and a_2 is the shear wave

$$a_1 = \left(\frac{3K + 4\mu}{3\varrho}\right)^{1/2}, \quad a_2 = \left(\frac{\mu}{\varrho}\right)^{1/2},$$

speed, k_0 is the yield limit for pure shear.

The system of the equations for the problem: constitutive equations (27.56), equations of motion (27.59), with equations (27.55) and (27.60) taken into account, can, after some algebra, be represented in terms of the introduced dimensionless magnitudes (27.66) in the following forms:

$$\begin{aligned}
U_{,\tau} &= -U_{,\alpha}U - U_{,\beta}V + \frac{1}{\cosh\alpha - \cos\beta}[(U^2 - V^2)\sinh\alpha \\
&\quad + 2UV\sin\beta] + \frac{1}{g_{11}^2}\left[\frac{\partial}{\partial\alpha}(g_{11}^2 P) + \frac{\partial}{\partial\beta}(g_{11}^2 T)\right] + \frac{\sinh\alpha}{\cosh\alpha - \cos\beta}(P + Q), \\
V_{,\tau} &= -V_{,\alpha}U - V_{,\beta}V - \frac{1}{\cosh\alpha - \cos\beta}[(U^2 - V^2)\sin\beta - 2UV\sinh\alpha] \\
&\quad + \frac{1}{g_{11}^2}\left[\frac{\partial}{\partial\alpha}(g_{11}^2 T) + \frac{\partial}{\partial\beta}(g_{11}^2 Q)\right] + \frac{\sin\beta}{\cosh\alpha - \cos\beta}(P + Q), \\
P_{,\tau} &= \left[c_3 U_{,\alpha} - c_5 V_{,\beta} - \frac{1}{\cosh\alpha - \cos\beta}(U\sinh\alpha + V\sin\beta)b_{21}\right]g^{11} \\
&\quad + \tilde{\gamma}\langle\tilde{\Psi}\rangle\left\{c_5c_6[(P + Q)c_7 + g^{11}c_8S] - 3[Pc_3 - Qc_5]\right\}, \\
Q_{,\tau} &= \left[c_3 V_{,\beta} - c_5 U_{,\alpha} - \frac{1}{\cosh\alpha - \cos\beta}(U\sinh\alpha + V\sin\beta)b_{21}\right]g^{11} \\
&\quad + \tilde{\gamma}\langle\tilde{\Psi}\rangle\left\{c_5c_6[(P + Q)c_7 + g^{11}c_8S] + 3[Pc_5 - Qc_3]\right\}, \\
T_{,\tau} &= \frac{g^{11}}{\Gamma^2}(U_{,\beta} + V_{,\alpha}) - \tilde{\gamma}\frac{6}{\Gamma^2}\langle\tilde{\Psi}\rangle T, \\
S_{,\tau} &= -b_7(P_{,\tau} + Q_{,\tau})g_{11} - \frac{1}{b_2}\tilde{\gamma}\langle\tilde{\Psi}\rangle[2S - (P + Q)g_{11}],
\end{aligned} \tag{27.67}$$

where

$$\begin{aligned}
&b_2 = \Gamma^2\frac{\Gamma^2 - 1}{3\Gamma^2 - 4}g_{11}, \quad b_3 = \frac{\Gamma^2}{2}\frac{2 - \Gamma^2}{3\Gamma^2 - 4}, \quad b_7 = \frac{b_3}{b_2}, \\
&c_1 = \frac{\varrho a_1^2}{2\mu}, \quad c_2 = (1 - b_7)b_3, \quad c_4 = 2c_1c_2\frac{1}{a^2}, \quad c_6 = \frac{c_1}{c_2}, \\
&c_3 = \frac{c_1 + c_2}{c_1(c_1 + 2c_2)}, \quad c_5 = \frac{c_2}{c_1(c_1 + 2c_2)}, \quad c_7 = 1 - b_7, \\
&c_8 = 1 + 2b_7, \quad b_{21} = c_3 - c_5, \quad \tilde{\gamma} = \frac{r_0\gamma}{3a_1},
\end{aligned} \tag{27.68}$$

and

$$\langle \widetilde{\Psi} \rangle = \langle \Phi \left(\frac{\sqrt{\widetilde{J}_2}}{W} - 1 \right) \frac{1}{\sqrt{\widetilde{J}_2}}, \tag{27.69}$$

$$\widetilde{J}_2 = \frac{1}{2} \left\{ g_{11}^2 [P^2 + Q^2 + 2T^2] + S^2 - \frac{1}{9} [P + Q + Sg^{11}] [2g_{11}(3 - g_{11}) - 1] \right\}. \tag{27.70}$$

The initial conditions (27.61) take the forms:

$$\begin{aligned} &P(\alpha, \beta, 0) = Q(\alpha, \beta, 0) = T(\alpha, \beta, 0) = S(\alpha, \beta, 0) \equiv 0, \\ &U(\alpha, \beta, 0) = V(\alpha, \beta, 0) \equiv 0. \end{aligned} \tag{27.71}$$

Now the boundary conditions (27.62)–(27.65) transform to the following forms:

Case I:

for $\alpha = 0$

$$P(0, \beta, \tau) = (1 - \cos \beta)^2 P_0(\beta, \tau), \quad T(0, \beta, \tau) = 0; \tag{27.72}$$

for $\alpha = \alpha_1$

$$P(\alpha_1, \beta, \tau) = 0, \quad T(\alpha_1, \beta, \tau) = 0.$$

Case II:

for $\alpha = 0$

$$P(0, \beta, \tau) = 0, \quad T(0, \beta, \tau) = 0, \quad T(0, \beta, \tau) = 0, \tag{27.73}$$

for $\alpha = \alpha_1$

$$P(\alpha_1, \beta, \tau) = (h_r - \cos \beta)^2 P_0(\tau), \quad T(\alpha_1, \beta, \tau) = 0.$$

The components of the strain tensor ϵ_{ij} can be determined from formulae (3.18). Since the displacement vector $\mathbf{u}$ only has the non-vanishing components u_α and u_β, we obtain the following expressions for the strain tensor components:

$$\begin{aligned} \epsilon_{\alpha\alpha} &= \frac{1}{a} \left[\frac{\partial u_\alpha}{\partial \alpha} (\cosh \alpha - \cos \beta) - u_\beta \sin \beta \right], \\ \epsilon_{\beta\beta} &= \frac{1}{a} \left[\frac{\partial u_\beta}{\partial \beta} (\cosh \alpha - \cos \beta) - u_\alpha \sinh \alpha \right], \\ \epsilon_{\alpha\beta} &= \frac{1}{2a} \left[\left(\frac{\partial u_\alpha}{\partial \beta} + \frac{\partial u_\beta}{\partial \alpha} \right) (\cosh \alpha - \cos \beta) + u_\alpha \sin \beta + u_\beta \sinh \alpha \right]. \end{aligned} \tag{27.74}$$

The contravariant components of the strain tensor ϵ^{ij} are determined from (3.19). We obtain:

$$\begin{aligned} \epsilon^{\alpha\alpha} &= g^{11} g^{11} \epsilon_{\alpha\alpha}, \\ \epsilon^{\beta\beta} &= g^{11} g^{11} \epsilon_{\beta\beta}, \\ \epsilon^{\alpha\beta} &= g^{11} g^{11} \epsilon_{\alpha\beta}. \end{aligned} \tag{27.75}$$

The physical components of stress tensor $\sigma^{(ij)}$, strain tensor $\epsilon^{(ij)}$, and velocity $v^{(i)}$ are deduced from (3.20) and (27.41). We obtain

(27.76)
$$\begin{gathered}
\sigma^{(\alpha\alpha)} = g_{11}\sigma^{\alpha\alpha}, \quad \sigma^{(\beta\beta)} = g_{11}\sigma^{\beta\beta}, \\
\sigma^{(zz)} = \sigma^{zz}, \quad \sigma^{(\alpha\beta)} = g_{11}\sigma^{\alpha\beta}, \\
\epsilon^{(\alpha\alpha)} = g_{11}\epsilon^{\alpha\alpha}, \quad \epsilon^{(\beta\beta)} = g_{11}\epsilon^{\beta\beta}, \quad \epsilon^{(\alpha\beta)} = g_{11}\epsilon^{\alpha\beta}, \\
v^{(\alpha)} = \sqrt{g_{11}}\, v^{\alpha}, \quad v^{(\beta)} = \sqrt{g_{11}}\, v^{\beta}.
\end{gathered}$$

A tilde over a symbol designates the physical components of the stress tensor or of the velocity vector, the notation (27.66) being used. We obtain the following relations:

(27.77)
$$\begin{gathered}
\widetilde{P} = g_{11}P, \quad \widetilde{Q} = g_{11}Q, \quad \widetilde{S} = S, \quad \widetilde{T} = g_{11}T, \\
\widetilde{U} = \sqrt{g_{11}}\, U, \quad \widetilde{V} = \sqrt{g_{11}}\, V.
\end{gathered}$$

The system of equations of the problem, (27.56) and (27.59), with expressions (27.55) and (27.60) taken into account, can be represented in a more concise, matrix notation:

(27.78)
$$L[\mathbf{u}] = \mathbf{A}\mathbf{u}_{,t} + \mathbf{B}\mathbf{u}_{,\alpha} + \mathbf{C}\mathbf{u}_{,\beta} + \mathbf{D} = 0,$$

where **u** is a vector; **A**, **B** and **C** are symmetric matrices of the dimension 6 × 6, and matrix **A** is positively definite; **D** denotes another vector:

$$\mathbf{A} = \begin{bmatrix}
\varrho & 0 & 0 & 0 & 0 & 0 \\
0 & \varrho & 0 & 0 & 0 & 0 \\
0 & 0 & \frac{1}{\mu} g_{11} & 0 & 0 & 0 \\
0 & 0 & 0 & b_2 & -b_3 g_{11} & -b_3 \\
0 & 0 & 0 & -b_3 g_{11} & b_2 & -b_3 \\
0 & 0 & 0 & -b_3 & -b_3 & b_2 g^{11}
\end{bmatrix},$$

$$\mathbf{C} = \begin{bmatrix}
\varrho V & 0 & -1 & 0 & 0 & 0 \\
0 & \varrho V & 0 & 0 & -1 & 0 \\
-1 & 0 & 0 & 0 & 0 & 0 \\
0 & 0 & 0 & 0 & 0 & 0 \\
0 & -1 & 0 & 0 & 0 & 0 \\
0 & 0 & 0 & 0 & 0 & 0
\end{bmatrix},$$

$$\mathbf{u}^{\mathrm{T}} = \left[\, U, V, T, P, Q, S \,\right],$$

$$\mathbf{B} = \begin{bmatrix} \varrho U & 0 & 0 & -1 & 0 & 0 \\ 0 & \varrho U & -1 & 0 & 0 & 0 \\ 0 & -1 & 0 & 0 & 0 & 0 \\ -1 & 0 & 0 & 0 & 0 & 0 \\ 0 & 0 & 0 & 0 & 0 & 0 \\ 0 & 0 & 0 & 0 & 0 & 0 \end{bmatrix}, \qquad \mathbf{D} = \begin{bmatrix} -\varphi_1 \\ -\varphi_2 \\ \Phi_{44} \\ \Phi_{11} \\ \Phi_{22} \\ \Phi_{33} \end{bmatrix},$$

where

$$\varphi_1 = \frac{1}{\cosh\alpha - \cos\beta}\left[\sinh\alpha\,(P + Q - U^2 + V^2) + 2\sin\beta\, UV\right],$$

$$\varphi_2 = \frac{1}{\cosh\alpha - \cos\beta}\left[\sin\beta\,(P + Q - U^2 + V^2) + 2\sinh\alpha\, UV\right],$$

$$\Phi_{11} = \frac{1}{\cosh\alpha - \cos\beta}\left[U\sinh\alpha + V\sin\beta\right] + \tilde{\gamma}\langle\tilde{\Psi}\rangle\left(P - \frac{1}{3}\sum g^{11}\right)g_{11},$$

$$\Phi_{22} = \frac{1}{\cosh\alpha - \cos\beta}\left[U\sinh\alpha + V\sin\beta\right] + \tilde{\gamma}\langle\tilde{\Psi}\rangle\left(Q - \frac{1}{3}\sum g^{11}\right)g_{11},$$

$$\Phi_{33} = (g^{11})^2\,\tilde{\gamma}\langle\tilde{\Psi}\rangle\left(S - \frac{1}{3}\sum g^{11}\right),$$

$$\Phi_{44} = 2\tilde{\gamma}\langle\tilde{\Psi}\rangle T g_{11},$$

$$\sum = g_{11}(P + Q) + S.$$

The elements of the matrices **A**, **B**, and **C** and of vector **D** are functions of α, β, and **u**. The system of differential equations (27.78) is quasi-linear and of hyperbolic type.

On account of the non-linearity of the problem the solution of the initial value problem is sought numerically. The system of (27.78) can be solved by the finite difference method along the bicharacteristics.

We solve the spatial boundary value problem formulated above, however, by the method of direct integration of the equations of the problem (27.78) [107], [108]. In order to integrate the equation of the problem with respect to time, we make use of Treanor's method [158], generalized to the case of a system of differential equations.

Let us consider, in the three-dimensional space (α, β, τ), a family of planes $\tau = n^*H$, where n^* is an integer and H is a positive number equal to the discretization interval with respect to time.

In order to obtain a numerical solution of the system of (27.79), we look for a difference operator $L\,[\mathbf{U}] = 0$ such that, at any point of the region of definition of the $n + 1$ discretization plane $\tau = \text{const}$, the solution is approximated by the operation $L\,[\mathbf{U}] = 0$ to the required accuracy. Also, **U** is prescribed on the nth plane $\tau = \text{const}$ (this is an explicit discretization scheme). At the inner points of the region the problem reduces to a solution of the Cauchy problem. At the points lying on the boundary of the region, for a given mixed initial boundary value problem, the vector **U** must satisfy additional boundary conditions.

For the sake of simplifying the numerical calculations we assume that the tractions on the boundary of half-space $x^1 = 0$ (Fig. 84) are symmetrical with respect to the Ox^1 axis. This assumption is not essential to the method; it only serves to limit the calculations to the region, $\beta \leqslant 0$ (Fig. 86), and to the numerical computation time.

Region D $(0 \leqslant \alpha \leqslant \alpha_1, -\pi < \beta < 0)$ is divided into a rectangular net of straight lines $\alpha =$ const and $\beta =$ const (Fig. 90). $\Delta h_\alpha = \alpha_1/n_1$ and $\Delta h_\beta = \pi/n_2$ are the division intervals where n_1

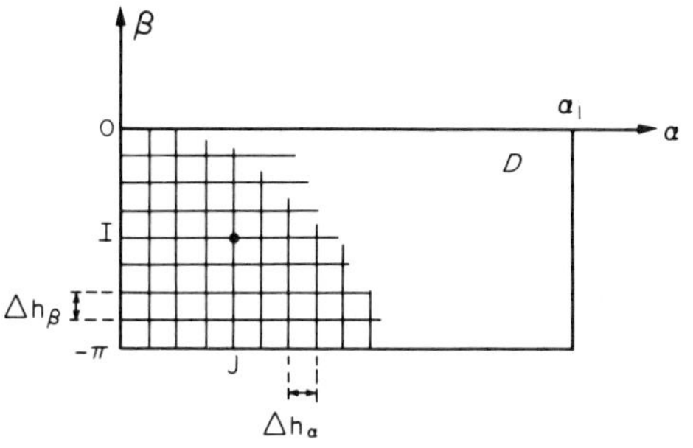

Fig. 90

and n_2 denote the numbers of intervals in the region D in the α and β directions respectively. Furthermore we assume that $n_\alpha = n_1 + 1, n_\beta = n_2 + 1$.

We determine the geometry of the problem at the nodal points of the net, i.e. we determine, at these points, the component of the metric tensor g^{11} (and g_{11}) and its derivatives in the directions $\alpha =$ const and $\beta =$ const.

We assume that the values of the vector **U** are known at the points of intersection of the net $(J, I), J = 1, \ldots, n_1, I = 1, \ldots, n_2$, on the nth plane of discretization with respect to time. Matrices **A**, **B**, and **C** and vector **D** in (27.78) are also known on the nth plane. We seek the operator $L[\mathbf{U}] = 0$ on the $n + 1$th plane.

At the inner points of region D (Fig. 90) we replace the partial derivatives of the components of vector **U** in the directions α and β by the corresponding finite differences (for a fixed time $\tau =$ const):

$$\frac{dU_i}{d\alpha} = \frac{1}{\Delta h_\alpha}[U_i(J+1, I) - U_i(J, I)],$$
$$(27.79) \qquad (i = 1, \ldots, 6)$$
$$\frac{dU_i}{d\beta} = \frac{1}{\Delta h_\beta}[U_i(J, I+1) - U_i(J, I)].$$

At the points on the boundary of region D, $\alpha = 0$ and $\alpha = \alpha_1$, we replace the partial derivatives of the components of the vector **U** in direction α (for fixed time $\tau =$ const) by the forward and backward differences

$$\frac{dU_i}{d\alpha} = \frac{1}{\Delta h_\alpha}[U_i(2, I) - U(1, I)],$$
$$(27.80)$$
$$\frac{dU_i}{d\alpha} = \frac{1}{\Delta h_\alpha}[U_i(n_\alpha, I) - U_i(n_\alpha - 1, I)].$$

The components of the vector **U** in the $(n+1)$th plane, i.e. for $\tau = (n+1)H$, have to satisfy the boundary conditions (27.72) for $\alpha = 0$ (the boundary of the half-space) and for $\alpha = \alpha_1$ (the boundary of the cylindrical cavity). Moreover, the components of vector **U** have to satisfy the conditions resulting from the symmetry of the problem (it was assumed that the loading of the boundary $x^1 = 0$ is symmetrical with respect to the Ox^1 axis). Thus the following conditions hold:

$$\left.\begin{aligned} &T(\alpha, 0, \tau) = 0, \\ &V(\alpha, 0, \tau) = 0, \end{aligned}\right\} \quad \text{for } \beta = 0,$$

(27.81)

$$\left.\begin{aligned} &T(\alpha, -\pi, \tau) = 0, \\ &V(\alpha, -\pi, \tau) = 0, \end{aligned}\right\} \quad \text{for } \beta = -\pi.$$

Also, on account of the fact that the following components of vector **U** – P, Q, S, and U – are symmetrical with respect to the Ox^1 axis while components V and T are antisymmetrical with respect to that axis, we obtain the following conditions:

(27.82)

$$\begin{aligned} P(\alpha, \pi - \Delta h_\beta, \tau) &= P(\alpha, -\pi + \Delta h_\beta, \tau), \\ Q(\alpha, \pi - \Delta h_\beta, \tau) &= Q(\alpha, -\pi + \Delta h_\beta, \tau), \\ S(\alpha, \pi - \Delta h_\beta, \tau) &= S(\alpha, -\pi + \Delta h_\beta, \tau), \\ U(\alpha, \pi - \Delta h_\beta, \tau) &= U(\alpha, -\pi + \Delta h_\beta, \tau), \\ V(\alpha, \pi - \Delta h_\beta, \tau) &= -V(\alpha, -\pi + \Delta h_\beta, \tau), \\ T(\alpha, \pi - \Delta h_\beta, \tau) &= -T(\alpha, -\pi + \Delta h_\beta, \tau), \end{aligned}$$

and

(27.83)

$$\begin{aligned} P(\alpha, \Delta h_\beta, \tau) &= P(\alpha, -\Delta h_\beta, \tau), \\ Q(\alpha, \Delta h_\beta, \tau) &= Q(\alpha, -\Delta h_\beta, \tau), \\ S(\alpha, \Delta h_\beta, \tau) &= S(\alpha, -\Delta h_\beta, \tau), \\ U(\alpha, \Delta h_\beta, \tau) &= U(\alpha, -\Delta h_\beta, \tau), \\ V(\alpha, \Delta h_\beta, \tau) &= -V(\alpha, -\Delta h_\beta, \tau), \\ T(\alpha, \Delta h_\beta, \tau) &= -T(\alpha, -\Delta h_\beta, \tau). \end{aligned}$$

The point O, singular in the (α, β) plane, corresponds to infinity, for the region in the x^1–x^2 plane: we assume that $\mathbf{U} = 0$ represents that part of the region. It means that since the disturbances generated by the external load do not reach the points infinitely remote from the origin of the coordinate system in the cartesian space x^i during the time in which we examine the motion of the medium, we therefore assume the condition

$$U_i = (0, 0, \tau) = 0. \tag{27.84}$$

The system of equations obtained can be written in the form

$$\frac{dU_i}{d\tau} = f_i(\tau, U_i), \quad i = 1, \ldots, 6, \tag{27.85}$$

with the initial condition for $\tau = \tau_n = n \cdot H$: $U_i = U_{i1}(\tau_n)$. We shall integrate the above system of equations by use of Treanor's algorithm [158], [174], generalized to the case of a system of differential equations. This algorithm resembles the Runge–Kutta method of the fourth order.

We assume that the function $f_i(\tau, U_i)$ is approximated in interval τ_n to $\tau_{(n+1)} \cdot H$ as follows (at an arbitrary point (J, I) of region D):

$$\frac{dU_i}{d\tau} \approx -P_n^i(U_i - U_{i1}) + a_n^i + b_n^i(\tau - \tau_n) + c_n^i(\tau - \tau_n)^2. \tag{27.86}$$

From (27.86) we deduce the approximate solution for vector $\mathbf{U}$ in the form:

$$\begin{aligned} U_i(\tau) \cong U_{i1}(\tau_n) &- \frac{1}{P_n^i}\left(a_n^i - \frac{b_n^i}{P_n^i} + 2\frac{c_n^i}{(P_n^i)^2}\right) \exp\left[-P_n^i(\tau - \tau_n)\right] \\ &+ \frac{1}{P_n^i}\left\{\left[a_n^i - \frac{b_n^i}{P_n^i} + 2\frac{c_n^i}{(P_n^i)^2}\right] + \left(b_n^i - 2\frac{c_n^i}{P_n^i}\right)(\tau - \tau_n) \right. \\ &\left. + c_n^i(\tau - \tau_n)^2\right\}. \end{aligned} \tag{27.87}$$

This solution depends on constants P_n^i, a_n^i, b_n^i, and c_n^i which can be determined on the chosen planes $\tau = \text{const}$, in the time interval τ_n to τ_{n+1}.

In the same way as before we consider the following expansion:

$$\frac{dU_i}{d\tau} = f_i(\tau, U_i) \approx -P_n^i(U_i - U_{i1}) + a_n^i + d_n^i(\tau - \tau_n), \tag{27.88}$$

and we obtain

$$\begin{aligned} U_i(\tau) \approx U_{i1}(\tau_n) &- \frac{1}{P_n^i}\left(a_n^i - \frac{d_n^i}{P_n^i}\right) \exp\left[-P_n^i(\tau - \tau_n)\right] \\ &+ \frac{1}{P_n^i}\left\{\left(a_n^i - \frac{d_n^i}{P_n^i}\right) + d_n^i(\tau - \tau_n)\right\}, \end{aligned} \tag{27.89}$$

where the solution depends on constants P_n^i, a_n^i, and d_n^i. The constants are determined on planes $\tau = \tau_n$, $\tau = \tau_{n+1/2} = \tau_n + 0.5H$, and $\tau = \tau_{n+1} = \tau_n + H$.

On plane $\tau = \tau_n$ we use the notation $U_i(\tau_n) = U_{i1}$ and $f_{i1} = U'_{i1}$, then we obtain on plane $\tau = \tau_{n+1/2}$:

$$U_{i2} = U_{i1} + \tfrac{1}{2}Hf_{i1} \quad \text{for} \quad \tau = \tau_n + \tfrac{1}{2}H.$$

Next on plane $\tau_{n+1/2}$ we write $f_{i2} = U'_{i2}$ and obtain on plane τ_{n+1}

$$U_{i3} = U_{i1} + \tfrac{1}{2} H f_{i2} \quad \text{for} \quad \tau = \tau_n + H.$$

In turn, on plane τ_{n+1} we have $f_{i3} = U'_{i3}$. Substituting the above expressions into (27.88) we find that

$$\begin{aligned} a^i_n &= f_{i1}, \\ P^i_n &= -\frac{f_{i2} - f_{i3}}{U_{i2} - U_{i3}}, \\ d^i_n &= \frac{1}{H}\left\{ f_{i2} + P^i_n(U_{i2} + U_{i3}) + f_{i2} + f_{i3} - 2\left(f_{i1} + P^i_n U_{i1}\right)\right\}. \end{aligned} \tag{27.90}$$

By virtue of $(27.90)_2$ we determine constants P^i_n in turn and assume in what follows that their maximum value is

$$P_n = \left\langle \sup_{i=1,\ldots,6} \left\{ P^i_n \right\} \right\rangle. \tag{27.91}$$

Now we denote

$$\begin{aligned} &F_1 = \frac{1 - \exp(-P_n H)}{P_n H}, \quad F_2 = \frac{1 - F_1}{P_n H}, \quad F_3 = \frac{0.5 - F_2}{P_n H}, \\ &V_1 = 4F_3 - F_2, \quad V_2 = 2(F_2 - 2F_3), \quad V_3 = 4F_3 - 3F_2. \end{aligned} \tag{27.92}$$

Assuming the approximate form (27.88) we obtain, using (27.89) on plane $\tau_{n+1} = \tau_n + H$,

$$U_{i4} = U_{i1} + H[2f_{i3}F_2 + f_{i2}P_n H F_2 + f_{i1}(F_1 - 2F_2)]. \tag{27.93}$$

Next, substituting the consecutive pairs $(\tau_n, U_{i1}), \ldots, (\tau_{n+1}, U_{i4})$ into (27.86) we determine a_n, b_n, and c_n.

From (27.87) we obtain, as a result, the expression for the vector $\mathbf{U}$ on plane $\tau = \tau_{n+1}$:

$$\begin{aligned} U_{i(n+1)} = U_i(\tau_n + H) = U_{i1} + H[&f_{i1}F_1 + HV_3(P_n U_{i1} + f_{i1}) \\ &+ HV_2(P_n U_{i2} + f_{i2}) + HV_2(P_n U_{i3} + f_{i3}) \\ &+ HV_1(P_n V_{i4} + f_{i4})]. \end{aligned} \tag{27.94}$$

It should be observed that in the case when the correction coefficient of the method P_n is equal to zero we have to assume in the calculations the following values (from (27.92)):

$$F_1 = 1, \quad F_2 = 0.5, \quad F_3 = \frac{1}{6} \quad \text{when} \quad P_n = 0. \tag{27.95}$$

If, in this case, we assume in (27.94) that $P_n \to 0$, then we obtain the same expressions as in the case of the fourth order method of Runge–Kutta [36].

So far we have assumed a constant integration step H. Now we assume that the relative error committed during the determination of the solution at point (J, I) of the region (Fig. 90) on plane $\tau_{n+1} = \tau_n + H$ is equal to

(27.96)
$$Er(J,I) = \sup_{i=1,\dots,6}\left\{Er_i(J,I)\right\} = \sup_{i=1,\dots,6}\left\{\frac{|U_{i(n+1)}(J,I) - U_{i4}(J,I)|}{U_{i(n+1)}(J,I)}\right\},$$
$$\text{if}\quad U_{i(n+1)}(J,I) \neq 0.$$

Next we assume that the error committed during the determination of the solution on plane τ_{n+1}, in the whole region D, is equal to

(27.97)
$$Er = \sum_{\substack{J=1,\,n_\alpha \\ I=1,\,n_\beta}} Er(J,I).$$

Assuming that Er_d is the admissible error of the method, for example that Er_d is of order 10^{-2}, we can determine the law for the change of integration step dependent on the ratio

(27.98)
$$B = \frac{Er}{Er_d}.$$

The law of the integration step change with time can be determined in the same way that it was described by Zarka and Frelat [174].

We assume that the loading on the boundary of the half-space monotonically increases from zero to a constant value within a time of order 10^{-6}. Such a time change of the loading of the bounding plane can be treated as an approximation for step loading. At the same time the case of the waves of strong discontinuity are neglected since the assumed model for the medium (27.51) does not admit the appearance of strong wave discontinuities in the case of continuous boundary conditions.

We shall present a few interesting results from the numerical calculations [108], under the assumption that function $\Phi(F)$ is determined by formula (27.58) and for the following numerical values for the physical constants (aluminium):

$$E_0 = 7.06 \times 10^{10}\ \text{Pa},\quad \varrho = 2.7 \times 10^{-3}\ \text{kg cm}^{-3},\quad k_0 = 4.9 \times 10^{7}\ \text{Pa},$$
$$\nu = 0.3,\quad \gamma = 500\ \text{s}^{-1}.$$

We take the boundary conditions of type (27.72), assuming the loading of the half-space $P_0(\beta,\tau)$ to be of the forms

$$P_0(\beta,\tau) = \left[1 - \exp - \left(\frac{\beta - \beta_k}{a_{10}}\right)^2\right] p_0(1 - e^{\tau}),\quad \text{for}\quad 0 < \tau \leqslant \tau_k,$$

$$P_0(\beta,\tau) = \left[1 - \exp - \left(\frac{\beta - \beta_k}{a_{10}}\right)^2\right] p_0(1 - e^{\tau_k}),\quad \text{for}\quad \tau \geqslant \tau_k,$$

where the following numerical values are used:

$$\beta_k = \frac{2}{3}, \quad \tau_k = 4.485, \quad a_{10} = 10, \quad p_0 = 0.02, 0.2.$$

The propagation of the plastic front with time, for the value $p_0 = 0.02$ and for the submersion of the cavity in the half-space of $h_r = 10$, is presented in Fig. 91 in the XY plane

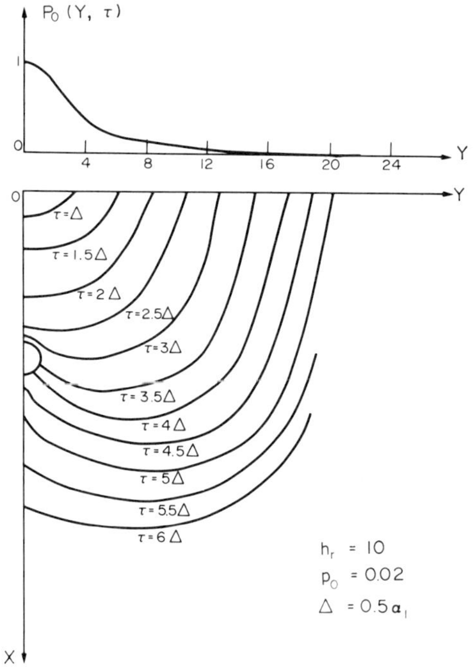

Fig. 91

for $Y \geqslant 0$. In the neighbourhood of the cylindrical cavity we find the distinct regions of smaller stress intensity. For time $\tau > 3\alpha_1$ the effect of the cylindrical cavity on the half-space starts declining.

The stress distributions Q and S on the cylindrical cavity, for the case discussed, are consecutively presented in Fig. 92a and b. Q and S represent the non-dimensionalized forms of the components of the stress tensor $\sigma^{\beta\beta}$ and σ^{zz}, expressed as contravariant components, in the natural basis. The diagrams are plotted for times $\tau_1 = 2\alpha_1$, $\tau_2 = 2.25\alpha_1$, and $\tau_3 = 2.5\alpha_1$. The variations of both of the stress components on the cavity contour and with time are alike. We observe that the maximum values of stress are attained on the side boundaries of the cylindrical cavity and diminish in the direction of the OX axis, changing sign and becoming positive in the vicinity of the OX axis.

The stress intensity diagrams, in this case, on the boundary of the cylindrical cavity for various times, from $\tau = 1.5\alpha_1$ to $\tau = 2.75\alpha_1$, are presented in Fig. 92c. These diagrams show at which points of the boundary of the cylindrical cavity the stress concentration appears, in the case when it is assumed that the cavity is free from normal stresses, i.e. for the boundary conditions (27.72).

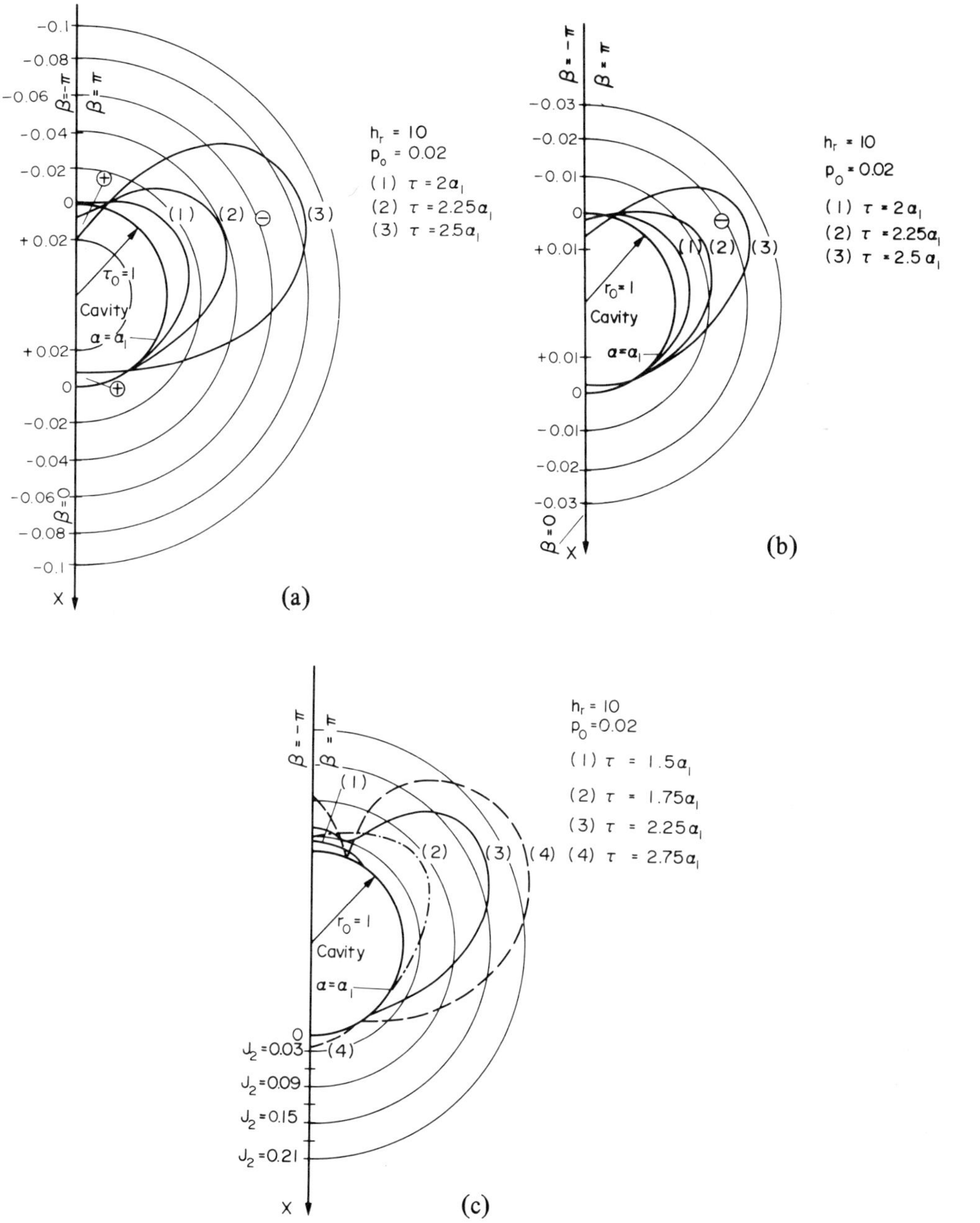

Fig. 92

Assuming ten times higher stress on the boundary of the half-space, i.e. taking $p_0 = 0.2$, we obtain, in the XY plane, the picture of the propagation of the plastic wave fronts as given in Fig. 93. The regions of smaller stress intensity are seen more clearly in this case in the vicinity of the cylindrical cavity as compared with the previous case. For time $\tau > 2\alpha_1$ we can assume

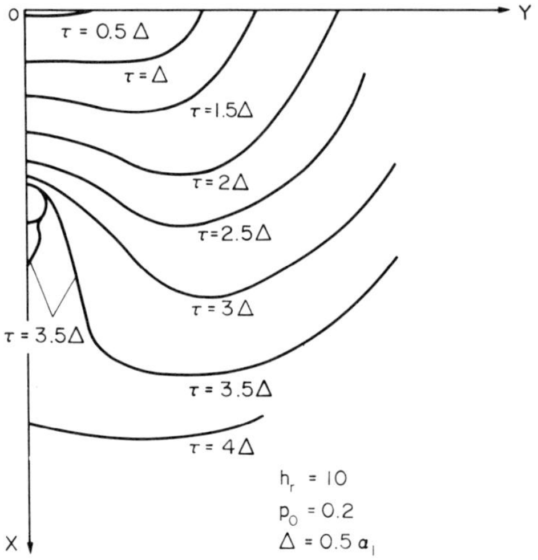

Fig. 93

that the effect of the cavity has already declined; compare the curves for $\tau = 4\Delta$. In the vicinity of the cylindrical cavity we obtain a similar form for the stress distribution Q and S with time as in the case of the assumption $p_0 = 0.02$ (see Fig. 92a and b). Also the stress intensity distribution on the contour of the cylindrical cavity has a similar shape as in Fig. 92c.

CHAPTER VI

THERMAL STRESS WAVES

28. Solution of the differential equations of the theory of thermal stresses

We present a certain method of obtaining a particular solution of the fundamental equations of thermoelasticity for a certain quite wide class of problems [101]. We neglect the existence of body forces and of heat sources and we assume homogeneous initial conditions for the temperature field and for the quantities characterizing the strain state.

Making use of an operator method we obtain a particular solution for an elastic medium [101]. The solution is distinguished by its simplicity and because it can be expressed by integrals of the temperature field and its gradients. Thus, using the proposed method, we are able to reduce a certain dynamic problem in the theory of thermoelasticity to a corresponding problem in the theory of elasticity. The methods usually applied require multiple integration and a knowledge of the general solution. The latter, however, is known only in the case of a few configurations of the body.

The solution of the fundamental equations of dynamic thermoelasticity with no body forces,

$$\nabla^2 \mathbf{u} + \frac{\lambda+\mu}{\mu} \operatorname{grad} \operatorname{div} \mathbf{u} = \frac{\varrho}{\mu} \ddot{\mathbf{u}} + \frac{3\lambda+2\mu}{\mu} \alpha \operatorname{grad} T, \tag{28.1}$$

is sought in the form of the sum of two functions

$$\mathbf{u} = \mathbf{u}^{\mathrm{I}} + \mathbf{u}^{\mathrm{II}}, \tag{28.2}$$

where function $\mathbf{u}^{\mathrm{I}}$ satisfies the inhomogeneous equation (28.1) independently of the assumed boundary and initial conditions while function $\mathbf{u}^{\mathrm{II}}$ satisfies the homogeneous equation

$$\nabla^2 \mathbf{u}^{\mathrm{II}} + \frac{\lambda+\mu}{\mu} \operatorname{grad} \operatorname{div} \mathbf{u}^{\mathrm{II}} = \frac{\varrho}{\mu} \ddot{\mathbf{u}}^{\mathrm{II}} \tag{28.3}$$

with appropriately modified boundary and initial conditions.

If the field of temperature T satisfies the homogeneous Fourier heat conduction equation, written in the following form:

$$\frac{a_1^2}{k} D_1 T - a_1^2 \nabla^2 T = 0, \tag{28.4}$$

where k is the thermal diffusion coefficient, $a_1 = \sqrt{(\lambda+2\mu)/\varrho}$ is the speed of propagation of the longitudinal waves and $D_1 = \partial/\partial t$, then the particular solution $\mathbf{u}^{\mathrm{I}}$ (for the conditions presented below) can be obtained by straightforward integration.

Solution $\mathbf{u}^{\mathrm{I}}$ of (28.1) is sought in the form

$$\mathbf{u}^{\mathrm{I}} = \operatorname{grad} \Phi. \tag{28.5}$$

Substituting (28.5) into (28.1) we obtain

$$\frac{2\mu+\lambda}{\alpha a_1^2(3\lambda+2\mu)}(D_2 - a_1^2 \nabla^2)\Phi + T = 0, \tag{28.6}$$

where $D_2 = \partial^2/\partial t^2$.

Applying to each side of (28.6) in turn the operators $-\dfrac{a_1^2}{k} D_1$ and D_2 and combining the equations obtained, we obtain, using (28.4),

$$(D_2 - a_1^2 \nabla^2)\left[D_2 \Phi - \frac{a_1^2}{k} D_1 \Phi + \alpha a_1^2 \frac{3\lambda+2\mu}{2\mu+\lambda} T\right] = 0. \tag{28.7}$$

Thus a particular solution of the last equation can be obtained if we assume that

$$\frac{\partial^2 \Phi}{\partial t^2} - \frac{a_1^2}{k}\frac{\partial \Phi}{\partial t} = -\alpha a_1^2 \frac{3\lambda+2\mu}{2\mu+\lambda} T(Q,t), \tag{28.8}$$

whence

$$\Phi = \frac{k}{a_1^2}\left(1 - e^{a_1^2 \frac{t}{k}}\right)\Phi_1 + \frac{3\lambda+2\mu}{2\mu+\lambda}\alpha k \int_0^t \left[1 - e^{a_1^2(t-\eta)/k}\right] T\, d\eta + \Phi_2. \tag{28.9}$$

By the direct substitution of function (28.9) into (28.6) we deduce that (28.9) is a particular integral of (28.6) provided that functions Φ_1 and Φ_2 satisfy the following equations:

$$\nabla^2 \Phi_1 - \frac{a_1^2}{k}\Phi_1 = -\frac{3\lambda+2\mu}{\lambda+2\mu}\frac{\alpha a_1^2}{k} T(Q,0), \qquad \nabla^2 \Phi_2 = \frac{1}{k}\Phi_1. \tag{28.10}$$

In the case of homogeneous initial conditions for T we can substitute $\Phi_1 = \Phi_2 = 0$, then we have

$$\Phi = \frac{3\lambda+2\mu}{2\mu+\lambda}\alpha k \int_0^t \left[1 - \exp\left(\frac{a_1^2(t-\eta)}{k}\right)\right] T(Q,\eta)\, d\eta. \tag{28.11}$$

Consequently, in the case of homogeneous initial conditions for $\mathbf{u}$ and under the above assumptions with respect to T, we have the following general solution of the initial value problem of the dynamic equations of the theory of thermal stresses:

for the displacement vector:

$$\mathbf{u} = \operatorname{grad} \Phi = \frac{3\lambda+2\mu}{2\mu+\lambda}\alpha k \int_0^t \left[1 - \exp\left(\frac{a_1^2(t-\eta)}{k}\right)\right] \operatorname{grad} T\, d\eta, \tag{28.12}$$

and for the strain components:

$$\varepsilon_{ij} = \frac{3\lambda+2\mu}{2\mu+\lambda}\alpha k \int_0^t \left[1 - \exp\left(\frac{a_1^2(t-\eta)}{k}\right)\right] T_{;ij}(x^m,\eta)\, d\eta = \varepsilon_{ij}^{\mathrm{I}}, \tag{28.13}$$

where the semicolon denotes covariant differentiation.

Making use of the Duhamel–Neumann law,

$$\sigma_{ij} = \lambda g_{ij}\,\theta + 2\mu\varepsilon_{ij} - (3\lambda+2\mu)\,g_{ij}\alpha T \tag{28.14}$$

where $\theta = g^{ij}\,\epsilon_{ij}, g^{ij}$ is the metric tensor, we obtain the following solution for the stresses:

$$\sigma_{ij} = \sigma^{\mathrm{I}}_{ij} = -\lambda\frac{3\lambda+2\mu}{2\mu+\lambda}\alpha\frac{a_1^2}{k}g_{ij}\int_0^t \exp\left(\frac{a_1^2(t-\eta)}{k}\right) T(Q,\eta)\,d\eta \tag{28.15}$$

$$+\,2\mu\frac{3\lambda+2\mu}{2\mu+\lambda}\alpha k\int_0^t \left[1-\exp\left(\frac{a_1^2(t-\eta)}{k}\right)\right] T_{;ij}(x^m,\eta)\,d\eta - (3\lambda+2\mu)\,g_{ij}\,\alpha T.$$

The relative change of volume at a given point of the body is equal to

$$\theta = g^{ij}\,\varepsilon_{ij} = -\frac{3\lambda+2\mu}{2\mu+\lambda}\frac{\alpha a_1^2}{k}\int_0^t \exp\left(\frac{a_1^2(t-\eta)}{k}\right) T(Q,\eta)\,d\eta. \tag{28.16}$$

This equation is obtained by use of the heat conduction equation (28.4) and of the condition $T(Q, 0) = 0$.

Integration with respect to time of the solution presented above can be readily performed provided we take into account that the general solution of the heat conduction equation (28.4) together with the conditions

$$M(P)\frac{\partial T}{\partial n} + N(P)(T-T_0) = 0, \qquad T(Q,0) = 0 \tag{28.17}$$

(where P is a point on the body surface) can be represented in the forms

$$T(Q,t) = T_0 - \sum_{n=1}^{\infty} a_n\,\varphi_n(Q)\,\mathrm{e}^{-k\alpha_n^2 t}, \tag{28.18}$$

$$a_n = \frac{T_0\int_D \varphi_n(Q)\,dV(Q)}{\int_D \varphi_n^2(Q)\,dV(Q)}, \tag{28.19}$$

where $\varphi_n(Q)$ and α_n^2 are the eigenfunctions and the eigenvalues of the corresponding eigenvalue problem.

The two remarks below refer to the general solution of the equations under the above assumptions and the assumption of homogeneous initial conditions. The remarks concern the time interval $0 < t < t^*$, where $t^* = r_Q/a_1$ denotes the time of arrival at a given point Q of the body of the first wave which carries the effect of the boundary, and r_Q denotes the distance between the point Q and the nearest point on the surface of the body.

From the above discussion we can draw the conclusions that:

(1) since the displacement vector field is a potential one, the body, at each point, is in a state of pure strain (i.e. no rotation);

(2) if the temperature at each point of the body surface $T(P, t) \geqslant 0$ then at each inner point of the body, within the time interval $0 < t < t^*$, the relative change of volume θ is always less than zero. This is undoubtedly the characteristic effect of the dynamic heat expansion of a body.

Making use of the method presented above we can reduce the solution of many one-dimensional dynamic problems in the theory of thermal stresses to the computation of a finite number of integrals of temperature and its derivatives.

The method presented here has been generalized to the case of dynamic problems in thermo-viscoelasticity by means of the elastic–viscoelastic analogy [102]. This method can also be adapted to cases of dynamic thermoplastic processes, namely in the range of the elastic deformation, in the unloading range and in the range of plastic deformation (when one assumes a sectionally linear material which is hardening or ideally viscoplastic).

29. Thermal shock on the boundary of an elastic/viscoplastic half-space

We shall consider the solution of the problem of the propagation of waves in a half-space generated by thermal shocks on its boundary [95], [96], [131]. The first attempts to solve dynamic problems in thermoplasticity belong to Suvorov [148], [149]. It was assumed there that the coefficient of heat conduction depends on temperature. A consequence of such an assumption is that thermal waves propagate with a finite velocity. This would be a good approximation to reality only if it matched the behaviour of a wide class of materials. The assumption of a non-linear conduction equation permitted the equations to be simplified considerably; however, it furnished artificial, dispersionless solutions of the conduction equation for a wave of weak discontinuity. The problem of the propagation of plane thermoplastic waves was studied in full details in a paper by Raniecki [131]. The problem of a thermal shock applied to the boundary of a half-space filled with an elastic–plastic medium was solved under the assumption that the physical constants do not depend on temperature. The solution depends on the method of characteristics applied in the solutions of problems of purely mechanical shocks. The thermal shock problem was then generalized to the case of elastic/viscoplastic media [95], [96]. In this chapter we confine ourselves to the case of the deformation of an elastic/viscoplastic medium governed by (3.31).

We assume the classical heat conduction equation which is solved independently of the equation of motion. Also we assume that the physical constants do not depend on temperature, that is we suppose the solutions refer to temperatures less than $T \leqslant 600$ K. In this range of temperature the changes in the plastic limit, the elastic modulus, the hardening parameter, and in the viscosity coefficient are not substantial for quite a wide class of metals. At the same time, for this temperature plastic deformation may occur due to thermal shocks.

The solutions of the problems in the phase plane will be derived by the use of the method of characteristics which is frequently applied in the solution of problems of the propagation of plastic waves generated by mechanical loads.

Let us consider, in a cartesian coordinate system x_i, an elastic/viscoplastic medium occupying the half-space $x_1 \geqslant 0$. The medium occupying region $x_1 > 0$, initially kept at temperature $T = 0$, is subjected, at instant $t = 0$, to a temperature which increases linearly from zero, and is then held constant:

$$T(0,t) = \begin{cases} 0 & \text{for} \quad t < 0, \\ T_0 \dfrac{t}{t_0} & \text{for} \quad 0 < t \leqslant t_0, \\ T_0 & \text{for} \quad t_0 \leqslant t. \end{cases} \tag{29.1}$$

The constitutive equations (3.31), for the case of a half-space in a uniaxial state of strain, reduce to:

$$\begin{aligned} \dot{\varepsilon}_{11} &= \frac{1}{2\mu}(\dot{\sigma}_{11}-\dot{\sigma}_{22}) + \sqrt{3}\,\gamma \left\langle \Phi\left[\frac{\sqrt{J_2}}{\varkappa}-1\right]\right\rangle, \\ \dot{\sigma}_{11}+2\dot{\sigma}_{22} &= 3K(\dot{\varepsilon}_{11}-3\alpha\dot{T}), \end{aligned} \tag{29.2}$$

where we have assumed that $f(\sigma_{ij}, \epsilon_{ij}^{p}) = \sqrt{J_2}$, $\sqrt{J_2} = \frac{1}{3}(\sigma_{11}-\sigma_{22})$.

The heat conduction equation (28.4), in the case of the half-space, takes the form

$$T_{,11} - \frac{1}{k}\dot{T} = 0. \tag{29.3}$$

The solution of the heat conduction equation for the boundary condition (29.1) and the initial condition

$$T/_{t=0} = 0 \tag{29.4}$$

and the condition that the solution is bounded at infinity takes the forms

for $0 \leqslant t < t_0$

$$T(x_1, t) = \frac{T_0}{t_0}\left[\left(t+\frac{x_1^2}{2k}\right)\operatorname{erfc}\left(\frac{x_1}{2\sqrt{kt}}\right) - x_1\sqrt{\frac{t}{k\pi}}\exp\left(-\frac{x_1^2}{4t}\right)\right] = \varphi_1(x_1, t), \tag{29.5}$$

for $t \geqslant t_0$

$$T(x_1, t) = \varphi_1(x_1, t) - \varphi_1(x_1, t-t_0).$$

The system of (29.2), together with the equation of motion (5.5) (for $X_i \equiv 0$), and with (5.7), which in the case of a uniaxial half-space take the forms

$$\frac{\partial \sigma_{11}}{\partial x_1} = \varrho \frac{\partial v}{\partial t}, \qquad \frac{\partial \varepsilon_{11}}{\partial t} = \frac{\partial v}{\partial x_1}, \tag{29.6}$$

constitutes the complete system of equations for the problem posed. To define the problem uniquely we need initial conditions

$$u/_{t=0} = v/_{t=0} = 0 \tag{29.7}$$

and a set of boundary conditions

$$\sigma_{11}/_{x_1=0} = 0. \tag{29.8}$$

The system of (29.2) and (29.6) is hyperbolic. It possesses the following families of characteristics:

(29.9) $$\xi = \pm\eta + \text{const}, \qquad \xi = \text{const},$$

where the following dimensionless variables have been introduced:

(29.10) $$\xi = \frac{ax_1}{k}, \qquad \eta = \frac{a^2 t}{k}, \qquad a^2 = \frac{3K+4\mu}{3\varrho}.$$

Along characteristics (29.9) the following relations hold under the assumption that the solution of the heat conduction equation is of the form (29.5):

along $\xi = \pm\eta + \text{const}$

(29.11) $$d\sigma_{11} \mp \varrho a\, dv \pm \left\{\frac{3K\alpha T_0}{\eta_0}\operatorname{erfc}\left(\frac{\xi}{2\sqrt{\eta}}\right) + \frac{4\sqrt{3}}{3a^2}\gamma\left\langle\Phi\left[\frac{\sqrt{J_2}}{\varkappa}-1\right]\right\rangle\right\} d\xi = 0,$$

along $\xi = \text{const}$

$$d\sigma_{11} - \varrho a^2\, d\varepsilon + \left\{\frac{3K\alpha T_0}{\eta_0}\operatorname{erfc}\left(\frac{\xi}{2\sqrt{\eta}}\right) + \frac{4\sqrt{3}}{3a^2}\mu k\gamma\left\langle\Phi\left[\frac{\sqrt{J_2}}{\varkappa}-1\right]\right\rangle\right\} d\eta = 0$$

for $0 < \eta \leqslant \eta_0$, whereas for $\eta \geqslant \eta_0$ the argument of the error function occurring in (29.11) changes. For $\eta > \eta_0$ we have to insert into (29.11) $\operatorname{erfc}(\xi/2\sqrt{\eta-\eta_0})$ instead of $\operatorname{erfc}(\xi/2\sqrt{\eta})$.

Now we present the solutions in the various regions of the (ξ, η) phase space (Fig. 94).

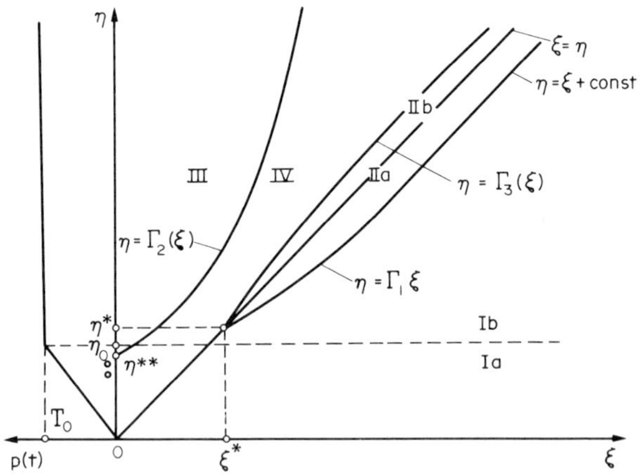

Fig. 94

Region I. This region is the one of elastic deformation. Because the solution of the conduction equation (29.5) is expressed in two parts, this region is split into two subregions a and b. The system of (29.2) and (29.6) for the elastic range (i.e. for $\gamma\Phi = 0$) can be reduced to equation (28.1) in terms of displacements. The solutions for the strain field, the stress field, and the velocity in region I are obtained by the method given in the preceding section from (28.13), (28.15), and (28.12) upon differentiating the last one with respect to time. We also make use of the solution of the conduction equation (29.5) and of the initial conditions (29.7) and (29.8). In region Ia we obtain:

$$(29.12)\quad \begin{aligned} \varepsilon &= 2\frac{\vartheta}{\eta_0}\left\{\left(1+\eta+\frac{\xi^2}{2}\right)\operatorname{erfc}\left(\frac{\xi}{2\sqrt{\eta}}\right)-\frac{1}{2}\left[e^{\eta-\xi}\operatorname{erfc}\left(\frac{\xi}{2\sqrt{\eta}}-\sqrt{\eta}\right)\right.\right.\\ &\quad\left.\left.+e^{\xi+\eta}\operatorname{erfc}\left(\frac{\xi}{2\sqrt{\eta}}+\sqrt{\eta}\right)\right]-\xi\sqrt{\frac{\eta}{\pi}}\exp\left(-\frac{\xi^2}{4\eta}\right)\right\}\equiv\psi_1(\xi,\eta),\\ \sigma_{11} &= 2\varrho a^2\frac{\vartheta}{\eta_0}\left\{\operatorname{erfc}\left(\frac{\xi}{2\sqrt{\eta}}\right)-\frac{1}{2}\left[e^{\eta-\xi}\operatorname{erfc}\left(\frac{\xi}{2\sqrt{\eta}}-\sqrt{\eta}\right)\right.\right.\\ &\quad\left.\left.+e^{\xi+\eta}\operatorname{erfc}\left(\frac{\xi}{2\sqrt{\eta}}+\sqrt{\eta}\right)\right]\right\}\equiv\psi_2(\xi,\eta),\\ v &= 2a\frac{\vartheta}{\eta_0}\left\{\frac{1}{2}e^{\eta-\xi}\operatorname{erfc}\left(\frac{\xi}{2\sqrt{\eta}}-\sqrt{\eta}\right)-\frac{1}{2}e^{\eta+\xi}\operatorname{erfc}\left(\frac{\xi}{2\sqrt{\eta}}+\sqrt{\eta}\right)\right.\\ &\quad\left.+\xi\operatorname{erfc}\left(\frac{\xi}{2\sqrt{\eta}}\right)\exp\left(-\frac{\xi^2}{4\eta}\right)\right\}\equiv\psi_3(\xi,\eta),\end{aligned}$$

for $0<\eta\leqslant\eta_0$, while in region Ib, for $\eta\geqslant\eta_0$, we obtain:

$$(29.13)\quad \begin{aligned}\varepsilon &= \psi_1(\xi,\eta)-\psi_1(\xi,\eta-\eta_0),\\ \sigma_{11} &= \psi_2(\xi,\eta)-\psi_2(\xi,\eta-\eta_0),\\ v &= \psi_3(\xi,\eta)-\psi_3(\xi,\eta-\eta_0),\end{aligned}$$

where we have put $\vartheta = 3K\alpha T_0/2\varrho a^2$.

We introduce a certain dimensionless magnitude e_s, dependent on the physical properties of the medium and temperature T_0 and defined by the formula $e_s=\varrho a^2\varepsilon_{is}/2\sqrt{3}K\alpha T$. We shall examine for which values of the parameter e_s and the dimensionless magnitude η_0 permanent strains occur in a half-space heated on the boundary according to (29.1) – in other words, for which values of e_s and η_0 a plastic wave starts propagating in the medium $x_1\geqslant 0$.

Examining the change in the strain ε with time at a fixed cross-section, we deduce from $(29.12)_1$ that it is a monotonically increasing function attaining a maximum value on the characteristic $\xi=\eta$. At a certain point (ξ^*,η^*) (Fig. 94) it reaches the value

$$(29.14)\quad \varepsilon(\xi,\eta)/_{\xi=\eta=\xi^*=\eta^*} = -\frac{\sqrt{3}}{2}\varepsilon_{is}.$$

From this condition we can determine the coordinates of the point of plastic wave $\eta=\Gamma_1(\xi)$ initiation (Fig. 94). For $\xi>\xi^*$ (and for $\xi>\eta$) the medium is deformed plastically.

We have the following condition for the generation of the plastic wave $\eta=\Gamma_1(\xi)$:

$$(29.15)\quad \lim_{\xi\to\infty}\varepsilon(\xi,\eta)/_{\xi=\eta}\leqslant-\frac{\sqrt{3}}{2}\varepsilon_{is},$$

where $\epsilon(\xi, \eta)$ is determined here from formula $(29.13)_1$, i.e. for $\eta \geqslant \eta_0$:

$$\begin{aligned}\varepsilon(\xi,\eta)/_{\xi=\eta} = \frac{\vartheta}{\eta_0}\Big\{&(2+2\xi+\xi^2)\operatorname{erfc}\left(\frac{1}{2}\sqrt{\xi}\right)-\operatorname{erfc}\left(-\frac{1}{2}\sqrt{\xi}\right)\\&-e^{-2\xi}\operatorname{erfc}\left(\frac{3}{2}\sqrt{\xi}\right)-2\xi^{3/2}\pi^{-1/2}\exp\left(-\frac{\xi}{4}\right)\\&-(2+2\xi-2\eta_0+\xi^2)\operatorname{erfc}\left(\frac{\xi}{2\sqrt{\eta-\eta_0}}\right)+e^{-\eta_0}\operatorname{erfc}\left(\frac{2\eta_0-\xi}{2\sqrt{\xi-\eta_0}}\right)\\&+e^{2\eta-\xi_0}\operatorname{erfc}\left(\frac{3\xi-2\eta_0}{2\sqrt{\xi-\eta_0}}\right)+\xi\sqrt{\frac{\xi-\eta_0}{\pi}}\exp\left(-\frac{\xi^2}{4(\xi-\eta_0)}\right)\Big\}.\end{aligned} \tag{29.16}$$

Making use of (29.16) in (19.15) and passing to the limit, we obtain

$$\lim_{\xi\to\infty} \varepsilon(\xi,\eta)/_{\xi=\eta} = \frac{\vartheta_0}{\eta_0}\{-1+e^{-\eta_0}\} \leqslant -\frac{\sqrt{3}}{2}\varepsilon_i, \tag{29.17}$$

hence, eventually we find the condition the generation of the plastic wave $\eta = \Gamma_1(\xi)$ in the form

$$e_s \leqslant \frac{1-e^{-\eta_0}}{\eta_0}. \tag{29.18}$$

The curve $e_s = (1-e^{-\eta_0})/\eta_0$ is shown in Fig. 95. Plastic strain arises in the half-space heated on the boundary according to (29.1) for arbitrary values of the coordinates (e_s, η_0) of the points lying in region R. As $\eta_0 \to 0$ we obtain the condition for plastic strain generation in a half-space under a thermal shock $e_s = 1$ [95], [131], i.e. when the temperature T_0 appears suddenly at instant $t = 0$ on the plane bounding the half-space. The boundary condition $T_{x_1=0} = T_0$, for $t > 0$, corresponds to the contact of the surface with a large quantity of flowing liquid of constant temperature T_0 under conditions of ideal heat transmission.

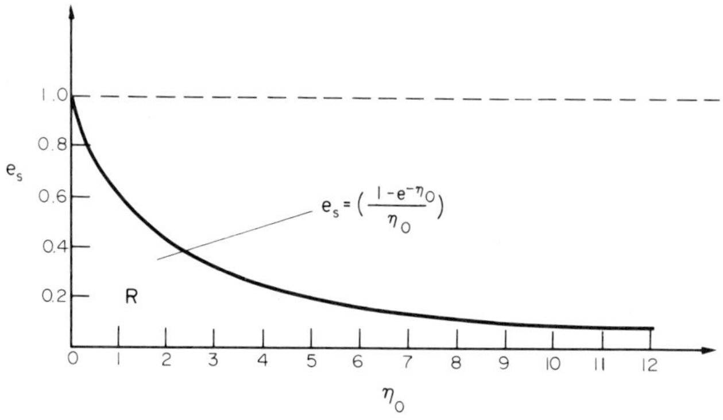

Fig. 95

The question arises as to whether the plastic strain appears on the boundary of the half-space under the condition (29.18). Is condition (29.18) sufficient for the existence of a plastic wave $\eta = \Gamma_2(\xi)$ on the boundary of the half-space (Fig. 94)?

From the constitutive equations (20.2), for $\gamma\langle\Phi\rangle = 0$ (elastic loading process), and from the boundary condition (29.8) we deduce that

$$\varepsilon(0,\eta) = 2\vartheta\frac{\eta}{\eta_0}. \tag{29.19}$$

The condition for the generation of the plastic wave $\eta = \Gamma_2(\xi)$ is as follows:

$$\varepsilon(0,\eta) = +\frac{\sqrt{3}}{2}\varepsilon_{is}, \tag{29.20}$$

whence

$$\frac{\eta}{\eta_0} = \frac{\sqrt{3}\,\varepsilon_{is}}{4\vartheta} = e_s. \tag{29.21}$$

But, by virtue of (29.18), $e_s \leqslant 1$, hence $\eta \leqslant \eta_0$. Accordingly, if condition (29.18) is satisfied, i.e. the condition for which plastic compressive strains appear in the medium – region II, likewise plastic tensile strains arise at the boundary $\xi = 0$ – region III. A plastic wave of weak discontinuity $\eta = \Gamma_2(\xi)$ is initiated on the boundary of the half-space $\xi = 0$ and appears at the instant $\eta = \eta^{**} \leqslant \eta_0$. The limiting case ($\eta = \eta^{**} = \eta_0$) occurs only when $e_s = 1$; this corresponds, from (29.18), to the value $\eta_0 = 0$ (thermal shock of constant temperature T_0).

Now we determine the plastic wave $\eta = F_1(\xi)$ and its asymptotes. Equation (29.14) constitutes the condition from which wave $\eta = \Gamma_1(\xi)$ is determined:

$$\varepsilon(\xi,\Gamma_1(\xi)) = -\frac{\sqrt{3}}{2}\varepsilon_{is}. \tag{29.22}$$

Making use of the solution in region I for the strain (formula $(29.12)_1$) we obtain, using (29.22), an implicit equation from which the form of wave $\eta = \Gamma_1(\xi)$ can be determined:

$$\begin{aligned} F[\xi,\Gamma_1(\xi)] = \Bigg\{&\left[1+\Gamma_1(\xi)+\frac{\xi^2}{2}\right]\operatorname{erfc}\left(\frac{\xi}{2\sqrt{\Gamma_1(\xi)}}\right)-\frac{1}{2}\Bigg[e^{\Gamma_1(\xi)-\xi}\\ &\times\operatorname{erfc}\left(\frac{\xi-2\Gamma_1(\xi)}{2\sqrt{\Gamma_1(\xi)}}\right)+e^{\Gamma_1(\xi)+\xi}\operatorname{erfc}\left(\frac{\xi+2\Gamma_1(\xi)}{2\sqrt{\Gamma_1(\xi)}}\right)\\ &-\xi\sqrt{\frac{\Gamma_1(\xi)}{\pi}}\exp\left(-\frac{\xi^2}{4\Gamma_1(\xi)}\right)\Bigg]\Bigg\}+e_s\eta_0 \equiv \psi_1[\xi,\Gamma_1(\xi)]+e_s\eta_0 = 0\\ &\text{for}\quad 0<\eta\leqslant\eta_0, \end{aligned} \tag{29.23}$$

and

$$F[\xi,\Gamma_1(\xi)] = \psi_1[\xi,\Gamma_1(\xi)]-\psi_1[\xi,F_1(\xi-\eta_0)]+e_s\eta_0 = 0 \quad \text{for} \quad \eta\geqslant\eta_0. \tag{29.24}$$

Now we examine the derivative of the wave $\eta = \Gamma_1(\xi)$:

$$\frac{d\Gamma_1}{d\xi} = -\frac{\dfrac{\partial F}{\partial\xi}}{\dfrac{\partial F}{\partial\Gamma_1}} \tag{29.25}$$

(29.25)
[*cont.*]

$$= -\frac{\xi\,\mathrm{erfc}\left(\frac{\xi}{2\sqrt{\Gamma_1}}\right)+\frac{1}{2}e^{\Gamma_1-\xi}\,\mathrm{erfc}\left(\frac{\xi-2\sqrt{\Gamma_1}}{2\sqrt{\Gamma_1}}\right)-\frac{1}{2}e^{\Gamma_1-\xi}\,\mathrm{erfc}\left(\frac{\xi+2\Gamma_1}{2\sqrt{\Gamma_1}}\right)-2\sqrt{\frac{\Gamma_1}{\pi}}\,e^{-\frac{\xi^2}{4\Gamma_1}}}{\mathrm{erfc}\left(\frac{\xi}{2\sqrt{\Gamma_1}}\right)-\frac{1}{2}e^{\Gamma_1-\xi}\,\mathrm{erfc}\left(\frac{\xi-2\Gamma_1}{2\sqrt{\Gamma_1}}\right)-\frac{1}{2}e^{\Gamma_1-\xi}\,\mathrm{erfc}\left(\frac{\xi+2\Gamma_1}{2\sqrt{\Gamma_1}}\right)}$$

$$= -\frac{\psi_1[\xi,\Gamma_1(\xi)]_{,\xi}}{\psi_1[\xi,\Gamma_1(\xi)]_{,\Gamma_1}} \quad \text{for} \quad 0<\eta\leqslant\eta_0,$$

and

(29.26) $$\frac{d\Gamma_1}{d\xi} = -\frac{\psi_1[\xi,\Gamma_1(\xi)]_{,\xi}-\psi_1[\xi,\Gamma_1(\xi-\eta_0)]_{,\xi}}{\psi_1[\xi,\Gamma_1(\xi)]_{,\Gamma_1}-\psi_1[\xi,\Gamma_1(\xi-\eta_0)]_{,\Gamma_1}} \quad \text{for} \quad \eta\geqslant\eta_0.$$

Comparing (29.25) and (29.26) with the solution in region I (formulae (29.12)) we see that

(29.27) $$\frac{d\Gamma_1}{d\xi} = -\varrho a\,\frac{v(\xi,\Gamma_1)}{\sigma_{11}(\xi,\Gamma_1)}.$$

From an analysis of formulae (29.12) we deduce, for arbitrary values of $\xi\geqslant\eta$, that

(29.28) $$v(\xi,\eta)\geqslant 0, \qquad \sigma_{11}(\xi,\eta)\leqslant 0 \qquad \text{for} \quad \xi\geqslant\eta,$$

and since relations (29.12) are valid only for $\xi\geqslant\eta$ we have

(29.29) $$\frac{d\Gamma_1}{d\xi}\geqslant 0,$$

where $\sigma_{11}(\xi,\Gamma_1)=0$ has to be excluded from (29.27) on physical grounds. Furthermore we can assert that

(29.30) $$\sigma_{11}(\xi,\Gamma_1)\geqslant -\varrho a\,v(\xi,\eta),$$

for any ξ and η with $\xi\geqslant\eta$, whence

(29.31) $$\frac{d\Gamma_1}{d\xi}\leqslant 1.$$

Consequently we find that the derivative of the plastic wave of weak discontinuity $\eta=\Gamma_1(\xi)$ is bounded by the interval

(29.32) $$0\leqslant\frac{d\Gamma_1}{d\xi}\leqslant 1.$$

We write down (29.27) in a slightly different form, namely

(29.33) $$\frac{d\Gamma_1}{d\xi} = 1-\frac{\varrho a v(\xi,\Gamma_1)+\sigma_{11}(\xi,\Gamma_1)}{\sigma_{11}(\xi,\Gamma_1)},$$

and we examine the limit of this derivative for $\eta\geqslant\eta_0$, as $\xi\to\infty$,

(29.34) $$\lim_{\xi\to\infty}\frac{d\Gamma_1}{d\xi} = 1-\lim_{\xi\to\infty}\frac{\varrho a\,v(\xi,\Gamma_1)+\sigma_{11}(\xi,\Gamma_1)}{\sigma_{11}(\xi,\Gamma_1)}.$$

Making use of formulae (29.12) and of the equation of the plastic wave (29.24) we deduce from (29.34) that

$$\lim_{\xi \to \infty} \frac{d\Gamma_1}{d\xi} = 1, \tag{29.35}$$

i.e. a straight line of equation

$$\eta = \xi + \beta, \qquad \beta = \text{const.} \tag{29.36}$$

constitutes the asymptote to the plastic wave $\eta = \Gamma_1(\xi)$. Substituting the equation of the asymptote (29.36) into the equation for the wave (29.24) and proceeding to the limit $\xi \to \infty$, we obtain

$$\lim_{\xi \to \infty} F(\xi, \xi + \beta) = -e^{\beta} + e^{-\eta_0 + \beta} + e_s \eta_0 = 0, \tag{29.37}$$

whence

$$\beta = -\ln \frac{1 - e^{-\eta_0}}{e_s \eta_0}, \tag{29.38}$$

i.e. the straight line of the equation

$$\xi = \eta + \ln \frac{1 - e^{-\eta_0}}{e_s \eta_0} \tag{29.39}$$

is the equation of the asymptote to the plastic wave $\eta = \Gamma_1(\xi)$.

If $\eta_0 \to 0$ we obtain the asymptote of the wave $\eta = \Gamma_1(\xi)$ for the case of a thermal shock of constant temperature T_0 [95], [131], namely $\xi = \eta + \ln(1/e_s)$.

The plastic loading wave $\eta = \Gamma_1(\xi)$ is presented in Fig. 96 for the numerical data: $\eta_0 = 1$ and $e_s = 0.5$. The straight line of the equation $\eta = \xi - 0.23$ is the asymptote to the wave in accordance with (29.39). As can be seen from Fig. 96, the plastic wave tends rapidly to its asymptote.

Now we pass to the solution of the problem in the remaining regions of the phase plane.

Region IIa. In this region we have to solve the Cauchy problem for the system of (29.2) and (29.6). Since we cannot integrate the relations along characteristics (29.11) for the system of equations of the problem (29.2) and (29.6), the solution in region IIa has to be constructed numerically. To this end we make use of the method of characteristic nets. The initial values on curve $\eta = \Gamma_1(\xi)$, for the Cauchy case, are determined from the solutions in region I.

Regions III, IV, and IIb. The solutions in these regions are performed simultaneously. The plastic wave of weak discontinuity $\eta = \Gamma_2(\xi)$, bounding region III – the region of tensile plastic strain – starts propagating from point $(0, \eta^{**})$ (Fig. 94).

The condition for the determination of wave $\eta = \Gamma_2(\xi)$ is as follows:

$$\varepsilon|_{\eta = \Gamma_2(\xi)} = +\frac{\sqrt{3}}{2} \varepsilon_{is} \qquad \text{for} \quad 0 \leqslant \xi \leqslant \xi^* \tag{29.40}$$

and

$$\varepsilon|_{\eta = \Gamma_2(\xi)} = +\frac{\sqrt{3}}{2} [2\varepsilon_{is} + \varepsilon_{i0}(\xi)] \qquad \text{for} \quad \xi > \xi^*, \tag{29.41}$$

where by $\epsilon_{io}(\xi)$ we denoted the strain intensity on the unloading wave $\eta = \Gamma_3(\xi)$ (Fig. 94). The existence of the unloading wave $\eta = \Gamma_3(\xi)$ has been demonstrated on the grounds of

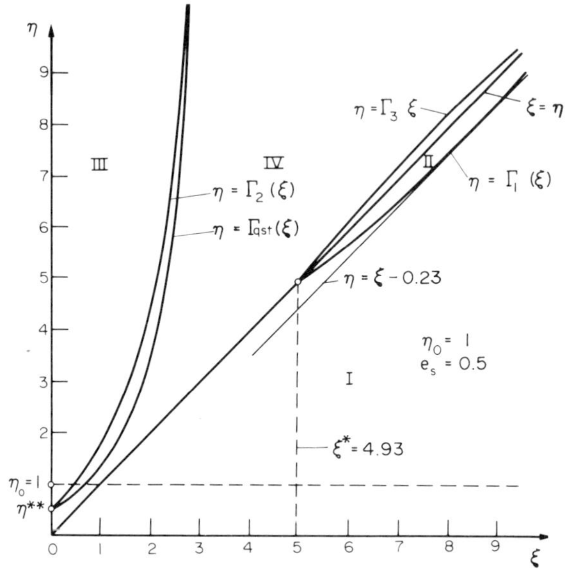

Fig. 96

computations which have been performed. This wave starts propagating from point $\xi = \xi^*$; the condition for its determination is the following one:

$$\sqrt{J_2} = \varkappa|_{\eta=\Gamma_3(\xi)} \tag{29.42}$$

or, in the case of the constitutive equations (3.13), referring to the strain theory, we obtain

$$\sigma_i = f(\varepsilon_i)|_{\eta=\Gamma_3(\xi)}, \tag{29.43}$$

and after expansion we have the condition in the form

$$(\sigma_{11} - \sigma_{22})|_{\eta=\Gamma_3(\xi)} = \sqrt{3}\, f\left(\frac{2}{\sqrt{3}}\,\varepsilon\right)\Big|_{\eta=\Gamma_3(\xi)} \tag{29.44}$$

As has already been said, the solutions in regions III, IIb, and IV (region IV, for $\xi \leqslant \xi^*$, is a region of elastic loading and for $\xi \geqslant \xi^*$ and unloading region) and the determination of waves $\eta = \Gamma_2(\xi)$ and $\eta = \Gamma_3(\xi)$ from conditions (29.40)–(29.44) are performed simultaneously by the method of characteristic nets, using relations (29.11).

The solution in phase plane, for values $\eta_0 = 1$, $e_s = 0.5$, is presented in Fig. 96. The diagrams of strain variations with time at fixed cross-sections $\xi = 1.0; 2.0; 3.0$ are shown in Fig. 97.

We shall solve the quasi-static problem for the temperature distribution on the boundary of the half-space given in form (29.1). Curve $\eta = \Gamma_{qst}(\xi)$ (Fig. 96) bounding the region of

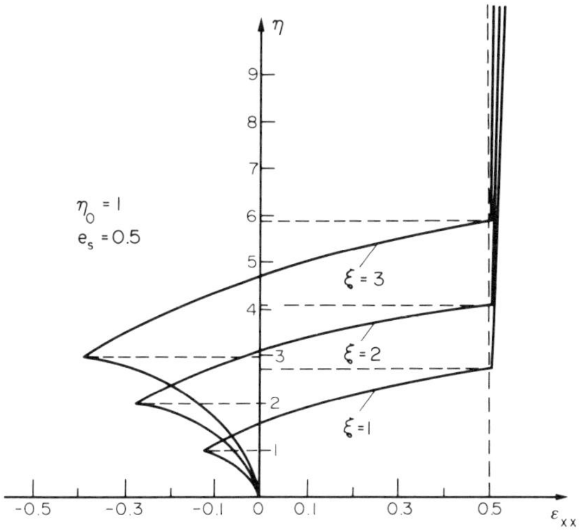

Fig. 97

viscoplastic deformation III is determined from the relation on the vertical characteristic ξ = const $(29.11)_2$:

$$
d\sigma_{11} - \varrho a^2 d\varepsilon + \frac{3K\alpha T_0}{\eta_0} \operatorname{erfc}\left(\frac{\xi}{2\sqrt{\eta}}\right) d\eta = 0 \qquad \text{for} \quad 0 < \eta \leqslant \eta_0,
$$

(29.45)

$$
d\sigma_{11} - \varrho a^2 d\varepsilon + \frac{3K\alpha T_0}{\eta_0} \left[\operatorname{erfc}\left(\frac{\xi}{2\sqrt{\eta}}\right) - \operatorname{erfc}\left(\frac{\xi}{2\sqrt{\eta - \eta_0}}\right)\right] d\eta = 0 \quad \text{for} \quad \eta \geqslant \eta_0.
$$

We have assumed that $\langle \Phi \rangle \equiv 0$ since wave $\eta = \Gamma_{qst}(\xi)$ is determined from the side of the elastic region.

In the quasi-static solution we have to assume that $d\sigma_{11} = 0$; this directly follows from the equation of motion $(29.6)_1$. Thus we deduce from (29.45) that:

$$
\varrho a^2 d\varepsilon = \frac{3K\alpha T_0}{\eta_0} \operatorname{erfc}\left(\frac{\xi}{2\sqrt{\eta}}\right) d\eta \qquad \text{for} \qquad 0 < \eta \leqslant \eta_0,
$$

(29.46)

$$
\varrho a^2 d\varepsilon = \frac{3K\alpha T_0}{\eta_0} \left[\operatorname{erfc}\left(\frac{\xi}{2\sqrt{\eta}}\right) - \operatorname{erfc}\left(\frac{\xi}{2\sqrt{\eta - \eta_0}}\right)\right] d\eta \qquad \text{for} \quad \eta \geqslant \eta_0.
$$

Integrating the above equations and making use of the initial condition $\epsilon(\xi, 0) = 0$, we obtain

$$
\varepsilon(\xi, \eta) = \frac{3K\alpha T_0}{\varrho a^2 \eta_0} \left[\left(\eta + \frac{\xi^2}{2}\right) \operatorname{erfc}\left(\frac{\xi}{2\sqrt{\eta}}\right) - \xi \sqrt{\frac{\eta}{\pi}} \exp\left(-\frac{\xi^2}{4\eta}\right)\right] \equiv \varphi(\xi, \eta)
$$

(29.47)

$$
\text{for} \quad 0 < \eta \leqslant \eta_0,
$$

(29.47) [cont.]

$$\varepsilon(\xi,\eta)=\frac{3K\alpha T_0}{\varrho a^2\eta_0}\left[\varphi(\xi,\eta)-\varphi(\xi,\eta-\eta_0)\right] \quad \text{for} \quad \eta\geqslant\eta_0.$$

Setting, in formulae (29.47), $\epsilon(\xi,\eta)=+\frac{\sqrt{3}}{2}\epsilon_{is}$, we determine the curve $\eta=\Gamma_{qst}(\xi)$. Thus we obtain an equation, in implicit form, from which the curve $\eta=\Gamma_{qst}(\xi)$ is determined, namely,

$$\left[\left(\Gamma_{qst}+\frac{\xi^2}{2}\right)\operatorname{erfc}\left(\frac{\xi}{2\sqrt{\Gamma_{qst}}}\right)-\xi\sqrt{\frac{\Gamma_{qst}}{\pi}}\exp\left(-\frac{\xi^2}{4\sqrt{\Gamma_{qst}}}\right)\right]-e_s\eta_0=0 \quad \text{for} \quad 0<\eta\leqslant\eta_0,$$

(29.48)

$$\varphi(\xi,\Gamma_{qst})-\varphi(\xi,\Gamma_{qst}-\eta_0)-e_s\eta_0=0 \quad \text{for} \quad \eta\geqslant\eta_0.$$

The curve $\eta=\Gamma_{qst}(\xi)$ for $e_s=0.5$, $\eta_0=1$, is also presented in Fig. 96. We can observe that already at small times the plastic wave of weak discontinuity $\eta=\Gamma_2(\xi)$ tends to the quasi-static solution $\eta=\Gamma_{qst}(\xi)$.

In conclusion we can state, from the results of the solutions obtained, that as $\eta_0\to 0$ the plastic wave of weak discontinuity $\eta=\Gamma_2(\xi)$ becomes the plastic wave of strong discontinuity; for $0\leqslant\xi\leqslant\xi^0$ (Fig. 98). Next, for $\xi^0\leqslant\xi\leqslant\xi^*$ (if $\xi^*>\xi^0$) an elastic wave of strong discontinuity will propagate; this wave coincides with characteristic $\xi=\eta$. The solution at the

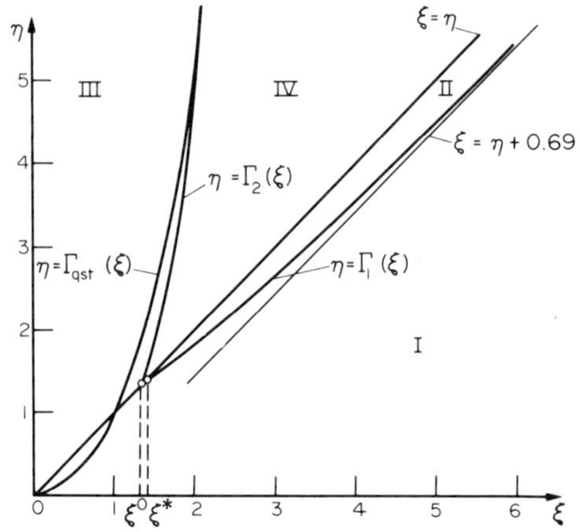

Fig. 98

front of the wave of strong discontinuity $\xi=\eta$ can be reduced to a Volterra integral equation of the second kind (similar to the way as was shown in section 15). We make the use of the dynamic continuity condition of the relation on the positive characteristic (29.11) and of the fact that the strain rate on the front of a wave of strong discontinuity is infinite. As a result we obtain the integral equation for strain ϵ in the form

(29.49)
$$\varepsilon = 2\vartheta + \vartheta\left[1 - 3\,\mathrm{erf}\left(\frac{1}{2}\sqrt{\xi}\right) - e^{2\xi}\mathrm{erfc}\left(\frac{3}{2}\sqrt{\xi}\right)\right] - \\ - \frac{2\mu k}{\sqrt{3}\,\varrho a^4}\int_0^{\xi} \gamma\Phi\left[\frac{(1-2\nu)\,\varrho a^2 \varepsilon(\bar{\xi})}{\sqrt{3}\,(1-\nu)\,\varkappa\,[\varepsilon(\bar{\xi}), \bar{\xi}]} - 1\right] d\bar{\xi} \qquad \text{for} \quad 0 \leqslant \xi \leqslant \xi^*.$$

The constant of integration is determined from the boundary condition $\sigma_{11}|_{\xi=0} = 0$. It was shown in [95] that the integrand is bounded and satisfies the Lipschitz condition (this also was shown in section 15 for the identical integrand).

We observe from (29.49) that the strain on the wave approaching from the side of region III (Fig. 98) on the boundary $\xi = 0$ is equal to 2ϑ and it then decreases with the increase of ξ. The point of the decay of the plastic wave of strong discontinuity ξ^0 is determined from the condition

(29.50)
$$\varepsilon_{\xi=\xi^0} = +\frac{\sqrt{3}}{2}\varepsilon_{is}.$$

For $\xi > \xi^0$ the plastic wave of strong discontinuity becomes a wave of weak discontinuity of equation $\xi = \Gamma_2(\xi)$ (Fig. 98). On the wave of strong discontinuity $\xi = \eta$ for $\xi^0 \leqslant \xi \leqslant \xi^*$ the solution is obtained from (29.49) by assuming that $\langle\Phi\rangle \equiv 0$, thus

(29.51)
$$\varepsilon = \vartheta\left[3\,\mathrm{erfc}\left(\frac{1}{2}\sqrt{\xi}\right) - e^{2\xi}\mathrm{erfc}\left(\frac{3}{2}\sqrt{\xi}\right)\right] \qquad \text{for} \qquad \xi^0 \leqslant \xi \leqslant \xi^*,$$

(29.52)
$$\varepsilon = \vartheta\left[3 - \mathrm{erf}\left(\frac{1}{2}\sqrt{\xi^*}\right) - 2\,\mathrm{erf}\left(\frac{1}{2}\sqrt{\xi}\right) - e^{2\xi^*}\mathrm{erfc}\left(\frac{3}{2}\sqrt{\xi^*}\right)\right] + \frac{1}{2a}[v_2 + a\varepsilon_2].$$

The function $[\nu_2 = a\epsilon_2]$ for $\xi \geqslant \xi^*$ cannot be determined from formulae (29.12). It has to be determined from the discrete solutions in region II. The solutions in regions III and IV are constructed as before in a numerical fashion. On the grounds of these numerical computations we can assert that the plastic wave of weak discontinuity of equation $\eta = \Gamma_2(\xi)$ after only a small lapse time approaches the curve bonding the region of plastic strain for the quasi-static problem $\eta = \Gamma_{qst}(\xi)$ (Fig. 98) determined from the formula

(29.53)
$$\xi = 2\sqrt{\eta}\,\mathrm{erf}^{-1}[1 - e_s],$$

where erf^{-1} is the inverse to the error function.

On the boundary of the half-space the solution can be obtained in an exact form, coinciding with the quasi-static solution. The strain on the boundary of the half-space increases from value $\epsilon(0, 0) = 2\vartheta$, attaining the asymptotic value

(29.54)
$$\lim_{\eta\to\infty} \varepsilon(0, \eta) = 2.22 \qquad \text{for} \qquad \xi = 0,$$

provided we assume that the linear hardening coefficient of material $\mu_1/\mu = 0.009$. In the case when the hardening of the material is disregarded, we obtain

(29.55)
$$\lim_{\eta\to\infty} \varepsilon(0, \eta) = 2.27 \qquad \text{for} \qquad \xi = 0.$$

The asymptotic difference between the strains on the boundary of the half-space is of the order 6% in both cases and it decreases with distance from the boundary. After a sufficiently long lapse of time the dynamic effect decays. Generally we can remark, comparing the results presented here for an elastic/viscoplastic medium, that these solutions, in the asumptotic state, are, in great measure, close to the solutions by Raniecki [131] based on the model of the strain theory of plasticity.

The thermal shock problem on the boundary of the half-space was presented above. The problem is much more complicated if the thermal shock is accompanied by mechanical dynamic loads. The picture of the solution to this problem in the phase plane is quite complicated. The mechanical load propagates with a finite velocity equal to the velocity of longitudinal waves into a medium deformed by the change of temperature on the boundary. For $t < x/a$ the picture of the solution will be, of course, the same as in the case of the thermal shock only. On the other hand, already on the wave of strong discontinuity $x = at$, carrying the disturbance from the boundary, interference of the mechanical loading boundary condition is observed.

If we assume the boundary condition for the pressure in the form $\sigma_{11} = -p(t) \geqslant 0$, then the mechanical loading will produce, in the half-space, compressive strain only. The thermal loading, however, will generate in the half-space tensile strain for times $t \geqslant x/a$. The value of the pressure applied to the boundary at the initial instant and the change of temperature on the boundary of the half-space will be decisive in the solution on the wave $x = at$ and for the configuration of the plastic strain regions in the phase plane for $t > x/a$. The solution in the regions of the phase plane lying above the characteristic $x = at$ offers great difficulty first of all on account of the necessity to examine a number of variants of the solutions (dependent on the value and the variations in time of the loads on the boundary). The application of the method of the characteristic net is also more complicated. This is due to the difficulty of selecting an appropriate value for the mesh size in the characteristic nets: the thermal effects decay very fast with the distance while the disturbances generated by the mechanical loads vanish in the medium very slowly. As $x \to \infty$ the stresses tend to the value corresponding to the yield limit. This is why the solution for $t > x/a$ has to be constructed in another way, for example by the method of successive iterations due to Courant.

30. Thermal shock on the boundary of a spherical cavity in an elastic/viscoplastic space

We consider an unbounded medium with a spherical cavity of radius r_0. The cavity surface is subjected at instant $t = 0$ to sudden, uniform heating up to temperature T_0 and then it is kept at that temperature. The constitutive equations (3.31) in the case of spherical symmetry (see the assumptions in section 16) and under the assumption that $f(\sigma_{ij}, \epsilon_{ij}^{p}) = \sqrt{J_2}$ take the forms [97]:

(30.1)
$$\dot{\varepsilon}_{rr} - \dot{\varepsilon}_{\varphi\varphi} = \frac{1}{2\mu}(\dot{\sigma}_{rr} - \dot{\sigma}_{\varphi\varphi}) + \sqrt{3}\,\gamma \left\langle \Phi\left[\frac{\sqrt{J_2}}{\varkappa} - 1\right]\right\rangle,$$
$$\dot{\sigma}_{rr} + 2\dot{\sigma}_{\varphi\varphi} = 3K(\dot{\varepsilon}_{rr} + 2\dot{\varepsilon}_{\varphi\varphi} - 3\alpha\dot{T}),$$

where
$$\sqrt{J_2} = \frac{1}{3}(\sigma_{rr} - \sigma_{\varphi\varphi}).$$

The heat conduction equation (28.4) for the prescribed boundary conditions

$$T|_{r=r_0} = T_0, \quad \lim_{r\to\infty} T = 0, \tag{30.2}$$

and the initial condition

$$T|_{t=0} = 0, \tag{30.3}$$

has a solution of the following form:

$$T(r,t) = T_0 \frac{r_0}{r} \operatorname{erfc}\left(\frac{r-r_0}{\sqrt{4kt}}\right). \tag{30.4}$$

The system of (30.1) and (30.4), together with the equation of motion (5.5) and the continuity condition (5.7),

$$\sigma_{rr,r} + \frac{2}{r}(\sigma_{rr} - \sigma_{\varphi\varphi}) = \varrho v_{,t}, \qquad v_{,r} = \varepsilon_{rr,t} \tag{30.5}$$

permits us to find a unique solution for the given boundary and initial conditions

$$\sigma_{rr}(r,t) = 0, \quad u(r,0) = v(r,0) = 0. \tag{30.6}$$

Using the dimensionless variables

$$\xi = \frac{ar}{k}, \quad \xi_0 = \frac{ar_0}{k}, \quad \eta = \frac{a^2 t}{k}, \tag{30.7}$$

we can show that the system of (30.1) and (30.5) has the following characteristics:

$$\xi = \xi_0 \pm \eta + \text{const}, \quad \xi = \text{const}. \tag{30.8}$$

If we take the solution of the heat conduction equation given by (30.4), we obtain the following relations [97] along the characteristics (30.8):

$$\pm 3a\,d\sigma_{rr} - (4\mu + 3K)\,dv + \left\{4\sqrt{3}\,\gamma\left\langle\Phi\left[\frac{\sqrt{J_2}}{\varkappa} - 1\right]\right\rangle + (4\mu - 6K)\frac{v}{\xi}\right.$$
$$\left. + \frac{9\xi_0 T_0 K\alpha}{2\sqrt{\pi\eta^3}}\left(1 - \frac{\xi_0}{\xi}\right)\exp\left[-\frac{(\xi-\xi_0)^2}{4\eta}\right] \pm \frac{6a}{\xi}(\sigma_{rr} - \sigma_{\varphi\varphi})\right\} d\xi = 0$$

(30.9) for $\xi = \xi_0 \pm \eta + \text{const}$,

$$\left(\frac{3K}{2\mu} + 2\right) d\sigma_{\varphi\varphi} - \left(\frac{3K}{2\mu} - 1\right) d\sigma_{rr} - 3K\left\{\sqrt{3}\,\gamma\left\langle\Phi\left[\frac{\sqrt{J_2}}{\varkappa} - 1\right]\right\rangle\right.$$
$$\left. + 3\frac{v}{\xi} - \frac{3\alpha T_0 \xi_0 K}{2\sqrt{\pi\eta^3}}\left(1 - \frac{\xi_0}{\xi}\right)\exp\left[-\frac{(\xi-\xi_0)^2}{4\eta}\right]\right\} d\xi = 0 \quad \text{for} \quad \xi = \text{const}.$$

Now let us pass to the solution in the phase plane (ξ, η) (Fig. 99). Region I is defined physically as the region of elastic deformation. Since the relations along the characteristics (30.9) cannot be integrated even in the range of the elastic deformation (for $\langle\Phi\rangle \equiv 0$), the solution in region I must therefore be determined in another way by means of the

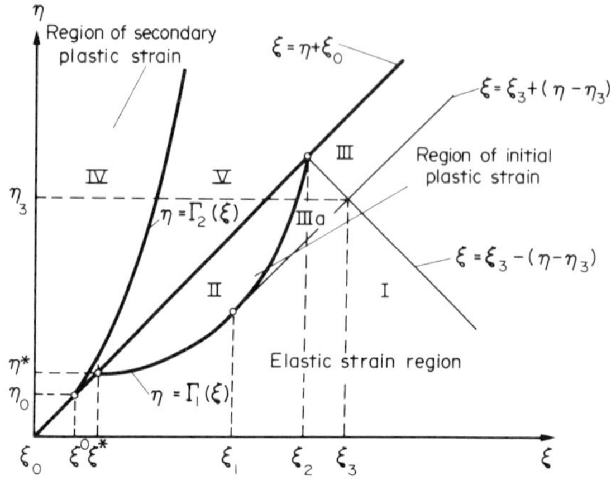

Fig. 99

method given in section 28. The system of (30.1) and (30.5) in the elastic strain range can be reduced by taking note that $\varepsilon_{\varphi\varphi} = u/r$, $\varepsilon_{rr} = \partial u/\partial r$, and $v = \partial u/\partial t$, to the form

$$L[u] = -\frac{3K\alpha}{\varrho} T_{,r}, \tag{30.10}$$

where

$$L[u] = \frac{\partial^2}{\partial t^2} - a^2\left(\frac{\partial^2}{\partial r^2} + \frac{2}{r}\frac{\partial}{\partial r} - \frac{2}{r^2}\right),$$

i.e. to the form (28.1).

The function

$$u(r,t) = \frac{3\lambda+2\mu}{2\mu+\lambda}\alpha k \int_0^t \left[1 - \exp\left(\frac{a^2(t-\zeta)}{k}\right)\right] T_{,r}\, d\zeta \tag{30.11}$$

is a particular integral of the equation for the temperature field for zero initial conditions (see section 28).

On account of the initial conditions $(30.6)_2$ the function (30.11) constitutes the solution of the initial problem in region I. Computing the gradient of the temperature field defined by formula (30.4) and substituting it in formula (30.11), we arrive at the expression for the displacements in region I which, in terms of the dimensionless variables (30.7), take the form

$$u(\xi,\eta) = \vartheta\frac{\xi_0}{\xi^2}\Bigg\{[\xi^2 - (2\eta + \xi_0^2 + 2)]\operatorname{erfc}\left(\frac{\xi-\xi_0}{2\sqrt{\eta}}\right) \tag{30.12}$$

$$-2\sqrt{\frac{\eta}{\pi}}(\xi+\xi_0)\exp\left[-\frac{(\xi-\xi_0)^2}{4\eta}\right] + (1-\xi)\operatorname{erfc}\left(\frac{\xi-\xi_0}{2\sqrt{\eta}} + \sqrt{\eta}\right)\exp(\eta+\xi-\xi_0)$$

$$+(1+\xi)\operatorname{erfc}\left(\frac{\xi-\xi_0}{2\sqrt{\eta}} - \sqrt{\eta}\right)\exp(\eta-\xi+\xi_0)\Bigg\}.$$

The expressions for the strain and velocity fields are obtained directly from (30.12):

$$\varepsilon_{\varphi\varphi}(\xi,\eta) = \vartheta\frac{\xi_0}{\xi^3}\Big\{[\xi^2-(2\eta+\xi_0^2+2)]\operatorname{erfc}\Big(\frac{\xi-\xi_0}{2\sqrt{\eta}}\Big)$$
$$-2\sqrt{\frac{\eta}{\pi}}(\xi+\xi_0)\exp\Big[-\frac{(\xi-\xi_0)^2}{4\eta}\Big]+(1-\xi)\operatorname{erfc}\Big(\frac{\xi-\xi_0}{2\sqrt{\eta}}+\sqrt{\eta}\Big)\exp(\eta+\xi-\xi_0)$$
$$+(1+\xi)\operatorname{erfc}\Big(\frac{\xi-\xi_0}{2\sqrt{\eta}}-\sqrt{\eta}\Big)\exp(\eta-\xi+\xi_0)\Big\},$$

(30.13)
$$\varepsilon_{rr}(\xi,\eta) = 2\vartheta\frac{\xi_0}{\xi^3}\Big\{(2\eta+\xi_0^2+2)\operatorname{erfc}\Big(\frac{\xi-\xi_0}{2\sqrt{\eta}}\Big)$$
$$+2\sqrt{\frac{\eta}{\pi}}(\xi+\xi_0)\exp\Big[-\frac{(\xi-\xi_0)^2}{4\eta}\Big]+\Big(\xi-\frac{1}{2}\xi^2-1\Big)\exp(\eta+\xi-\xi_0)$$
$$\times\operatorname{erfc}\Big(\frac{\xi-\xi_0}{2\sqrt{\eta}}+\sqrt{\eta}\Big)-\Big(1+\xi+\frac{1}{2}\xi^2\Big)\exp(\eta-\xi+\xi_0)\operatorname{erfc}\Big(\frac{\xi-\xi_0}{2\sqrt{\eta}}-\sqrt{\eta}\Big)\Big\},$$
$$v(\xi,\eta) = \vartheta\frac{\xi_0}{\xi^2}\Big\{(1-\xi)\exp(\eta+\xi-\xi_0)\operatorname{erfc}\Big(\frac{\xi-\xi_0}{2\sqrt{\eta}}+\sqrt{\eta}\Big)$$
$$+(1+\xi)\exp(\eta-\xi+\xi_0)\operatorname{erfc}\Big(\frac{\xi-\xi_0}{2\sqrt{\eta}}-\sqrt{\eta}\Big)-2\operatorname{erfc}\Big(\frac{\xi-\xi_0}{2\sqrt{\eta}}\Big)\Big\}.$$

The stress tensor components are determined from the Duhamel–Neumann law (28.14). In region I we have

(30.14)
$$\sigma_{rr}(\xi,\eta) = \varrho a^2\varepsilon_{rr}(\xi,\eta)+2(\varrho a^2-2\mu)\varepsilon_{\varphi\varphi}(\xi,\eta)-3K\alpha T_0\frac{\xi_0}{\xi}\operatorname{erfc}\Big(\frac{\xi-\xi_0}{2\sqrt{\eta}}\Big),$$
$$\sigma_{\varphi\varphi}(\xi,\eta) = (\varrho a^2-2\mu)\varepsilon_{rr}(\xi,\eta)+2(\varrho a^2-\mu)\varepsilon_{\varphi\varphi}(\xi,\eta)-3K\alpha T_0\frac{\xi_0}{\xi}\operatorname{erfc}\Big(\frac{\xi-\xi_0}{2\sqrt{\eta}}\Big).$$

A plastic loading wave of weak discontinuity will appear in the medium if the strain intensity ϵ_i in region I reaches the value ϵ_{is}. In Fig. 99 it is represented by the curve $\eta = \Gamma_1(\xi)$, and is determined from the condition

(30.15)
$$\varepsilon_i = \frac{2}{\sqrt{3}}(\varepsilon_{rr}-\varepsilon_{\varphi\varphi})|_{\eta=\Gamma_1(\xi)} = -\varepsilon_{is}.$$

Substituting $(30.13)_{1,2}$ into (30.15) we obtain the equation, in an implicit form, which determines the shape of the wave $\eta = \Gamma_1(\xi)$. In this way we determine wave $\eta = \Gamma_1(\xi)$ only up to a certain cross-section $\xi = \xi_1$ for which $\partial\Gamma_1/\partial\xi|_{\xi=\xi_1} = 1$. On account of geometrical dispersion the strain intensity decreases to zero along the characteristics, as $\xi \to \infty$. Therefore condition (30.15) is satisfied on the wave $\eta = \Gamma_1(\xi)$ provided $\partial\Gamma_1/\partial\xi > 1$ for $\xi > \xi_1$ (Fig. 99). For $\xi > \xi_1$ the wave $\eta = \Gamma_1(\xi)$ is determined numerically together with the solutions in regions II and IIIa by the method of characteristic nets, use being made of the relations along characteristics (30.9) and of condition (30.15) on the boundary of regions.

In region III (region of elastic deformation) the motion of the medium is governed by (30.10). In this region we have to pose a Darboux problem with the conditions given on characteristics $\xi = \xi_3 \pm (\eta - \eta_3)$:

(30.16)
$$u = u_0', \quad \text{for} \quad \xi = \xi_3-(\eta-\eta_3),$$
$$u = u_0'', \quad \text{for} \quad \xi = \xi_3+(\eta-\eta_3).$$

The solution of (30.10) by the method discussed in section 28 is presented in the form of the sum

$$u = u^{\text{I}} + u^{\text{II}}, \tag{30.17}$$

where u^{I} is a particular integral of (30.10), while u^{II} is the solution of the homogeneous equation

$$L[u^{\text{II}}] = 0, \tag{30.18}$$

with the conditions

$$\begin{aligned} u^{\text{II}} &= u_0' - u_{10}' = u_{20}' \quad \text{for} \quad \xi = \xi_3 - (\eta - \eta_3), \\ u^{\text{II}} - u_0'' - u_{10}'' &= u_{20}' \quad \text{for} \quad \xi = \xi_3 + (\eta - \eta_3). \end{aligned} \tag{30.19}$$

Functions u_{10}' and u_{10}'' are determined from the particular integral (30.11) of (30.10) on characteristics $\xi = \xi_3 \mp (\eta - \eta_3)$, respectively. Since the particular integral (30.11) for the initial problem (Cauchy's problem in region I) is simultaneously the general integral of (30.10), consequently, also along characteristic $\xi = \xi_3 + (\eta - \eta_3)$, the second condition (30.19) is simplified to give

$$\begin{aligned} u^{\text{II}} &= u_0' - u_{10}' = u_{20}' \quad \text{for} \quad \xi = \xi_3 - (\eta - \eta_3), \\ u^{\text{II}} &= 0 \quad \text{for} \quad \xi = \xi_3 + (\eta - \eta_3). \end{aligned} \tag{30.20}$$

The function u_{10}' is determined from solution (30.12) for $\eta = \eta_3 - \xi + \xi_3$,

$$u_{10}' = \frac{3\lambda + 2\mu}{2\mu + \lambda} \frac{\alpha k}{a} \int_0^{\eta_3 - \xi + \xi_3} [1 - \exp(\eta - \zeta)]\, T_{,\xi}|_{\eta = \eta_3 - \xi + \xi_3}\, d\zeta; \tag{30.21}$$

u_0' is determined from the solution in region IIIa for $\xi = \xi_3 - (\eta + \eta_3)$.

The solution in region II has therefore eventually been reduced to that of the homogeneous equation (30.18) with conditions (30.20). The solution of (30.18) can be represented in the form

$$u^{\text{II}}(\xi, \eta) = \frac{\varphi'(\xi - \eta - \xi_0)}{\xi} - \frac{\varphi(\xi - \eta - \xi_0)}{\xi^2} + \frac{\psi'(\xi + \eta - \xi_0)}{\xi} - \frac{\psi(\xi + \eta - \xi_0)}{\xi^2}. \tag{30.22}$$

Substituting the above result into the initial conditions we obtain

$$\begin{aligned} &\psi(\xi + \eta - \xi_0) \equiv 0, \\ &\varphi(\xi - \eta - \xi_0) = (\xi_0 + \xi_3)[\xi_3 + \xi - \eta - \eta_3]\left[1 - \frac{\xi_3 + \xi - \eta - \eta_3}{2(\xi_0 + \xi_3)}\right] u_{20}' \\ &\qquad + \frac{1}{2}[\xi_3 + \xi - \eta - \eta_3] \int_{\xi_3 - \eta_3}^{\xi - \xi_0 - \eta} \frac{u_{20}'}{z + \xi_3 + \eta_3 + \xi_0}\, dz. \end{aligned} \tag{30.23}$$

Finally, the solution in region III in terms of strain (and in the dimensionless variables ξ, η) will take the form

$$u(\xi, \eta) = \frac{3\lambda + 2\mu}{2\mu + \lambda} \frac{\alpha k}{a} \int_0^{\eta} \left[1 - \exp\left(\frac{\eta - \zeta}{k}\right)\right] T_{,\xi}\, d\zeta + \frac{\varphi'(\xi - \eta - \xi_0)}{\xi} - \frac{\varphi(\xi - \eta - \xi_0)}{\xi^2}. \tag{30.24}$$

The remaining parameters are readily obtainable.

We pass in turn to the solutions in the next regions IV and V. On account of the loading on the boundary $\xi = \xi_0$, a wave of strong discontinuity of equation $\eta = \xi - \xi_0$ starts propagating into the medium $\xi > \xi_0$. It can be proved by analogy with the half-space case that, if condition (30.15) holds, the wave $\eta = \xi - \xi_0$ over a certain segment $(0, \xi^0)$ will be a plastic wave.

The solution on the wave $\xi = \eta + \xi_0$ is obtained from relation (30.9) on the positive characteristic. We have taken into account relations (30.14) which are valid across the front of a wave of strong discontinuity for an elastic/viscoplastic medium. Furthermore, we have made use of the condition of the kinematic continuity

$$[v] = -a\,[\varepsilon_{rr}], \tag{30.25}$$

and of the condition of displacement continuity from which the condition of the continuity of strain component $\varepsilon_{\varphi\varphi}$ results.

By means of the above relations we obtain the following singular integral equation for the strain component $\epsilon_{rr}(\xi)$ on the front of wave $\xi = \xi_0 + \eta$:

$$\varepsilon_{rr}(\xi) = F(\xi) + \int\limits_{\xi_0}^{\xi} \psi[\varepsilon_{rr}(\bar{\xi}), \bar{\xi}]\, d\bar{\xi} \qquad \text{for} \quad \eta = \xi - \xi_0, \tag{30.26}$$

where

$$\begin{aligned} F(\xi) = \vartheta \left[1 + \frac{\xi_0}{\xi} \operatorname{erfc}\left(\frac{1}{2} \sqrt{\xi - \xi_0} \right) \right] + \frac{1}{2} \left(\frac{v_1}{a} + \varepsilon_{rr1} \right) \\ -\left(1 - \frac{2\mu}{\varrho a^2} \right) \varepsilon_{\varphi\varphi 1} + \int\limits_{\xi_0}^{\xi} \frac{1}{\bar{\xi}} \left[\left(1 - \frac{2\mu}{\varrho a^2} \right) \left(\frac{v_1}{a} + \varepsilon_{rr1} \right) + \frac{2\mu}{\varrho a^2} \varepsilon_{\varphi\varphi 1} \right. \\ \left. - \frac{1}{2} \vartheta \frac{\xi_0}{\sqrt{\pi(\bar{\xi} - \xi_0)}} \exp\left(-\frac{\bar{\xi} - \xi_0}{4} \right) \right] d\bar{\xi}, \end{aligned} \tag{30.27}$$

$$\psi[\varepsilon_{rr}(\xi), \xi] = -\frac{\varepsilon_{rr1}}{\xi} - \frac{2\mu k}{\sqrt{3}\,\varrho a^4} \gamma \left\langle \Phi \left[\frac{2\mu(\varepsilon_{rr} - \varepsilon_{\varphi\varphi 1})}{\sqrt{3}\,\sigma_{is}} - 1 \right] \right\rangle.$$

The constant of integration in (30.27) is determined from boundary condition $(30.6)_1$: $C = \vartheta$. This results from the assumption of zero initial conditions. Functions v_1, ϵ_{rr1} and $\varepsilon_{\varphi\varphi 1}$ are determined by formulae (30.13) for $\eta = \xi - \xi_0$. The integrals of these functions are not elementary and therefore some of them have to be computed by expanding them into series and taking into account the first few terms in the expansion.

Examining the convergence of the above solution we can state that

$$|\varepsilon_{rr}| \leqslant |F(\xi) + \int\limits_{\xi_0}^{\xi} \psi[\varepsilon_{rr}(\bar{\xi}), \bar{\xi}]\, d\bar{\xi}| \leqslant |F(\xi)| + |\int\limits_{\xi_0}^{\xi} \psi[\varepsilon_{rr}(\bar{\xi}), \bar{\xi}]\, d\bar{\xi}|. \tag{30.28}$$

Since the function $F(\xi)$ is bounded ($|F(\xi)| \leqslant 2\vartheta$) we obtain

$$|\varepsilon_{rr}| \leqslant 2\vartheta + |\int\limits_{\xi_0}^{\xi} [\varepsilon_{rr}(\bar{\xi}), \bar{\xi}]\, d\bar{\xi}|. \tag{30.29}$$

Then the estimation of the solution is identical with that of (20.10).

The end of the plastic wave of strong discontinuity is determined from the condition

$$\frac{2}{\sqrt{3}}(\varepsilon_{rr}-\varepsilon_{\varphi\varphi 1}) = +\varepsilon_{is}. \tag{30.30}$$

At the point ($\xi = \xi^0$) (Fig. 99) the plastic wave of strong discontinuity becomes a wave of weak discontinuity $\eta = \Gamma_2(\xi)$ similarly to the case of plane waves. The solution on the wave of strong discontinuity $\eta = \xi - \xi_0$, for $\xi > \xi^*$, is also obtained from (30.26) using the assumption $\gamma\langle\Phi\rangle = 0$. Simultaneously, the functions ε_{rr1}, $\varepsilon_{\varphi\varphi 1}$ and v_1 occurring in (30.27) have to be determined from the solutions in regions II and III.

Region IV is the region of viscoplastic deformation bounded by the wave of equation $\eta = \Gamma_2(\xi)$ (Fig. 99). Region V, for $\xi^0 \leqslant \xi \leqslant \xi^*$ and $\xi > \xi_2$, is a region of unloading. The solutions in regions IV and V are found simultaneously with the determination of the curve $\eta = \Gamma_2(\xi)$ by the method characteristic nets. The wave $\eta = \Gamma_2(\xi)$ is determined from the condition

$$(\varepsilon_{rr}-\varepsilon_{\varphi\varphi})|_{\eta=\Gamma_2(\xi)} = +\frac{\sqrt{3}}{2}\varepsilon_{is} \quad \text{for} \quad \xi^0 \leqslant \xi \leqslant \xi^* \quad \text{and} \quad \xi \geqslant \xi_2, \tag{30.31}$$

$$(\varepsilon_{rr}-\varepsilon_{\varphi\varphi})_{\eta=\Gamma_2(\xi)} = +\frac{\sqrt{3}}{2}[2\varepsilon_{is}+\varepsilon_{io2}(\xi)] \quad \text{for} \quad \xi^* \leqslant \xi \leqslant \xi_2, \tag{30.32}$$

where $\epsilon_{io2}(\xi)$ denotes the strain intensity on the characteristic $\eta = \xi - \xi_0$ approached from the side of region II.

A numerical example in [97] is computed for the same numerical data as in the preceding section with the additional assumption that the radius of the spherical cavity is equal to $\xi_0 = 10^7$. Quantities of smaller order are disregarded in relations (30.13), (30.14), and (30.27) and as a result, the calculations are considerably simplified (terms of order ξ_0^{-1} are neglected). The results of the calculations are presented in Fig. 100.

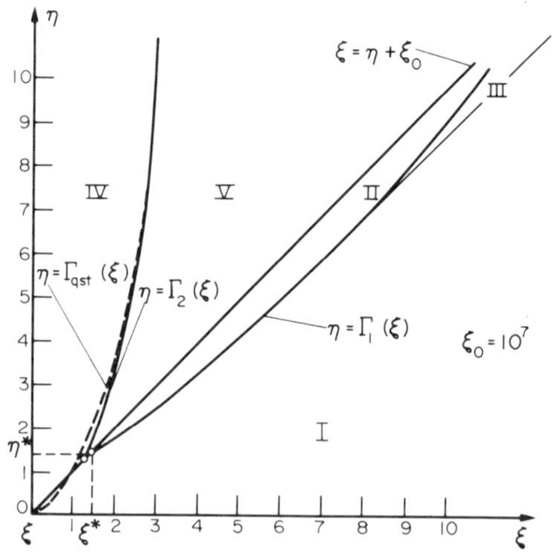

Fig. 100

We can assert that the plastic wave $\eta = \Gamma_2(\xi)$, bounding the region of viscoplastic deformations, tends, even for very small times, to the curve $\eta = \Gamma_{qst}(\xi)$ which bounds the region of plastic deformation in the quasi-static treatment of the problem. The solution of the quasi-static problem was given by Raniecki [130] for the case of spherical symmetry.

The problem of a thermal shock applied to the boundary of the spherical cavity in an unbounded elastic–plastic space was solved in [132].

It should also be stated that, in the case of spherical waves generated by a thermal shock, the distribution of the permanent strain in the medium is almost identical for the case of the constitutive equations of the theory of viscoplasticity and for the Hencky–Ilyushin strain theory.

Concluding this chapter we can state that, in cases when a body is heated from the outside, the influence of the temperature on the strain field should be taken into account. In the case of a thermal shock an increment of temperature on the boundary of order 150 K produces plastic strain of the order 2%. On the other hand, when heat is not supplied to the body from outside and if we take into account the coupling with the temperature field in the constitutive equations and in the heat conduction equation the coupling with the strain field, then we find negligibly small changes of temperature in the body and insignificant changes in the strain field generated by thermal effects. In the case of the coupled equations of thermo-viscoplasticity [106], it was found, for the adiabatic process of an aluminium bar striking an undeformable obstacle with a speed of approximately 53 m s^{-1}, that the mean temperature of the bar increases by about 6 K while the strain changes generated by the increment of temperature are unnoticeable. Similar phenomena have been observed also in certain quasi-static loading processes.

BIBLIOGRAPHY

1. AGGARWAL, H. R., SOLDATE, A. M., HOOK, J. F., and MIKLOWITZ, J., Bilinear theory in plasticity and an application to two-dimensional wave propagation, *J. Appl. Mech.*, June 1964.
2. BALTOV, A., The plane problem for elastic/viscoplastic bodies, *Archwm Mech. stosow.*, 1967, **18**, 2.
3. BALTOV, A., PhD thesis, IFTR, Pol. Acad. Sci., Warsaw, 1967.
4. BALTOV, A., Tuong Vinh, *CR Acad. Sci. Paris*, **275**, Série A, 291, 1972.
5. BEJDA, J., Propagation of two-dimensional stress waves in elastic/viscoplastic material, *Proceedings of the 12th International Congress of Applied Mechanics* (ed. M. Hetency), Springer-Verlag, 1969.
6. BEJDA, J., *Spatial Wave Problems in Inelastic Media* [in Polish], IFTR, Pol. Acad. Sci., Warsaw, 1972, 27.
7. BEJDA, J., A difference method for plane problems in dynamic viscoelasticity, *Partial Differential Equations and Waves* (ed. Froissart), Springer-Verlag, 1969.
8. BEJDA, J., Dynamic response of elastic/viscoplastic and viscoelastic–plastic elements, *3rd European Symposium on Earthquake Engineering, Sofia, 1970.*
9. BEJDA, J., Propagation of plane and axially symmetric two-dimensional elastic/viscoplastic waves, *Revue roum. Sci., Tech., Série Appl. Mech.*, 1970, **15**, 2.
10. BEJDA, J. Propagation of two-dimensional strong discontinuity waves in elastic/viscoplastic material, *Archs. Mech.*, 1972, **24**, 3.
11. BEJDA, J. and PERZYNA, P., The propagation of stress waves in a rate sensitive and work-hardening plastic medium, *Archwm Mech. stosow.*, 1964, **16**, 5.
12. BERTHOLF, L. D., Numerical solution for two-dimensional elastic wave propagation in finite bars, *J. Appl. Mech.*, 1967, **34.**
13. BLEICH, H. H., MATHEWS, A. T., and WRIGH, J. P., Moving step load on the surface of a half-space of granular material, *Int. J. Solids Struct.*, 1968, **4**, 2.
14. BLEICH, H. H. and NELSON, I., Plane waves in an elastic–plastic half-space due to combined surface pressure and shear, *J. Appl. Mech.*, March 1966.
15. BLEICH, H. H. and MATHEWS, A. T., Step load moving with superseismic velocity on the surface of an elastic–plastic half-space, *Int. J. Solids Struct.*, 1967, **3**, 5.
16. BUI, H. D., Dissipation d'énergie dans une déformation plastique, *Cah. Rhéologie*, 1965, **1**, 1.
17. BURAVTSEV, A. N., Analytical method of construction of unloading wave [in Russian], *Vest. Leningr. gos. Univ.*, 1970, **1**.
18. BURKE, J. J. and WEISS, V. (eds.) *Shock Waves and the Mechanical Properties of Solids, Syracuse* University Press, 1971.
19. BURNAT, M., KIELBASINSKI, A., and WAKULICZ, A., The method of characteristics for a multi-dimensional gas flow, *Archwm Mech. stosow.*, 1964, **16**, 3.
20. BUTLER, D. S., The numerical solution of hyperbolic systems of partial differential equations in three independent variables, *Proc. Soc., London*, 1962.
21. CAMPBELL, J. D. and COOPER, R. H., Yield and flow low carbon steel at medium strain rates, *Proceedings of the Physics Conference on Basil Yield Fracture, 1967.*
22. CAMPBELL, J. D. and FERGUSON, W. G., The temperature and strain rate dependence of shear strength of mild steel, *Phil. Mag.*, 1970, **21**, 169.
23. CAMPBELL, J. D. and MARSH, K. J., The effect of strain on the post-yield flow of mild steel, *J. Mech. Phys. Solids*, 1963, **11**, 1.
24. CHIDDISTER, J. L. and MALVERN, L. E., Compression impact testing of aluminium at elevated temperature, *Expl. Mech.*, 1963, 3.
25. CLARK, D. S. and DUWEZ, P. E., The influence of strain rate on some tensile properties of steel, *Proc. Am. Soc. Test. Mater.*, 1950, 50.

26. CHOU, P. C. and KARP, R. R., Two-dimensional dynamic elasticity problems by the numerical method of characteristics, *Proceedings of the Army Symposium on Solid Mechanics, September 1968.*
27. CLIFTON, R. J., *An Analysis of Combined Longitudinal and Torsional Waves in a Thin-walled Tube*, Report No. 5, Brown Univ., Providence, May 1960.
28. CLIFTON, R. J., Unloading waves for combined longitudinal and torsional plastic waves, *J. Appl. Mech.*, 1968, **35**, 4.
29. CLIFTON, R. J. and LIPKIN, J., Plastic waves of combined stresses due to longitudinal impact of a pretorqued tube: Parts I and II, *Qt. Appl. Math.*, 1967, **25**, 1.
30. CLIFTON, R. J., A difference method for plane problems in dynamic elasticity, *Q. Appl. Math.*, 1967, **25**, 1.
31. CLIFTON, R. J., Analysis of dynamic deformation of elastic–plastic solids under condition of plane strain, PhD thesis, Inst. Techn., Pittsburg, 1963.
32. COLLOQUES IRIA, *Méthodes de calcul scientifique et technique*, Rocquencourt, 17–21 décembre 1973.
33. COURANT, R. and HILBERT, D., *Partial Differential Equations*, New York, 1962.
34. CRISTESCU, N., *Dynamic Plasticity*, North-Holland, 1967.
35. CRISTESCU, N., On the propagation of elastic–plastic waves for combined stresses, *Prikl. Mat. Mekh.* 1959, **23**.
36. CESCHINO, F. and KUNTZMANN, J., *Problèmes differentiels de conditions initiales,* Dunod, Paris, 1963.
37. DŻYGADLO, Z., KALISKI, S., SOLARZ, L., and WLODARCZYKE, E., *Vibrations and waves* [in Polish], PWN, Warsaw, 1966.
38. ESTRIN, M. I., On the equations of the dynamics of compressible plastic medium [in Russian], *Dokl. Akad. Nauk SSSR*, 1960, **135**, 1.
39. FERGUSON, W. G., KUMAR, A., and DORN, J. E., Dislocation damping in aluminium at high strain rate, *J. Appl. Phys.*, 1967, 38.
40. FONG, J. T., *Elastic–Plastic Waves in a Half-space of Linearly Work-hardening Material for Coupled Loadings*, Rep. No. 161, Standard Univ., May 1966.
41. FREUDENTHAL, A. M., The mathematical theories of the inelastic continuum, *Encyclopedia of Physics*, vol. VI, Springer-Verlag, Berlin, 1958.
42. FRELAT, J., NGUYEN, Q. S., and ZARKA, J., Some remarks about classical problems in plasticity and viscoplasticity. Application to their numerical resolution. Lecture at Univ. of Wales, Swansea, May 1974.
43. GERMAIN, P., *Cours de mécanique des milieux continus*, vol. 1, *Théorie générale*, Masson, Paris, 1973.
44. GERMAIN, P. and LEE, E. H., *International Symposium, Foundations of Plasticity* (ed. A. Sawczuk), Warsaw, 1972.
45. GOEL, R. P. and MALVERN, L. E., Biaxial plastic simple waves with combined kinematic and isotropic hardening, *Trans. ASME*, December 1971.
46. GOLDSMITH, W., *Impact: The Theory and Physical Behaviour of Colliding Solids*, London, 1960.
47. GRYBOS, R., *Theory of Impact in Discrete Mechanical Systems* [in Polish], PWN, Warsaw, 1969.
48. GRIGORIAN, S. S., On basic notions of soil dynamics [in Russian], *Prikl. Mat. Mekh.*, 1960, **24**, 6.
49. GUTOWSKI, R., KALISKI, S. and OSIECKI, J., Propagation of plane unloading wave in a non-homogeneous soil [in Polish], *Biul. Wojsk. Akad. Tech.*, 1959, 2/85/.
50. HADAMARD, J., *Cours d'Analyse*, Paris, 1927.
51. HAUSER, F. E., SIMMONS, J. A., and DORN, J. E., Strain rate effects in plastic wave propagation: Response of metals to high velocity deformation, *Interscience*, New York, 1961.
52. HOHENEMSER, K. and PRAGER, W., Über die Ansätze der Mechanik isotroper Kontinua, *Z. angew, Math. Mech.*, 1932, vol. 12.
53. HSU, J. C. C. and CLIFTON, R. J., Plastic waves in a rate sensitive material: II, Waves of combined stress, *J. Mech. Phys. Solids*, 1974, **22**.
54. JEFFERY, G. B., Plane stress and plane strain in bipolar coordinates, *Phil. Trans. R. Soc.*, Série A, 1921.
55. KALISKI, S., On certain equations of dynamics of an elastic/viscoplastic body: The strain hardening properties and the influence of strain rate, *Bull. Acad. Pol. Sci., Série Techn.*, 1963, **11**, 7.
56. KALISKI, S., The unloading wave for a body with rigid unloading characteristic in layered media, *Proc. Vibr. Probl.*, 1961, 3.
57. KALISKI, S., NOWACKI, W. K., and WLODARCZYK, E., Propagation and reflection of a spherical wave in an elastic/viscoplastic strain hardening body, *Proc. Vibr. Probl.*, 1964, **5**, 1.
58. KALISKI, S., NOWACKI, W. K., and WLODARCZYK, E., On certain closed solution for the shockwave with rigid unloading, *Bull. Acad. Pol. Sci., Série Sci. Techn.*, 1967, **15**, 5; and *Biul. Wojsk. Akad. Tech.*, 1967, **16**, 3.
59. KALISKI, S., NOWACKI, W. K., and WLODARCZYK, E., Plane biwaves in an elastic/viscoplastic semi-space, *Proc. Vibr. Probl.*, 1967, **8**, 2.

60. KALISKI, S., NOWACKI, W. K., and WLODARCZYK, E., Propagation of plane loading and unloading biwaves in an elastic/viscoplastic semi-space; Part I, Theory, *Proc. Vibr. Probl.*, 1967, **8,** 3.
61. KALISKI, S., NOWACKI, W. K., and WLODARCZYK, E., Propagation of plane loading and unloading biwaves in an elastic/viscoplastic semi-infinite body; Part II, Numerical analysis, *Proc. Vibr. Probl.*, 1967, **8,** 3.
62. KALISKI, S., NOWACKI, W. K., and WLODARCZYK, E., The influence of strain hardening in the problem of propagation of plane loading and unloading biwaves in an elastic/viscoplastic semi-infinite body, *Proc. Vibr. Probl.*, 1967, **8,** 4.
63. KALISKI, S. and OSIECKI, J., Unloading wave for a body with rigid unloading characteristic, *Proc. Vibr. Probl.*, 1959, **1,** 1.
64. KALISKI, S. and OSIECKI, J., The problem of reflection by a rigid or elastic wall of an unloading wave in a body with rigid unloading characteristic, *Proc. Vibr. Probl.*, 1959, **1.**
65. KALISKI, S. and WLODARCZYK, E., On certain closed-form solution of the propagation and reflection problem of an elastic/viscoplastic wave in a bar, *Archwm. Mech. stosow.*, 1967, **19,** 3.
66. KALISKI, S. and WLODARCZYK, E., Resonance of a longitudinal elastic/viscoplastic wave in a finite bar, *Proc. Vibr. Probl.*, 1967, **8,** 2.
67. KALISKI, S. and WLODARCZYK, E., The problem of resonance for longitudinal elastic–plastic waves in a finite bar, *Proc. Vibr. Probl.*, 1967, **8,** 1.
68. KLEPACZKO, J., *Experimental Investigations of Elastic–Plastic Wave Processes in Metals* [in Polish] IFTR, Pol. Acad. Sci., 1970, 61.
69. KOLAROV, D. and BALTOV, A., Dynamical problems of viscoplastic continuum in a temperature field [in Russian], *Conference on Mechanics, Bucharest, 1969.*
70. KOEHLER, J. S. and SEITZ, F., *On the Propagation of the Plastic Deformation Produced by an Expanding Cylinder*, NDRC-AOR, Report, No. A-139.
71. KOLSKY, H., *Stress Waves in Solids*, Oxford, 1953.
72. KRZYŻAŃSKI, M., *Partial Differential Equations of the Second Order* [in Polish], vol. 2, PWN, Warsaw, 1962.
73. KUKUDZHANOV, V. N., On numerical solution of elastic–plastic wave propagation problems [in Russian], *Symposium Alma-Ata, 1971.*
74. KUMAR, A., HAUSER, F. E., and DORN, J. E., Viscous drag on dislocations in aluminium at high strain rates, *Acta metall.*, 1968, 16.
75. LINDHOLM, U. S., Some experiments in dynamic plasticity under combined stress, *Symposium on the Mechanical Behavior of Materials under Dynamic Loads, San Antonia, Texas, 6–8 September, 1967.*
76. LINDHOLM, U. S. and YEAKLEY, L. M., Dynamic deformation of simple and polycristalline aluminium, *J. Mech. Phys. Solids*, 1965, 13.
77. LINDHOLM, U. S., Dynamic deformation of metals, in *Behavior of Metals under Dynamic Loading*, New York, 1965.
78. LINDHOLM, U. S., Some experiments in dynamic plasticity under combined stress, *Symposium on the Mechanical Behavior of Materials under Dynamic Loads, San Antonio, Texas, 6–8 September, 1967.*
79. LIONS, J. L., *Cours d'analyse numérique*, Ecole Polytechnique, Paris, 1973.
80. LIONS, J. L., *Quelques méthodes de résolution des problèmes aux limites non linéaires*, Dunod, Gauthier-Villars, 1969.
81. LIPKIN, J. and CLIFTON, R. J., An experimental study of combined longitudinal and torsional plane waves in a thin-walled tube, *Twelfth International Congress of Applied Mechanics, Stanford University, August 1968.*
82. LUNTS, I. L., Propagation of spherical elastic–plastic waves [in Russian], *Prikl. Mat. Mekh.*, 1949, **13,** 1.
83. MALVERN, L. E., The propagation of longitudinal waves of plastic deformations in a bar of material exhibiting a strain rate effect, *J. Appl. Mech.*, 1951, **18.**
84. MANJOINE, M., Influence of rate of strain and temperature on yield stresses of mild steel, *J. Appl. Mech.*, 1944, **11.**
85. MANDEL, J., *Cours de mécanique des milieux continus*, Paris, Gauthier-Villars, 1966.
86. MANDEL, J. and others, *Etude des cavités souterraines*, Rapport CEA, Saclay, 1973.
87. MANDEL, J., *Introduction à la mécanique des milieux continus déformables*, PWN, Warsaw, 1974.
88. MARSH, K. J. and CAMPBELL, J. D., The effect of strain rate on the post-yield flow of mild steel, *J. Mech. Phys. Solids*, 1963, **11,** 1.
89. MINDLIN, D., Stress distribution around a tunnel, *Am. Soc. Civ. Eng.*, April 1939.
90. MURAKAMI, S. and BEJDA, J., Two-dimensional cylindrical problem of elastic/viscoplastic wave propagation, *Archs. Mech.*, 1971, **23,** 2.
91. NAGHDI, P. M. and MURCH, S. A., *On the Mechanical Behavior of Viscoelastic/Plastic Solids*, Techn. Report, Univ. California, 1963.
92. NAN NING, Elastic–plastic waves for combined stress, Ph.D. thesis, Stanford Univ., 1968.

93. NIKITIN, L. V., Propagation of shear elastic/viscoplastic waves in beams and plates [in Russian], *Inzh. Sbornik*, 1960, 30.

94. NOWACKI, W. K., Propagation and reflection of plane stress waves from a deformable support in an elastic/viscoplastic strain hardening body, *Proc. Vibr. Problems*, 1964, **5**, 4.

95. NOWACKI, W. K., Thermal shock on the boundary of an elastic/viscoplastic semi-infinite body: Part I, *Bull. Acad. Pol. Sci., Série Sci. Techn.*, 1965, **13**, 2.

96. NOWACKI, W. K., Thermal shock on the boundary of an elastic/viscoplastic semi-infinite body: Part II, *Bull. Acad. Pol. Sci., Série Sci. Techn.*, 1965, **13**, 7.

97. NOWACKI, W. K., The problem of a thermal shock on the boundary of a spherical cavity in an elastic/ viscoplastic space, *Proc. Vibr. Probl.*, 1965, **6**, 3.

98. NOWACKI, W. K., The unloading wave in an elastic–plastic semi-space with a rigid unloading in the case of two-parameter loads, *Archwm Mech. stosow.*, 1969, **21**, 4.

99. NOWACKI, W. K., On certain closed form solution for the plane shear-pressure wave in elastic–plastic half-space, *Archwm Mech. stosow.*, 1968, **20**, 5.

100. NOWACKI, W. K., Elastic/viscoplastic plane waves with combined compressive and two shear stresses in a half-space, *Bull. Acad. Pol. Sci., Série Sci. Techn.*, 1974, **22**, 3.

101. NOWACKI, W. K., and RANIECKI, B., Note on the propagation of thermoelastic (non-coupled) waves, *Proc. Vibr. Probl.*, 1967, **8**, 2.

102. NOWACKI, W. K. and RANIECKI, B., Remarks on the solution for some dynamic problems of thermo-viscoelasticity, *Archwm Mech. stosow.*, 1968, **20**, 3.

103. NOWACKI, W. K. and RANIECKI, B., Note on the propagation of thermoelastic and thermovisco-elastic (non-coupled) waves, *Progress in Thermoelasticity*, PWN, Warsaw, 1967.

104. NOWACKI, W. K. and ZARKA, J., Etude de la frontière élastique des monocristaux d'aluminium, *Int. J. Solids Struct.*, 1971, **7**, 1277–87.

105. NOWACKI, W. K., *Stress Waves in Plastic Bodies*, Ed. Inst. of Technology, Poznań, 1974.

106. NOWACKI, W. K. and ZARKA, J., Sur le champ des températures obtenues en thermoviscoplasticité, *Archwm Mech. stosow.*, 1974, **21**, 4.

107. NOWACKI, W. K., Comportement dynamique d'une cavité dans le semi-space, *Bull. Acad. Pol. Sci, Série Sci. Techn.*, 1975, **23**, 1.

108. NOWACKI, W. K., Sur le comportement dynamique d'une cavité dans un massif semi-infini élasto/ viscoplastique, *Polish–French Symposium, Nicea, 1974.*

109. OSIECKI, J., Propagation of plane stress waves in a non-homogeneous solid medium, *Proc. Vibr. Probl.*, 1961, **2**, 2.

110. OSIECKI, J., Reflection of plane stress wave in a non-homogeneous solid medium, *Proc. Vibr. Probl.*, 1961, **2**, 2.

111. OLSZAK, W. and PERZYNA, P., The constitutive equations of the flow theory for a non-stationary yield condition, *International Congress of Applied Mechanics, Munich*, Springer-Verlag, 1966.

112. OLSZAK, W. and PERZYNA, P., Propagation of spherical waves in a non-homogeneous elastic/visco-plastic medium, *Bull. Acad. Pol. Sci., Série Sci. Techn.*, 1961, **9**, 9.

113. OLSZAK, W. and PERZYNA, P., On elastic/viscoplastic soil, *Proceedings of the Symposium on Rheology and Mechanics of Soils, Grenoble, April 1964.*

114. PERZYNA, P., *Theory of Viscoplasticity* [in Polish], PWN, Warsaw, 1966.

115. PERZYNA, P., The constitutive equations for rate sensitive plastic materials, *Appl. Math.*, 1963, **20**.

116. PERZYNA, P., The constitutive equations for work-hardening and rate sensitive plastic materials, *Proc. Vibr. Probl.*, 1963, **4**, 4.

117. PERZYNA, P., Théorie physique de la viscoplasticité, Conférences, fasc. 104, *Acad. Polon. Sci., Paris*, 1974.

118. PERZYNA, P., KLEPACZKO, J., BEJDA, J., NOWACKI, W. K., and WIERZBICKI, T., *Applications of Viscoplasticity* [in Polish], Ossolineum, 1971.

119. PERZYNA, P., Description of thermo-mechanical behavior of irradiated materials, *Polish–French Symposium, Nicea, 1974.*

120. PIELORZ, A., On approximate methods of solution of boundary value problems in inelastic media [in Polish], *Eng. Reports* [*Rozpr. Inz.*], 1969, **17**, 3.

121. PODOLAK, K., Propagation of plane stress waves produced by a moving load in an elastic-plastic medium, *Proc. Vibr. Probl.*, 1967, **8**, 3.

122. PODOLAK, K., Propagation of one-dimensional stress waves in an elastic–plastic medium subject to moving heat sources, *Proc. Vibr. Probl.*, 1968, **10**, 3.

123. PODOLAK, K., Reflection of stress plane wave in an elastic–plastic medium of variable yield limit, [in Polish], *Mech. Teor. Stosow.*, 1972, **10**, 3.

124. RAKHMATULIN, Kh. A., On propagation of unloading waves [in Russian], *Prikl. Mat. Mekh.*, 1945, **9**, 1.

125. RAKHMATULIN, Kh. A. and DEMIANOV, Iu. A., *Strength at Intensive Loadings of Short Duration* [in Russian], Moscow, 1961.

126. RAKHMATULIN, Kh. A., SAGOMONIAN, A. Ia., and ALEKSEYEV, N. A., *Problems of Soil Dynamics* [in Russian], MGU, Moscow, 1964.
127. RAKHMATULIN, Kh. A., On propagation of elastic–plastic waves at complex loadings [in Russian], *Prikl. Mat. Mekh.*, **22**, 6.
128. RAFA, J. and WLODARCZYK, E., Penetration of a moving pressure front into a semi-space filled with elastic/viscoplastic medium [in Polish], *Biul. Wojsk. Akad. Tech.*, 1969, **18**, 10.
129. RANIECKI, B., On the coupled equations for isotropic rate sensitive elastic–plastic materials, *Problèmes de la Rhéologie*, PWN, Warsaw, 1973.
130. RANIECKI, B., A quasi-static spherically symmetric problem of thermo-plasticity, *Bull. Acad. Pol. Sci., Série Sci. Techn.*, 1965, **13**, 2.
131. RANIECKI, B., Thermal shock on the boundary of an elastic–plastic semi-infinite body, *Proc. Vibr. Probl.*, 1964, **5**, 4.
132. RANIECKI, B., Spherical thermoplastic stress waves, *Proc. Vibr. Probl.*, 1965, **6**, 4.
133. RANIECKI, B., On collision of a cold plate with a hot elastic–plastic plate, *Proc. Vibr. Probl.*, 1972, **13**, 2.
134. RYCHMYER, R. D. and MORTON, K. W., *Difference Method for Initial-value Problems*, Wiley, New York, 1967.
135. RECKER, W. W., A numerical solution of three-dimensional problems in dynamic elasticity, *J. Appl. Mech.*, 1970, **37**, 1.
136. RICHARDSON, D. J., The solution of two-dimensional hydrodynamic equations by the method of characteristics, *Methods in Computational Physics*, Academic Press, 1964.
137. RUBINE, E. (Ed.), *Mathematics Applied to Physics*, Springer-Verlag, 1970.
138. RUSANOW, W. W., Difference method of constant direction, *Archwm Mech. stosow.*, 1968, **20**, 6.
139. SABODASH, P. F. and CHEREDNICHENKO, R. A., Numerical solution of the diffraction and wave propagation problem by the method of spatial characteristics [in Russian], *5th Soviet Symposium on Wave Diffraction*, Izd. LGU, 1970.
140. SAUERWEIN, H., Anisotropic waves in elastoplastic soils, *Int. J. Eng.*, 1967, **5**, 5.
141. SCHRFINER, K. E., Bending waves in impulsively loaded elastic–plastic plates, *Int. J. Solids Struct.*, 1969, **5**, 4.
142. SHIEH, R. C., On certain closed-form solutions to problems of wave propagation in a strain-hardening rod, *Q. Appl. Math.*, 1970, **27**, **4**.
143. SHIEH, R. C., HEGEMIER, G. A., and PRAGER, W., Closed-form solutions to problems of wave propagations in a rigid workhardening, locking rod. *Int. J. Solids and Struct.*, 1969, **5**.
144. SKOBIEYEV, A. M., On the theory of unloading waves [in Russian], *Prikl. Mat. Mekh.*, 1962, **26**, 6.
145. SKOBIEYEV, A. M., On some problems of soil dynamics [in Russian], *Inzh. Zhurnal*, 1966, 1.
146. SKOBIEYEV, A. M. and FLITMAN, L. M., Moving loading on an inelastic plane [in Russian], *Prikl. Mat. Mekh.*, 1970, **34**, 1.
147. SOKOLOVSKII, V. V., Propagation of elastic–viscoplastic waves in bars [in Russian], *Dokl. Akad. Nauk SSSR*, 1948, 60.
148. SUVOROV, Iu. P., Propagation of temperature loadings in elastic–plastic bar [in Russian], *Prikl. Mat. Mekh.*, 1963, **27**, 2.
149. SUVOROV, Iu. P., On propagation of elastic–plastic waves due to heating of a semi-infinite bar [in Russian], *Prikl. Mat. Mekh.*, 1964, **28**, 1.
150. SHEVCHENKO, Iu. I., Unloading theorems in the theory of elastic–plastic strain due to uneven heating [in Russian], *Prikl. Mat. Mekh.*, 1966, **30**, 7.
151. *Theory of Plasticity* [in Polish] (eds. OLSZAK, W., PERZYNA, P., and SAWCZUK, A.), PWN, Warsaw, 1965.
152. THOMAS, T. Y., *Plastic Flow and Fracture in Solids*, Academic Press, New York, 1961.
153. TING, T. C. T., On the initial slope of elastic–plastic boundaries in combined longitudinal and torsional wave propagation, *J. Appl. Mech.*, June, 1969.
154. TING, T. C. T., Interaction of shock waves due to combined loading, *Int. J. Solids Struct.*, 1969, **5**, 5.
155. TING, T. C. T., Elastic–plastic boundaries in the propagation of plane and cylindrical waves of combined stress, *Q. Appl. Math.*, 1970, **27**, **4**.
156. TING, T. C. T. and NAN NING, Planes waves due to combined compressive and shear stresses in half space, *J. Appl. Mech.*, June 1969.
157. TING, T. C. T. and Nan NING, Planes waves due to combined compressive and shear stresses in half space, *Am. Soc. Mech. Eng.*, No. APM, 12, 1969.
158. TREANOR, Ch. E., A method for the numerical integration of coupled first-order differential equations with greatly different time constants, *Math. Comp.*, 1966, **20**.
159. VERNER, E. A. and BECKER, E. B., Finite element stress formation for wave propagation, *Int. J. Numerical Meth. Eng.*, 1973, 7.
160. WILKINS, M. L., Calculation of elastic–plastic flow, *Methods in Comp. Physics*, Academic Press, 1964.

161. WIERZBICKI, T., Dynamics of rigid-viscoplastic circular plates [in Polish], PhD thesis, IFTR, Pol. Acad. Sci., Warsaw, 1965.

162. WLODARCZYK, E., Propagation and reflection of a plane and spherical shock-wave in an elastic–plastic body and a barotropic liquid, *Proc. Vibr. Probl.*, 1964, **5**, 4.

163. WLODARCZYK, E., Propagation of longitudinal–transverse loading and unloading waves in a non-homogeneous elastic/viscoplastic medium, *Pro. Vibr. Probl.*, 1968, **9**, 2.

164. WLODARCZYK, E., Propagation of longitudinal–transverse radial cylindrical wave in a non-homogeneous elastic/viscoplastic medium, *Proc. Vibr. Probl.*, 1968, **9**, 3.

165. WLODARCZYK, E., Propagation and reflection of one- and two-dimensional stress waves in plastic media [in Polish], DSc thesis, WAT, Warsaw, 1969.

166. WLODARCZYK, E., On a certain class of closed-form solutions of the propagation problems of an elastic/viscoplastic wave in a non-homogeneous rod. *Proc. Vibr. Probl.*, 1966, **7**, 4.

167. WLODARCZYK, E., On a certain closed-form solution of the propagation problem of a spherical wave in an elastic/viscoplastic medium, *Proc. Vibr. Probl.*, 1968, **9**, 3.

168. WLODARCZYK, E., Propagation of a longitudinal–transverse radial cylindrical wave in a non-homogeneous elastic/viscoplastic medium, *Proc. Vibr. Probl.*, 1968, **9**, 3.

169. YANG, C. Y., Strain hardening effects on unloading spherical waves in an elastic–plastic medium, *Int. J. Solids Struct.*, 1970, **6**, 6.

170. ZVOLINSKII, N. V. and SHKHINEK, K. N., On the determination of the unloading wave in a particular case [in Russian], *Prikl. Mat. Mekh.*, 1967, **31**, 1.

171. ZARKA, J., *Sur la viscoplasticité des métaux*, Art. Fr. 2ème fascicule, 1970, 223.

172. ZARKA, J., *Etude du comportement des monocristaux métalliques, Application à la traction du monocristal*, CFC, Cahier Spécial Rhéologie, 1973, **1**.

173. ZARKA, J., Modèle phénoménologique unidimensionnel pour l'étude du comportement viscoplastique du polycristal en grande déformations, *J. Méc.*, 1973, **12**, 2.

174. ZARKA, J. and FRELAT, J., Applications de l'algorithme de Treanor pour les problèmes en visco-plasticité. *Colloque: Méthodes Numériques en Calcul Scientifique et Technique, Paris, November 1974.*

175. ZIV, M., Two-spatial dimensional wave propagation by the theory of characteristics, *Int. J. Solids Struct.*, 1969, **5**.

176. ZIV, M., Generalised characteristics method for elastic wave propagations problems, *Israel. J. Techn.*, 1970, **8**, 1–2.

177. ZIV, M., The decay of loading elastic waves by the theory of characteristics, *Int. J. Engng Sci.*, 1970, **8**.

NAME INDEX

SUBJECT INDEX